D1418613

Defending The Indefensible: The Global Asbestos Industry and its Fight for Survival

Defending The Indefensible: The Global Asbestos Industry and its Fight for Survival

Jock McCulloch and Geoffrey Tweedale

Great Clarendon Street, Oxford OX2 6DP
United Kingdom

Oxford University Press is a department of the University of Oxford.
It furthers the University's objective of excellence in research, scholarship,
and education by publishing worldwide. Oxford is a registered trade mark of
Oxford University Press in the UK and in certain other countries

© Oxford University Press 2008

The moral rights of the author have been asserted

First published 2008
Reprinted 2013

All rights reserved. No part of this publication may be reproduced, stored in
a retrieval system, or transmitted, in any form or by any means, without the
prior permission in writing of Oxford University Press, or as expressly permitted
by law, by licence or under terms agreed with the appropriate reprographics
rights organization. Enquiries concerning reproduction outside the scope of the
above should be sent to the Rights Department, Oxford University Press, at the
address above

You must not circulate this work in any other form
and you must impose this same condition on any acquirer

Published in the United States of America by Oxford University Press
198 Madison Avenue, New York, NY 10016, United States of America

British Library Cataloguing in Publication Data
Data available

Library of Congress Cataloging in Publication Data
Data available

ISBN 978-0-19-953485-2

This book is dedicated to the memory of
Irving J. Selikoff (1915–1992)

Contents

List of Figures

List of Tables

List of Abbreviations

a/c	asbestos cement
ABREA	Brazilian Association of Workers Exposed to Asbestos
ACC	Accident Compensation Corporation
ACGIH	American Conference of Governmental Industrial Hygienists
ACPA	Asbestos Cement Producers' Association
AFA	[Belgian] Asbestos Fund
AHF	Air Hygiene Foundation
AIA	Asbestos International Association
AIA/NA	Asbestos Information Association/North America
AIC	Asbestos Information Committee/Centre
AIHA	American Industrial Hygiene Association
AJIM	*American Journal of Industrial Medicine*
ANDEVA	Association Nationale de Defence des Victimes de L'Amiante
ANYAS	*Annals of New York Academy of Sciences*
ARC	Asbestosis Research Council
ARD	Asbestos-related disease
ATI	Asbestos Textile Institute
BG	Berufsgenossenschafts
BJIM	*British Journal of Industrial Medicine*
CAAJ	Comité Anti-Amiante de Jussieu
CPA	Comité Permanent Amiante
CSR	Colonial Sugar Refining Company
EPA	Environmental Protection Agency
ERCA	Environmental Restoration & Conservation Agency
EEC	European Economic Community
EU	European Union

FDA	Food & Drug Administration
FIVA	Fond D'Indemnisation des Victimes de L'Amiante
FMP	Belgian Occupational Diseases Fund
FM/T&N	Federal Mogul/T&N
GEFCO	Griqualand Exploration & Finance Company
HEI	Health Effects Institute
HSE	Health & Safety Executive
IARC	International Agency for Research on Cancer
IAV	Dutch Institute for Asbestos Victims
IBAS	International Ban Asbestos Secretariat
ICOH	International Commission on Occupational Health
IHF	Industrial Health/Hygiene Foundation
IJHS	*International Journal of Health Services*
IJOEH	*International Journal of Occupational & Environmental Health*
ILO	International Labour Organization
INAIL	Italian National Institute of Insurance for Professional Illness & Injury
IOEH	Institute of Occupational & Environmental Health
IPCS	International Programme on Chemical Safety
JAMA	*Journal of the American Medical Association*
MetLife	Metropolitan Life Insurance Company
MRC	Medical Research Council
NIOH	National Institute of Occupational Health
NEJM	*New England Journal of Medicine*
NIOSH	National Institute for Occupational Safety & Health
NUM	National Union of Mineworkers
OECD	Organisation for Economic Cooperation & Development
OSHA	Occupational Safety & Health Administration
PRU	Pneumoconiosis Research Unit
PUA	polyvinyl acetate
QAMA	Quebec Asbestos Mining Association
QAPA	Quebec Asbestos Producers' Association
SAAPAC	South African Asbestos Producers' Advisory Committee
SACSIR	South African Council for Scientific & Industrial Research

SAIAC	Sociétés Associés d'Industries Amiante-Ciment
SBA	Safe Buildings Alliance
SUVA	Swiss Accident Insurance Organization
T&N	Turner & Newall
TLV	Threshold Limit Value
TUC	Trades Union Congress
UNARCO	Union Asbestos & Rubber Company
WHO	World Health Organization
WTO	World Trade Organization

1

Introduction

It is now clear . . . that the asbestos industry, internationally, is literally engaged in a struggle for survival. Recent developments, an increasing concern with potential environmental hazards (i.e. risk to the general public) and association with cancers other than lung cancer and mesothelioma have greatly increased the scope of the problem.

John H. Marsh (Raybestos Manhattan)
Report to Executive Council, AIA Conference, London
1 February 1977

All you really need to know about the root causes, the cover-ups, and the human impact of occupational cancer you can learn from the example of asbestos. It also tells you everything you need know about the reality of our economic system, what it values and what it fails to protect. It teaches about the collusion between government and industry. It addresses the issue of so-called 'junk science' and how the powers-that-be control information and public health policy. It reveals the hidden injuries of class. . . . And it drives home the old axiom of working class history–that everything you get in this society you must fight for.

Jim Brophy, 'Carcinogens at Work'
Conference on Everyday Carcinogens: Stopping Cancer Before It Starts
McMaster University, Hamilton, Ontario
27 March 1999

Few who watched the Twin Towers of the World Trade Center collapse in Manhattan after the terrorist attack of 9/11 would have had asbestos in mind. Yet within days of that tragic event, asbestos had become part of the drama. The thick dust clouds that had rolled menacingly towards spectators and the cameras, enveloping survivors and surrounding buildings, contained asbestos fibres that had been sprayed onto the girders of the Twin Towers during construction in the early 1970s. Not every floor in the Towers had been spray-coated with asbestos fire-proofing—the erection of the Towers coincided with a New York City ban on spray—but there was enough to deposit a measurable layer of asbestos dust on the surrounding district.

Within weeks of the collapse of the Towers, asbestos was in the headlines. Some wondered whether *enough* asbestos fireproofing had been used in the buildings: but the chief concern was asbestos exposure suffered by rescue workers and the future consequences of asbestos particles remaining in Manhattan office and residential buildings. A debate raged about the safety of working near Ground Zero, with the authorities providing reassurances, but some scientists noting that the future risk of asbestos cancers may be increased.[1] The episode, with its distant echoes of past controversies, was a graphic illustration of the ubiquity of the material and the problems posed by its use.

That the asbestos in the Towers survived one of the best-publicized conflagrations in history, then rose from the ashes to cause anxiety amongst the survivors, is entirely typical of this strange material. Asbestos, even from ancient times, was prized for its indestructibility and fire resistance—to say nothing of its high tensile strength, high electrical resistance, and flexibility. It was these qualities that from the late nineteenth century spawned a major global industry that produced thousands of different asbestos products in textiles, cement, and insulation. The elemental character of asbestos has ensured that asbestos remains the most controversial and feared of the industrial minerals.

A Deadly and Dangerous Commodity

'Asbestos' is a generic name given to a group of fibrous minerals. Put simply, asbestos is a rock that is dug from the ground. Two types exist: serpentine and amphibole. Chrysotile (or white asbestos) is the only member of the serpentine group and is mined mainly in Russia, Canada, China, Brazil, and Zimbabwe. The amphibole grouping includes, *inter alia*, two important commercial grades—amosite (brown) and crocidolite (blue)—which in the twentieth century were mainly mined in South

[1] P.J. Landrigan et al., 'Health and Environmental Consequences of the World Trade Center Disaster', *Environmental Health Perspectives* 112 (May 2004), pp. 731–9. See also M. Bowker, *Fatal Deception* (2003), pp. 277–91.

Africa.[2] Chrysotile has been the most widely used commercial grade, accounting for over 90 per cent of asbestos usage in the twentieth century.

All types of asbestos split longitudinally into fibres. It was this propensity to fiberize that (combined with its heat resistance and toughness) made asbestos so useful: it could be spun, mixed with cement, and used for insulation. But mining, crushing, working, and spinning a material that was basically a rock was not surprisingly very dusty. In fact, it was uniquely dusty, because fibres of asbestos can continue to break down almost to molecular level. Even rubbing asbestos between the fingers can make it 'smoke' and produce a small dust cloud of fibres of unimaginable fineness. Most can only be seen under an electron microscope. Such fibres are easily inhaled: even in the thickest clouds, the dust is breathed in with impunity, without irritating the airways—at least in the short term. Over the long term, the impact of inhalation is damaging and insidious. All types of asbestos are capable of producing three incurable and potentially fatal asbestos-related diseases (ARDs): asbestosis (lung scarring), lung cancer (originating in the lining of the airways of the lungs), and mesothelioma (a cancer of the lining of the chest or of the abdomen). Asbestos may also cause other cancers, such as gastrointestinal tumours.

The lungs have defences to deal with irritant particles, so that they are expelled or dissolved. Tiny hairs called cilia regularly sweep the airways, while scavenger cells (macrophages) try to digest the invaders. But asbestos fibres, especially amphiboles, are so durable and microscopically thin that they can persist in the lungs until death. Moreover, billions of fibres can be inhaled even in a single day and this continued assault on the lungs can cause inflammation and eventual scarring of the tissue, which compromises the ability of the lungs to perform effectively. Much of the damage remains hidden for years, because the lungs have spare capacity, but at some point breathlessness appears. This can be relatively minor, so that regular activities are not curtailed too much (in fact, many workers with asbestosis can continue working for years): but some sufferers eventually become incapacitated. It can be a progressive condition and sufferers are prone to lung infections and heart failure.

In the early twentieth century, asbestosis could appear within five years (even if the exposure had been only a matter of months) and symptoms could be severe. As working conditions improved, it took longer for workers to develop asbestosis and by the end of the twentieth century

[2] There are six types of asbestos: chrysotile (serpentine) and crocidolite, amosite, anthophyllite, tremolite, and actinolite (all amphiboles). See H.C.W. Skinner, M. Ross, and C. Frondel, *Asbestos and Other Fibrous Minerals* (1988).

it was no longer regarded as a death sentence. Nevertheless, even as late as the 1950s, in so-called advanced countries such as Britain, workers could be exposed to enough dust within six months to produce asbestosis (and cancer) a decade down the line.[3] Nor should one underestimate the suffering caused by asbestosis. A Japanese worker has described the symptoms:

> I always bring tissues and trash bags in my briefcase for my cough and phlegm.... I usually have four coughing fits every night. Since my coughs are very close together, I can't ingest any medicines. My wife rubs my back and sometimes asks me if we should call an ambulance, but I am not able to reply during a fit. During my two-hour fit, she just watches me and I feel more dead than alive.... When I have a light fit while riding on a train, my neighbours sometimes change their seats. This makes me feel very lonely.... And, I cannot forget even for an instant that I also carry a time bomb of asbestos in my lungs.[4]

The reference to a time bomb is apt, because asbestos is also a carcinogen. Individuals who develop asbestosis have an increased risk of developing lung cancer (and scientists now accept that asbestos can cause the latter disease, even without first producing asbestosis). Lung cancer needs little description: it is now a major killer in the developed world, largely due to cigarette smoking. Lung cancers caused by asbestos are no different in pathology than those due to any other cancer agent, such as tobacco, but how asbestos itself causes such tumours is still unclear. In the early 1940s, pathologists explained asbestos-related lung cancers in terms of 'irritation' caused by the harsh fibres, and even today with all our knowledge of DNA and cellular biology no better insight has been found. What is known, however, is that smoking and asbestos interact, so that individuals who smoke and are exposed to asbestos have a much greater chance of developing lung cancer than those who are exposed to only one of these carcinogens. Thus, there is a synergy between tobacco and asbestos, which again is not fully understood.

For most of the twentieth century, asbestos ranked as a dangerous industrial material. In 1945, one London physician described it as a 'deadly and dangerous commodity' that should probably be banned.[5] However,

[3] *West Norwood & Dulwich News*, 15 December 1967. Re. inquest on George Mittins (53), former asbestos grinder at Raybestos, who died of asbestosis and lung cancer after six months' exposure twelve years previously.

[4] R. Uno [former Yokosuka shipyard worker], 'Asbestosis: My Pain from Asbestosis', *Annals of the Global Asbestos Congress, Tokyo, Japan*, 19–21 November 2004.

[5] *Annual Report of [Barking] Medical Officer of Health* (1945), p. 23.

asbestos was not unique in killing workers. Asbestosis 'competed' with other industrial dusty-lung diseases (or *pneumoconioses*): these included coal miners' black lung, rock workers' silicosis, and cotton-mill byssinosis. At that time, the mortality or prevalence of asbestosis did not seem particularly striking. Nor was asbestos unique in producing tumours. Chemicals and oils had been causing industrial cancers in European workers since the late nineteenth century. Government and factory doctors did not ring any alarm bells, because industrial diseases and deaths were readily accepted and absorbed by a system that regarded them as an inevitable and necessary by-product of industrialization.

What undermined this complacency was mesothelioma—the deadliest ARD. As one French medical professor put it: '[Mesothelioma] is perhaps the most terrible cancer known, in which the decline is the most spectacular, the most cruel.'[6] Surrounding the lung is a paper-thin membrane called the pleura (a similar membrane surrounds the peritoneum or abdominal cavity). The pleura lies against the chest wall: and between the membrane and the lung is a lubricant which allows the lung to move as we breathe. Pleural mesothelioma occurs in that membrane, which few of us even know exists. Mesotheliomas usually form a white or grey-white sheet which can envelop the lungs or peritoneum (even sometimes the heart). The tumours are thick, hard, and so aggressive that some patients die from suffocation. No effective treatment exists. Chemotherapy and heroic surgery can sometimes offer hope and occasionally there are reports of sufferers surviving for years against the odds.[7] But most are dead within eighteen months of diagnosis (some survive less than six months). The symptoms of pleural mesothelioma are difficult to describe temperately: they are shortness of breath, a persistent cough, weight loss, anorexia, vomiting, and lassitude. Peritoneal mesothelioma is characterized by constipation and weight loss. A characteristic feature of both diseases is effusions, which produce either litres of fluid around the lungs that trigger breathlessness or fill the abdomen to produce a fearful bloating (ascites). The only relief for patients is draining the fluid at hospital at regular and increasing intervals—a miserable experience. Both types of mesothelioma are accompanied by intense pain, which can be unmanageable.

[6] A. Vircondelet, *Mortel Amiante* (1998), p. 15.

[7] These include American paleontologist Stephen Jay Gould, who died in 2002—twenty years after a diagnosis of peritoneal mesothelioma; and Margie Levine, an American school social worker, who died in 2004—fourteen years after she was diagnosed with mesothelioma at the age of 43. See M. Levine, *Surviving Cancer: One Woman's Story* (2001).

Mesothelioma has some unusual features. First, its only known cause is asbestos and in the majority of mesotheliomas evidence of past exposure can be found. It has been said that '[f]ew common cancers have such a direct causal relation with an exposure to a defined carcinogen as mesothelioma has with asbestos exposure—even lung cancer with cigarette smoking'.[8] Second, it can result from relatively trivial exposure. In some cases, individuals have developed the disease after exposure to asbestos of less than a day. One study recorded that the shortest duration of exposure was sixteen hours in a docker loading blue fibre.[9] Environmental (i.e. non-occupational) exposure has also been known to trigger the disease, leading to mesotheliomas amongst those who have never worked with asbestos: building workers, home renovators, housewives, office workers, nurses, and schoolteachers. Third, the disease has a long latency. Some mesotheliomas appear to have developed within fifteen years of exposure: more typically, the period between exposure and first symptoms is about thirty years and can be as long as sixty years. If there is such a thing as a typical mesothelioma victim, it would be a man who had worked with asbestos in mining, shipping, construction, or student jobs—'tough-guy' types, who are struck down in their forties or fifties, or as their retirement beckons; in other words, in their prime.

Exactly how asbestos produces a mesothelioma is, again, unclear, but asbestos fibres can penetrate the lungs to reach the pleura (in fact, pleural plaques—areas of calcification—are a common feature in those exposed to asbestos). Asbestos can also reach the abdomen via the diaphragm or bloodstream. Once there, the fibres can produce prolonged cycles of damage, repair, and local inflammation in the tissues, which can eventually trigger the tumour.

One can read about mesothelioma and even describe it without being able to appreciate the horror of the disease. Not surprisingly, mesothelioma sufferers do not make ideal autobiographers—time is too short. But the wives and offspring of such individuals have found the impact and consequences of living with such a disease so cathartic that a memoir has been the result. A handful of accounts of mesothelioma now exist, usually written by women. They are not well publicized and they have not been

[8] B.W.S. Robinson, A.W. Musk, and R.A. Lake, 'Malignant Mesothelioma', *The Lancet* 366 (30 July 2005), pp. 397–408.

[9] J. Leigh and T. Driscoll, 'Malignant Mesothelioma in Australia, 1945–2002', *IJOEH* 9 (July/September 2003), pp. 206–17.

bestsellers, but for those who can endure reading them they offer key insights into the problems caused by this disease.[10]

For example, Nancy Rossi wrote about her husband John, a New York tax lawyer, who developed peritoneal mesothelioma in the summer of 1978, when he was aged 33. John had been exposed to asbestos for only two weeks during a college job, when he had unloaded 'Flexboard' building sheets made by America's premier asbestos company, Johns-Manville. Nancy noted how he 'seemed to get thinner every day, especially his back. Every vertebra was so prominent, every rib so pronounced, bulging the thin dry skin that covered his back.... Unspoken, riveting pain produced by even slight movement made up [most] of his day.'[11] After twelve weeks he was dead from the Flexboard fibres later found in his lungs. Nancy was pregnant, but John did not survive to see his baby born.

From Discovery to Epidemic

The 'discovery' of the three key ARDs has attracted enormous attention from lawyers, the industry, scientists, and historians—so much so that there is now little left to discover about their chronology. Nevertheless, there is still room for debate. In each instance, the growth of evidence linking asbestos and disease was marked by the play of competing interests which shaped both the research process and the reception of knowledge. To protect itself, industry has always emphasized the slowness of medical discovery, pointing out that contests for certitude about the dangers of fibre often dragged on for decades. While that contains some truth, it should be remembered that industry interference in medical discovery has been so pronounced that in addition to the known medical literatures there are also 'secret' histories for each of the ARDs.

Medical discoveries—like most advances in knowledge—are the result of the accumulation of basic facts, until a flash of insight provides a new breakthrough or orthodoxy. The development of knowledge about ARDs is set out in Table 1.1.[12] The 'evil effects' of asbestos and the way in which the mineral killed workers by causing pulmonary fibrosis were

[10] M.J. Grove, *Asbestos Cancer: One Man's Experience* (1995); B. McKessock, *Mesothelioma: The Story of an Illness* (1995); A. Vircondelet, *Mortel Amiante* (1998); L. Kember, *Lean on Me: Cancer Through a Carer's Eyes* (2003); R. Albarado, *A Story Worth Telling: An Asbestos Tragedy* (2005).

[11] N. Rossi, *From This Day Forward: A True Love Story* (1983), pp. 152–3.

[12] G. Tweedale, 'Asbestos and Its Lethal Legacy', *Nature Reviews Cancer* 2 (April 2002), pp. 311–15.

Table 1.1. Asbestos-Related Disease Chronology, 1898–1964

1898—Factory Inspectors in the UK identify the 'evil effects' of asbestos and the 'easily demonstrated' danger to the workers' health.

1899—Asbestos worker admitted to a London hospital, suffering from pulmonary fibrosis. He dies the following year.

1924—In Britain the first inquest on an asbestos worker leads to the first medical description of asbestosis.

1931—British government introduces dust control regulations in the asbestos industry.

1935—Asbestosis and lung cancer cases appear in the medical literature in the US and Britain. A pathologist suggests to British government medical officers that the diseases might be linked.

1938—German pathologist declares for the first time that lung cancer is an occupational disease of asbestos workers. In Britain, the government notes a significant rise in lung cancers in asbestosis cases.

1943—German government recognizes asbestos-induced lung cancer as a compensable occupational disease. One German pathologist links asbestos with rare pleural cancers.

1947—British government statistics note a high percentage of asbestosis cases with lung (including pleural) cancers.

1955—First epidemiological study of a group of British asbestos workers confirms a lung cancer risk.

1960—Landmark study published by South African researchers shows a linkage between mesothelioma and both occupational and non-occupational exposure to asbestos.

1964—Catastrophic cancer mortality demonstrated among American insulation workers and publicized at a conference in New York.

known in Europe by about 1900. In the 1920s, asbestosis was described in the medical literature. By then its associated mortality had become so pronounced that the British government ordered an inquiry and in 1930 introduced government safety measures and workmen's compensation.[13] In the 1930s, the first medical reports appeared in the US and the UK that suggested a link between asbestosis and lung cancer.[14] Because the latter disease was not so prevalent in the 1930s, it was easier for physicians to 'see' the connection. The most sophisticated understanding of the asbestos lung cancer problem developed in Germany and in the UK. Germany made asbestos-related lung cancer a compensable disease in 1943; and although Britain did not follow this lead, data accumulated by the Factory Inspectorate provided compelling evidence by the late 1940s that asbestos could cause lung cancer. By the mid-1950s—when the first epidemiological study of lung cancer in asbestos workers was published—no one could doubt the connection.[15]

[13] E.R.A. Merewether and C.W. Price, *Report on Effects of Asbestos Dust on the Lungs and Dust Suppression in the Asbestos Industry* (1930).

[14] B.I. Castleman, *Asbestos: Medical and Legal Aspects* (2005), pp. 42–7.

[15] R. Doll, 'Mortality from Lung Cancer in Asbestos Workers', *BJIM* 12 (1955), pp. 81–6.

By then, the first mesotheliomas amongst asbestos workers had been described. It was such a rare and unusual tumour that the link with asbestos was even easier to see: 'Few [medical] authors ever expressed doubt about the relationship between malignant mesotheliomas and asbestos exposure, and by 1953, the issue seemed to be fairly well resolved.'[16] However, mesothelioma excited little interest, because it was still a rare disease. That situation changed dramatically in the late 1950s, when a team of South African pathologists uncovered a mesothelioma 'epidemic' around the asbestos mines in Cape Province. The research, published in 1960, connected asbestos with mesothelioma and also suggested that non-occupational (i.e. neighbourhood/environmental) exposure might cause the disease.[17] In 1964, a major American study of insulation workers gave worldwide publicity to the dangers of asbestos and the widespread nature of the risk.

The development of ARDs can best be viewed as a series of waves: 1930s (asbestosis); 1940s (asbestos-related lung cancer); and 1960s (mesothelioma). Besides the impact on individuals, each wave had a characteristic impact as it rippled through the industrial and sociopolitical systems in various countries. The impact of the final wave (the result of three ARDs combined) has been the most profound—making the period after 1960 of great significance. Until then, asbestos was virtually unknown to the wider public as a health hazard. But since the 1960s, the incidence of ARDs in many countries has increased dramatically, making the final wave of ARDs a tsunami of occupational and environmental disease.

Mesothelioma, in particular, is an increasing problem worldwide. At the start of the twentieth century, the disease was extremely rare, with perhaps an annual incidence of no more than one or two cases per million. However, that incidence has risen to over forty per million in some countries (while in some of the worst-affected regions around shipyards and asbestos factories the incidence can be as much as eighty-five per million).[18] At Prieska, a blue asbestos producing centre in South Africa, between 8 per cent and 15 per cent of people born in the 1940s has died (or will die) of mesothelioma. Over the last thirty years,

[16] P.E. Enterline, 'Changing Attitudes and Opinions Regarding Asbestos and Cancer 1934–1965', *AJIM* 20 (1991), pp. 685–700, 687.

[17] J.C. Wagner, C.A. Sleggs, and P. Marchand, 'Diffuse Pleural Mesotheliomas and Asbestos Exposure in the North-Western Cape Province', *BJIM* 17 (1960), pp. 260–71.

[18] Aborigines transporting raw crocidolite in Western Australia (site of the infamous Wittenoom mine) have reported rates of 250 per million per year, and some occupational cohorts have rates in the thousands of cases per million.

mesothelioma has achieved parity with other male cancers and is now roughly as common as cancers of the liver, bone, and bladder, especially in Europe and Australia. The disease is increasingly seen as an epidemic in the developed world and as a legacy of asbestos use that continued far too long. Because of the long incubation period between first exposure and the onset of disease, that legacy is only now being realized. It is disturbing that in some developed countries asbestos mortality is yet to peak and for the thousands of individuals with asbestos fibres in their lungs no help can be offered. In these countries, mesotheliomas may continue to appear until at least the middle of the present century. Meanwhile in the developing world, the asbestos problem is only now emerging.

Snapshots can be provided from epidemiological surveys and by looking at the situation in individual countries. In the US, one study in 2004 claimed that 10,000 Americans die each year of ARDs—a rate approaching thirty deaths per day.[19] Asbestos in the US thus kills thousands more people than skin cancer each year, and nearly the number that are slain with firearms. The epidemic is national in scope, affecting every state. In Canada, the home of supposedly 'safe' white asbestos, 343 persons died from mesothelioma in 2003—a 17 per cent increase from only four years previously.[20] In the UK, about 4,000 individuals die each year from asbestos—more than ten times the number of workers killed in industrial accidents. In Britain, the number of mesotheliomas has increased 40 per cent in the last eight years (to 2,037 in 2005). But this percentage rise is not unusual in industrialized countries. Every day in France, accidents kill on average two people, while asbestos kills eight. Japan's mesothelioma rate has almost doubled: from about 500 in 1995 to 953 in 2004. Australia (which has the highest reported incidence of mesothelioma in the world) saw the disease jump 60 per cent between 1995 and 2003 (to about 700 cases). But Germany, with about 900 cases in 2004 also saw a rise of 80 per cent since the mid-1990s.[21]

This inevitably begs the question: how many more deaths? It is difficult to know for sure (since many ARDs are poorly recorded and epidemiology is an inexact science), but wherever one looks the picture is dismal and

[19] Environmental Working Group. *Asbestos: Think Again* (2004).

[20] A. Sharp and J. Hardt, *Five Deaths a Day: Workplace Fatalities in Canada 1993–2005* (2006), p. 64.

[21] Data on the asbestos crisis are continually updated in medical articles and the press. Useful single surveys are: Special Issues on asbestos, *IJOEH* 9 (July/September 2004) and *IJOEH* 10 (April/June 2004); and *Annals of Global Asbestos Congress*, Tokyo.

worrying. ARDs over the next decade could claim the lives of over 100,000 Americans. Looking ahead to the year 2030, similar numbers have been projected for Britain, Japan, and France. In Australia the future death toll could reach 45,000, and in Germany more than 20,000. In Europe, it has been estimated that between 1995 and 2030 up to half a million deaths could occur from ARDs, especially mesothelioma.[22]

No one knows the final global death toll, but the International Labour Organization (ILO) and the World Health Organization (WHO) have recently stated that asbestos kills at least 90,000 workers worldwide each year *at present*. According to one report, the asbestos cancer epidemic could take at least five million (and possibly as many as ten million) lives before asbestos is banned worldwide and exposures cease.[23] Of course, one might argue that the chances of anyone contracting an ARD—especially a person born today—are slight. It is also true that other threats to health are equally, if not more, pressing: for example, HIV has already killed some twenty-five million, and tobacco kills nearly five million each year. On the other hand, no other industrial agent—not lead, not benzene, not vinyl-chloride, not chromium—even approaches the burden of disease that asbestos has caused worldwide. Moreover, it is the tragedy of asbestos that so much of this illness and death could—as we shall show—have been avoided.

Past, Current, and Future Issues

Not surprisingly, asbestos has been widely discussed and publicized since the 1960s. Following the discovery of the link between asbestos and mesothelioma, the mineral has been the subject of innumerable television and radio broadcasts. At the scientific level, an explosion in medical research has occurred that, according to Pub Med (the National Library of Medicine's online catalogue in the US), has produced over 9,000 publications since 1960 alone—a tally that is certainly an underestimate. Social scientists, journalists, and historians have been busy, too. Following the publication of Paul Brodeur's classic exposés of the American

[22] J. Peto et al., 'The European Mesothelioma Epidemic', *British Journal of Cancer* 79 (1999), pp. 666–72.

[23] J. LaDou, 'The Asbestos Cancer Epidemic', *Environmental Health Perspectives* 112 (March 2003), pp. 285–90.

asbestos industry in the 1970s and 1980s,[24] a steady stream of books has been published on asbestos—a flow that has quickened recently with the appearance of a number of histories and company studies (see Sources).

We ourselves have published three books on asbestos, which have covered the history of asbestos manufacture and mining in Britain, Australia, and South Africa.[25] Nevertheless, we found that asbestos continued to command our attention. As the world's asbestos health problem continued to develop, we saw that our texts were becoming (at least as regards contemporary events) obsolete. We found ourselves living through a scenario that was changing by the year—indeed by the month—and that we needed to keep ourselves up to date.

Our research had demonstrated our own ignorance of many aspects of asbestos, especially its global dimension. By looking at a single company, or even single countries, we had been unable to link crucial parts of the bigger picture. In particular, we had often failed to show how the asbestos industry operated as a worldwide business mechanism. Again, this has contemporary relevance, for although it is little publicized, asbestos is still mined and used in the developing world, where the problems that were experienced in America and Europe in the twentieth century are now being duplicated in China, Russia, India, and other countries in the Far East. Our books, by concentrating on countries that had either banned asbestos or ceased production, had only told part of the story. However, getting a handle on events as they unfold on a worldwide canvas is difficult. Conferences have brought together interested parties and have published useful proceedings, but these are not always easily available. We felt that the global aspects of asbestos and ARDs demanded another study.

This is especially so, because although journalists and academic historians have been almost unanimous in their condemnation of the industry, the historiography of asbestos is now mature enough to have spawned several attempts at revisionism. Jacqueline Corn's industry-sponsored work on the asbestos problem in American schools has attacked government policy and the asbestos industry's critics.[26] Rachel Maines (from an avowedly pro-business perspective) has emphasized the merits of asbestos,

[24] P. Brodeur, *Expendable Americans* (1974); Brodeur, *Outrageous Misconduct* (1985).

[25] J. McCulloch, *Asbestos: Its Human Cost* (1986); McCulloch, *Asbestos Blues* (2002); G. Tweedale, *Magic Mineral to Killer Dust* (2001).

[26] J.K. Corn, *Environmental Public Health Policy for Asbestos in Schools: Unintended Consequences* (1999).

absolved the American asbestos companies of conspiracy against the public interest, and placed the blame for litigation on journalists and on opportunistic lawyers.[27] In Britain, the asbestos industry's apologist has been Peter Bartrip, whose authorized histories of the British and American asbestos industries have depicted the leading companies as blameless and traduced by historians with an aversion to capitalism.[28] In the face of these attempts to revise the consensus, we have felt that the experience of asbestos and the lessons it teaches needed restating.

The opportunity to write a fresh study was facilitated by new documents and the availability of more detailed information. Asbestos had been one of the most inscrutable industries, until the tidal wave of litigation after the 1970s cracked its carapace of secrecy. In the US, legal discovery has been used to dramatic effect, bringing into the public domain a vast quantity of documents that were never intended for public scrutiny. Because the litigation has not stopped—indeed it has increased—the steady flow of incriminating documents has shown no slackening. Shortly after we completed our books, a treasure trove of unpublished asbestos documentation was generously donated to us by an American physician, Dr David Egilman. In addition to this material, much of which we had never seen, we found ourselves reading an endless stream of company correspondence, published government reports, epidemiological studies, think-tank analyses, and new historical articles.

However, the catalyst that pushed us towards writing this book was not new documentation, the industry's apologists, or gaps in the historiography. Instead, it was the need to explain one of the industry's greatest paradoxes—one of which most people remain unaware. The full extent of this paradox was revealed with the publication of a time-series, which synthesized all the available data on asbestos production (and use) worldwide between 1900 and 2003. This pioneering survey, which was unavailable for our previous work, was compiled by Robert Virta of the US Geological Survey.[29] We examined this windfall with considerable interest, especially Virta's tabulation of world production data. We did not regard the list of tonnage figures as particularly noteworthy, but when we began using it to compose a graph (see Figure 1.1), we got a shock.

To understand why, one needs to reflect on the big picture. Until the 1960s, no one seriously questioned the use of asbestos as a major

[27] R. Maines, *Asbestos and Fire: Technological Trade-offs and the Body at Risk* (2005).

[28] P. Bartrip, *The Way from Dusty Death* (2001); Bartrip, *Beyond the Factory Gates* (2005).

[29] R.L. Virta, *Worldwide Asbestos Supply and Consumption Trends from 1900 through 2003* (2006).

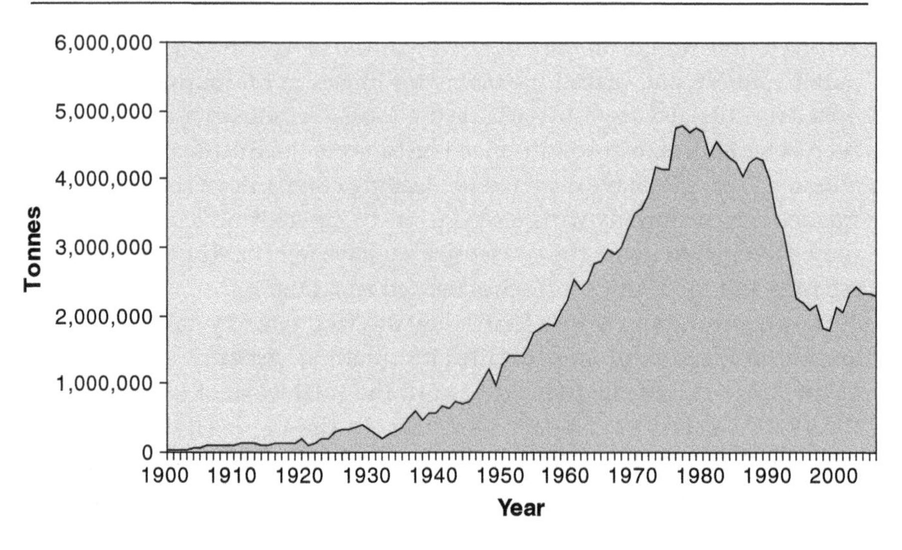

Figure 1.1. World Asbestos Production, 1900–2006

industrial product. To be sure, the mineral was associated with serious and debilitating industrial diseases: however, this was never serious enough to threaten the multinational asbestos companies. Society uses dangerous materials whenever the benefits apparently outweigh the risks. However, the publicity about mesothelioma in 1960 should have marked a sea-change. After all, mesothelioma demonstrated that occupational and environmental exposure to even relatively low doses of asbestos could be fatal. In other words, compelling scientific reasons existed for asbestos use to be severely restricted or perhaps even banned. This was particularly so, because during the 1960s and 1970s substitute materials were increasingly available for many asbestos products. Indeed, some have pointed out that many of these substitutes—such as fibreglass—had been available throughout the twentieth century. They did not cause fibrosis of the lungs and were not associated with mesothelioma or lung cancer. In short, one might have expected world asbestos production to decline rapidly after 1960.

Virta's figures show the opposite. Asbestos production *increased* after 1960. Between 1900 and 2004, world asbestos production was approximately 182 million tonnes. Of this total, 143 million tonnes were produced after 1960. Put another way, nearly 80 per cent of world asbestos production in the twentieth century was produced *after* the world learned that asbestos could cause mesothelioma! Of course, asbestos production did decline eventually and these are aggregate figures which do not show differentials between countries. Virta delineates greater rates of decline in asbestos use and production in certain countries. However, even in such medically sophisticated countries as Britain and the US, asbestos use continued to climb in the first decade or two after the discovery of mesothelioma. In some countries in the developing world, asbestos use is increasing even today.

What accounts for such a severe dislocation between medical knowledge and economic development? This burgeoning use of asbestos after 1960 has not escaped everyone's attention and Virta comments on it himself. He offers an explanation that is couched entirely in technological imperatives and medical uncertainties. Virta says that there was a 'lack of knowledge about the risks posed by exposure to high levels of airborne asbestos dust', and claims that the 'asbestos health issue did not arise until long after asbestos had been used by society for a long period of time'.[30] However, this is hardly convincing and is the kind of apologia that has always been offered by industry or government (or their supporters).

We believe that the reasons for this dislocation must be looked for elsewhere: first, in the history of asbestos before 1960, but particularly in events thereafter when the asbestos industry reacted to mesothelioma. That reaction was not simply dictated by a lack of knowledge or a failure to find adequate substitutes, but was often driven by purely political and commercial considerations. The argument that is advanced in this book is that asbestos has proved so enduring, because the industry was able to mount a successful defence strategy for the mineral—one that still operates in some parts of the world. Central to this strategy was a policy of concealment and, at times, misinformation that often amounted to a conspiracy to continue selling asbestos fibre irrespective of the health risks. We unravel that conspiracy to highlight how the industry censored scientific research; used reputable scientists to elide the health hazards and nurture scientific uncertainty; denied basic compensation (and sometimes human rights) to victims; and colluded with governments and scientific bodies.

In doing so, we uncover in detail some of the most interesting aspects of the asbestos story: the attempts to suppress the research on mesothelioma, the problem of asbestos in the environment, the inadequacies of compensation for industrial disease, and the fight over the continued use

[30] Virta, *Worldwide Asbestos* (2006), p. 16.

of asbestos in the developing world. Our canvas is global. The final picture shows how asbestos and the problems that surrounded it can be seen (as Jim Brophy suggests) as a paradigm for many industrial diseases—indeed for a whole range of hazards produced by our industrialized society. Not only is the asbestos story not over, but the same battles between science, corporate interests, and the public are now being played out in the debates over global warming and global inequalities of wealth. Asbestos therefore remains disturbingly relevant for our era of contested science and product defence.[31]

[31] D. Michaels, 'Manufactured Uncertainty: Protecting Public Health in the Age of Contested Science and Product Defense', *ANYAS* 1076 (September 2006), pp. 149–62.

2

Making the World Safe

Whoever you are, whatever your business, a Raybestos-Manhattan product touches your life.

Raybestos-Manhattan Inc.
Annual Report (1963)

Every time you walk into an office building, a home, a factory; every time you put your foot on the brake, ride in a train, see a bulldozer at work. Every time you see or do any of those things, the chances are that a product from the James Hardie (Asbestos) Group of Companies has a part in it.

John Reid, chairman of James Hardie Asbestos Ltd
In *Hardie Ferodo 1000: A James Hardie Group and Activity Report* (1978)

The global asbestos industry was born in the second half of the nineteenth century. The first modern mines, which were in Italy's alpine valleys, began production in the 1850s and for twenty years they enjoyed a monopoly supplying fibre to a small manufacturing market in Western Europe. In 1876, large chrysotile deposits were discovered at Thetford, in Canada. Two years later mining began and from the turn of the century Canada was the world's largest producer of chrysotile or white asbestos, a position it maintained for almost a hundred years. South Africa supplied virtually all of the world's blue and brown asbestos (crocidolite and amosite).[1]

The industry was so successful in marketing what became known as the 'magic mineral' that by 1950 most European and North American

[1] During the 1960s and 1970s, the world's major producers were the USSR with 45%, Canada with 35%, and South Africa with 10% of global output. For an overview of the global industry in that period see 'Asbestos Outlook', *The South African Mining and Engineering Journal* (October 1974), pp. 18–34.

households had incorporated some asbestos-based products into their daily lives. Asbestos was present in cement sheeting which clad many homes and in the pipes which supplied water; asbestos fibre was blended into vinyl floor tiles and in insulation designed to make oil refineries, hospitals, warships, cinemas, and domestic dwellings safe. It was used in rubber and plastic products, mixed with adhesives, cements, paints, and sealants. In automobiles asbestos was blended into gaskets, cylinder heads, spark plug washers, exhaust pipe insulation, radiator blankets, and brake linings. Some of its more exotic uses included cigarette filters, dish towels, surgical thread, banknotes, piano felts, ironing boards, berets, aprons, carpets, tampons and filters for rice, salt, beer, and fruit juice. In the 1920s, one British manufacturer sold mattresses filled with crocidolite or blue asbestos promising that their customers could with safety smoke in bed. The US postal service used asbestos mail bags. At one time toothpaste was fortified with fibre. The burning broomstick of the Wicked Witch in *The Wizard of Oz* was made from asbestos and the artificial snow in *Citizen Kane* was probably chrysotile. Asbestos was also a common ingredient in talcum powders, where it found its way into condoms. During the twentieth century, sufficient asbestos cement (a/c) pipe was manufactured in the US alone to circle the earth eight times and still run to the moon and back.[2]

Asbestos occurs widely in the earth's crust and many countries including the US, Australia, Italy, Brazil, India, and Kazakhstan have mined chrysotile. Mining began in Vermont in 1899 where the deposits are part of the serpentine belt which runs through Quebec. Mining took place in another dozen US states including Arizona and North Carolina. Between 1963 and 1985, Union Carbide, a company best remembered for the Bhopal tragedy, operated a number of chrysotile mines in California.[3] Only rarely did US output reach 100,000 tonnes per annum, with imports providing 90 per cent of domestic needs.[4] Most of the asbestos used in the US came from Quebec, which once boasted ten of the thirteen mines in Canada. The US also imported chrysotile from Zimbabwe and amphiboles from South Africa and Australia.

[2] J.E. Alleman and B.T. Mossman, 'Asbestos Revisited', *Scientific American* (July 1997), pp. 54–7.

[3] In 1970, US domestic production reached 125,314 tonnes or 3% of world output. See D.H.K. Lee and I.J. Selikoff, 'Historical Background to the Asbestos Problem', *Environmental Research* 18 (April 1979), ii, pp. 300–14.

[4] R.L. Virta, *Worldwide Asbestos Supply and Consumption Trends from 1900 through 2003* (USGS, 2006), pp. 5–6.

The relative cost of fibre in the manufacture of asbestos products encouraged vertical integration. The big companies such as Raybestos Manhattan Inc., Philip Carey, and Turner & Newall (T&N) owned mines in Canada and Southern Africa, which supplied fibre to their metropolitan factories. Prior to its retreat into bankruptcy in 1982, Johns-Manville controlled over 40 per cent of Canadian output and its holdings included Jeffrey, which was the world's largest asbestos mine. The industry's peak in North America and Western Europe coincided with what some economists have called the Golden Age of Capital (1945–72) and in that sense asbestos is symbolic of modern industrial production and its attendant global divisions of labour.

Asbestos was also important in the Soviet Union where the state was committed to industrialism and chrysotile enjoyed the patronage of political elites and official science. The Soviets had three asbestos producing regions. The major mines were at Bazhenovskoye (Uralasbest) in the Urals, which remains the world's largest mining and milling complex. Its centre, Asbest City, is less than 100 kilometres from Ekaterinburg. The other two mining centres were Dzhetygara and Kiembay in Kazakhstan and at Akdovurak in the Tuva Republic. During the 1960s and 1970s, the Soviet mines lay at the heart of Comecon and were able to export fibre into the captive markets of Eastern Europe. The break-up of the Soviet empire in 1991 brought an immediate, if temporary, decline in output. The industry has since recovered, due partly to markets in the developing world. Today Russia produces over half of the world's asbestos and is the world's second largest user (behind China).

In the US, asbestos use followed the pattern found elsewhere in the industrialized world. Consumption increased from less than 100,000 tonnes in 1912 to 750,000 tonnes in 1942.[5] The most rapid rise took place during the Second World War, which gave the industry an impetus it was able to maintain for more than three decades. In all, over 30 million tonnes of asbestos were consumed in the US during the twentieth century with half of that amount used after 1960. By the mid-1970s, in the US alone 200,000 people were employed in the manufacture of asbestos products, but many more worked in secondary industries and millions more were employed in downstream manufacturing, including 185,000 in shipyards and almost two million in automobile sales and repairs. On the global stage the pattern was similar. During the twentieth century, total

[5] J. Corn and J. Starr, 'Historical Perspective on Asbestos: Policies and Protective Measures in World War II Shipbuilding', *AJIM* 11 (1987), pp. 359–73.

world production was around 173 million tonnes, with the bulk being consumed after 1976.[6]

Inventing an Industry

The manufacture of asbestos products in the US began just before the Civil War with simple versions of asbestos paper, textiles, and packings being made in small workshops. Prior to the 1890s, asbestos was a novelty and its ability to resist corrosion, acids, alkalies, electricity, vibration, frost, and vermin was little understood. Its resistance to high temperatures, which made it so useful in any process involving the conversion or preservation of heat, was not utilized until the revolution in manufacturing and transport, which occurred towards the end of the nineteenth century. Once that happened, consumption rose inexorably.

The industry leaders gradually developed a range of building and insulation products. Johns-Manville first entered the market with stove and boiler insulation and later branched into theatre curtains and movie-projection booths. Most fire curtains were clearly labelled 'asbestos' to reassure audiences about their safety.[7] In 1899 an Austrian chemist, Ludwig Hatschek, discovered how to combine asbestos with cement and the mineral soon became popular as a reinforcing agent. The next major innovations were brake linings and friction materials for automobiles. Scottish industry was among the first in Western Europe to adopt asbestos for shipbuilding and heavy engineering. The mineral was initially imported from Canada in 1871. Before the end of that decade, the Scottish Asbestos Company, which made engine packing and insulation, was operating its own mines in Quebec. By 1885 there were nineteen manufacturers and distributors in Glasgow alone; a decade later there were fifty-two. In Scotland, as elsewhere, work conditions in factories and shipyards were shaped by rapacious employers, weak trade unionism, and negligent regulatory authorities.[8]

Ludwig Hatschek franchised his invention as 'Eternit' and over the next sixty years, licenses were granted to companies in Belgium, Switzerland, Italy, France, the UK, Germany, Chile, the Netherlands, Argentina, Hong Kong, Uruguay, China, Nigeria, and India.[9] Hatschek often took payment

[6] Virta, *Worldwide Asbestos*, p. 3.

[7] For a history of asbestos and theatre insulation, see R. Maines, *Asbestos and Fire* (2005).

[8] For a history of asbestos in shipbuilding, see R. Johnston and A. McIvor, *Lethal Work: A History of the Asbestos Tragedy in Scotland* (2000).

[9] R.F. Ruers and N. Schouten, *The Tragedy of Asbestos* (2005), p. 19.

in the form of share options and as a result Eternit became a vast enterprise. The Swiss and Austrian subsidiaries had a monopoly in a/c production while branches in Belgium, France, the Netherlands, and Germany controlled the local markets.[10] From 1920, the Swiss family headed by Ernst Schmidheiny and his company Holderbank played a key role in Eternit.[11] In France asbestos manufacture was dominated by only three companies: Eternit, which began manufacturing a/c products in Paris in 1922 using fibre from its mine in Corsica; Saint Gobain, which also produced a/c; and the T&N subsidiary Ferodo, which made asbestos textiles and friction products in Normandy.

In addition to the empires of Hatschek and Schmidheiny, numerous small manufacturers made asbestos-based products. One of the first factories to open in Prague was the Herman & Company Asbestos Works. The firm, which employed twenty-five textile workers, was managed by Karl Hermann, who shared ownership with his brother-in-law, Franz Kafka. For a brief period, Kafka managed the plant. He later wrote of that experience: 'The torment the factory brings me. Why did I not protest when I was obliged to work afternoons there. Now, nobody is violently forcing me, but Father with reproaches, Karl with silence and my conscience'.[12] Herman & Company shut down at the beginning of the First World War. Discouraged by that failure Kafka turned to writing.

By 1925, the Johns-Manville Corporation was one of the world's leading building materials and insulating companies. It had major contracts with the automobile, oil, and railroad industries and it enjoyed strong profits from a rapidly expanding market. Johns-Manville was in every respect a modern enterprise. It invested heavily in research, and as early as 1927 its New Jersey laboratory was developing new products.[13] In various ways Johns-Manville resembled that other conglomerate, US Steel. The latter owned coal mines and iron ore deposits, while Johns-Manville operated mines in Canada. By the early 1930s, US Steel controlled 40 per cent of American ingot capacity. Johns-Manville had about half of Canadian asbestos output and ran factories in a dozen states, with its

[10] For details of Eternit in Belgium, see S. Nay, *Mortel Amiante* (1997); for France, see O. Hardy-Hemery, *Eternit et L'Amiante 1922–2000: Aux Sources du Profit, une Industrie du Risque* (2005).

[11] M. Roselli, *Die Asbestlüge* (2007); Ruers and Schouten, *Tragedy of Asbestos*, pp. 7–8; W. Catrina, *Der Eternit-Report: Stephan Schmidheinys Schweres Erbe* (1989); R.F. Ruers, N. Schouten, and F. Iselin, *Eternit le Blanchiment de L'Amiante Sale: Les Consequences Tragiques de 100 Ans d'Amiante-Ciment* (CAOVA, 2006). Posted at: http://caova.ch/

[12] H. Salfellner, *Franz Kafka and Prague* (2003), pp. 121–2.

[13] Johns-Manville Corporation, *Annual Report* (1937), p. 3.

biggest plants at Manville, New Jersey, and Waukegan, Illinois. Johns-Manville and US Steel were major suppliers to the building, automobile, and railroad industries and both sold most of their products to buyer specifications. Both shared a close relationship with J.P. Morgan & Company. In 1927, J.P. Morgan bought over half of the shares in Johns-Manville.[14]

The leading British asbestos manufacturer T&N was founded in 1878 in Rochdale, near Manchester, by Samuel Turner. The company, which had its origins in the Lancashire cotton industry, was much larger than its competitors British Belting & Asbestos and the Cape Asbestos Company. By the 1920s, T&N accounted for nearly half of UK asbestos manufacture and its Rochdale factory was the world's largest asbestos textile plant.[15] T&N was the only company of comparable size and influence to Johns-Manville and, like its US competitor, T&N relied upon vertical integration to guarantee a supply of cheap fibre. It operated mines in Quebec and southern Africa.

The asbestos industry was too large and diverse to characterize easily. There was more than one type of fibre, and there were several asbestos markets. The longest fibres were used in textiles, woven brake linings, clutch pads, and marine insulation. Shorter fibres were combined with cement to make prefabricated wall panels, corrugated roof sections, and waters pipes or mixed with vinyl and asphalt to manufacture tiles. Asbestos was also a common filler to bolster cements, putties, and paints. Fibre quality was as important as type. Normally, if a pure chrysotile blend was used in cement sheets or pipes, the final product was likely to warp. To overcome that problem, manufacturers used a mixture of crocidolite and chrysotile. That was necessary with all white asbestos except for the fibre from Havelock mine in Swaziland, Cassiar in British Colombia, and from Shabanie and Gaths mines in Southern Rhodesia.[16]

The widespread use of asbestos in commercial, industrial, and domestic settings and the mobility of fibre through the cycles of mining, transport, and manufacture meant that during the second half of the twentieth century almost every citizen in the US or Western Europe was exposed in some way or other to airborne fibre. That is surprising, as the fire-resistant properties of asbestos are not unique, and when the mineral first appeared on the world market in the 1880s, there were already other products

[14] See 'Management by Morgan', *Fortune* (Magazine) 9 (1934), pp. 82–9.

[15] For a history of T&N, see G. Tweedale, *Magic Mineral to Killer Dust* (2001).

[16] Jock McCulloch interview with Pat Hart, CEO of Gefco, Braamfontein, Johannesburg, 6 July 2001.

which could serve the same purpose. Mineral wool was first manufactured in the 1840s for thermal insulation and rock wool was available before the end of the nineteenth century. In cement manufacture asbestos was merely a reinforcing agent which could have been replaced with any number of substitutes. The major appeal of asbestos to manufacturers lay in its low cost and that in turn derived from the conditions under which it was mined.

Making New Products

The US industry was structured around the specializations of paper, cement, asbestos magnesia, textiles, and brake linings. That diversity was repeated within each of those five divisions. Asbestos paper, for example, included all types of paper, millboard, pipe covering, cell pipe, and insulating cement. The bigger companies, such as Johns- Manville, Philip Carey, and Raybestos-Manhattan operated factories that straddled all five divisions. They also made non-asbestos products.[17] Raybestos-Manhattan (later Raymark), for example, was a leading manufacturer of bowling balls. The industry's diversity was reflected in its workforce and, besides the men and women who worked in factories, large numbers of jobbers, dealers, and contractors came into daily contact with the mineral.[18] Prior to the Second World War, more than 10 per cent of the US asbestos workforce was female and the tendency to pay women less than men was consistent with wider market conventions.

The manufacture of asbestos products required heavy machinery for carding, picking, spinning, and mixing and in each phase of the production process workers faced some risk of traumatic injury. All asbestos manufacture shared another common denominator—high levels of dust. This was a particular problem in asbestos textile plants where raw asbestos was blended manually. It was then disintegrated or beaten by hand to open up the fibres. During carding, asbestos was fed over rotating cylinders then transferred to spinning frames, which twisted and spun the fibre into yarn. The looms were poorly ventilated and mill workers were often knee-deep in dust.[19] The fibre was screened on open trays and the residue

[17] *Hearing on Code of Fair Competition for the Asbestos Industry: Transcripts of the Hearing and Appendix*, Washington, DC: National Recovery Administration (19 October 1933), p. 46.

[18] See Statement of Mr George Crabbs, President, the Philip Carey Manufacturing Company in *Hearing on Code of Fair Competition: Transcripts*, pp. 69–78.

[19] See D.W. Hills, 'Economics of Dust Control', *ANYAS* 132 (31 December 1965), i, pp. 322–37.

cleared into settling chambers which were cleaned manually at the end of each week.[20] Suppressing dust at carding machines was difficult. Carding material is easily disturbed by exhaust fans, which can damage the quality of the final product. To maintain the integrity of the asbestos–cotton mix requires still air.[21] In British and US factories the only concessions to safety were some basic ventilation and the use of respirators for the cleaning of the dust-settling chambers.

Nothing illustrates better the versatility of asbestos than the shipbuilding industry. The introduction of steamships in the latter half of the nineteenth century created a need for packing and insulation to make engines more efficient. In addition to running turbines, the steam generated by boilers was piped throughout ships for heating, cooking, and washing. Before the introduction of asbestos, such pipes were often uncovered. The resulting high temperatures from radiant heat could make work conditions below decks intolerable. Serious accidents were common on rolling ships as sailors were thrown against hot pipes. Amosite and crocidolite were light- and heat-resistant, and by the 1920s asbestos cladding had become a standard feature on ships. The new insulation was so popular that anyone employed in a shipyard or who worked on a ship after 1930 was exposed to asbestos.

Asbestos imports into Scottish docks came directly from South Africa and the biggest shipyards had their own sheds where fibre was unpacked and prepared by hand.[22] During construction the trades within a ship's hull included joiners, shipwrights, plumbers, electricians, smiths, and boiler makers. Laggers were mostly employed by outside contractors and in general they were not covered by shipyard regulations. Asbestos was sprayed onto bulkheads and deckheads and moulded sections made from almost pure amosite were used to insulate pipes and machinery and to fireproof living quarters. Heavy asbestos blankets covered boilers and steam lines were lagged with asbestos paste. In addition to the normal vibration which broke down insulation, constant refittings released dust into confined spaces. Ventilation systems circulated the dust, thereby contaminating sleeping quarters, messes, and lounges.

The most expensive items in the manufacture of asbestos brake linings and insulation were the raw materials, with wages and labour amounting

[20] Hills, 'Economics of Dust Control', p. 323.

[21] W.C. Dreessen et al., A *Study of Asbestosis in the Asbestos Textile Industry. Public Health Bulletin No 241, August 1938*, p. 21.

[22] Johnston and McIvor, *Lethal Work*, pp. 16–18.

to less than 40 per cent of the cost of production.[23] That cost structure, which changed little over time, made the price of raw asbestos the key to the industry's profitability. To ensure a supply of cheap fibre the major US manufacturers operated their own mines, but the problem of sudden fluctuations in output and price persisted. During the early 1920s, a price war broke out between Canadian, South African, and Soviet producers which led to the closure of a number of Quebec mines. Johns-Manville responded by increasing its investments in Canada, while on the domestic front it bought out a number of competitors. In 1928 Johns-Manville took over Banner Rock and Celite, both of which made non-asbestos insulation, asphalt roofing, decorative tiles, and bridge construction materials. T&N followed suit by purchasing seven insulation companies in an attempt to control the market.[24]

To solve the problem of supply, in 1929 Johns-Manville, T&N, Eternit, and Philip Carey arranged a cartel, which in effect divided the global market into spheres of influence. In Western Europe, firms from Belgium, France, Germany, Austria, Hungary, Britain, Spain, the Netherlands, Italy, and Switzerland participated in the Sociétés Associés d'Industries Amiante-Ciment (SAIAC) which had its secretariat in Switzerland.[25] According to T&N's management, the agreement was designed to standardize the product and secure fibre on the best terms. 'This miniature League of Nations', T&N told its shareholders, 'has a great future before it, for it is based upon the principle of mutual help, which now replaces the previous atmosphere of distrust and suspicion'.[26] In practice, the arrangement created a framework which enabled the big companies to dominate mining and manufacture.[27]

In the 1930s, Eternit bought a number of small mines in South Africa to form Everite, which grew into one of the region's biggest mining and manufacturing companies. T&N also acquired a controlling share in the Rhodesian & General Asbestos Corporation in Southern Rhodesia (Zimbabwe). With the benefit of rail subsidies and tax concessions, its output rose from about 8,000 tonnes in 1924 to over 36,000 tonnes in 1930, placing Southern Rhodesia in second place behind

[23] See Testimony of Mr Lewis H. Brown, President, Johns-Manville Corporation and President of Asbestos Institute in *Hearing on Code of Fair Competition: Transcripts*, p. 57.

[24] B.I. Castleman, *Asbestos: Medical and Legal Aspects* (2005), p. 29.

[25] Ruers and Schouten, *Tragedy of Asbestos*, p. 8.

[26] Turner & Newall, *Annual Report* (September 1929), p. 9.

[27] On the workings of that arrangement, see Monopolies Commission, *Asbestos and Certain Asbestos Products: A Report on the Supply of Asbestos and Certain Asbestos Products* (1973), p. 98. See also Castleman, *Asbestos*, pp. 29–30.

Canada.[28] That dramatic rise gave T&N a monopoly over high-quality Southern Rhodesian fibre and with it a more prominent voice within the global industry. In 1932, T&N began developing the Havelock mine in Swaziland. Having established a cartel and absorbed companies making asbestos-free substitutes the industry was able to survive the Great Depression. During testimony before the National Recovery Administration in 1933, the President of Johns-Manville, Lewis Brown, estimated the capital investment in US manufacture at around $50 million, with fifty-eight companies operating seventy-one plants. In addition, many of the bigger producers owned subsidiaries.[29] Despite the industry's top heavy structure, many companies were relatively small. Thirty-eight employed fewer than 100 workers and in the brake-lining division fifteen of the twenty-three plants employed fewer than 100 men and women.[30]

Chrysotile was first incorporated into brake linings in the UK in 1908, and it was soon taken up by American automobile manufacturers. By 1930, brake linings made from woven fibre constituted around a third of US industry revenue. Ten feet of lining was needed for each new car and the typical linings used in family automobiles were 80 per cent asbestos and 20 per cent cotton.[31] Leading manufacturers such as Ford and Chrysler had short lead times, with raw materials going in one door and finished products out the other. Such hectic production schedules had an impact upon work conditions.

Asbestos was a useful additive for industrial paper to enhance the retention of moisture. After pulping, beating, screening, and cleaning, the fibre was mixed with other wet-end materials, formed into sheets, pressed, rolled, and then dried. Millboard was made in a similar way, except it was passed through a hydraulic press and after drying, cut into sheets. Shingles, which consisted of around 30 per cent asbestos, were also subjected to tremendous hydraulic pressure. Magnesia insulation used for pipes and boilers was a standard Johns-Manville line and by the early 1930s it was the company's biggest seller. The product consisted of 85 per cent magnesia and 15 per cent asbestos. Magnesia contains masses of minute air cells which make it a poor conductor of heat, but it lacks strength. That came with the addition of chrysotile. The mix was shaped into strips to encase pipes or cut into blocks to cover boilers. The railways were the largest market for '85 per cent Magnesia',

[28] N.H. Wilson, *Notes on the Mining Industry of Southern Rhodesia* (1932), p. 40.
[29] Testimony of Brown, Hearing on Code of Fair Competition: Transcripts, pp. 4–69.
[30] Hearing on Code of Fair Competition: Transcripts, p. 60.
[31] 'Management by Morgan', p. 87.

as the product was generally known. The US Navy also used it in large quantities.

The industry's recovery from the Great Depression was helped by the development of new products. The sprayed asbestos insulation known as Limpet was devised in 1932 by J.W. Roberts, a T&N subsidiary. A spray made from blue asbestos and cement was used initially on railway coaches to reduce noise and condensation. Limpet was also later adapted for fireproofing and insulation.[32] The fireproofing blends were 5–30 per cent fibre, while the thermal insulation used on turbines was often pure asbestos. In power stations, spraying allowed turbines to be completely encased, thereby improving efficiency and safety. Sprayed insulation was introduced into the US in 1935. The material could be easily shaped and it became popular for the decoration of night clubs, restaurants, and hotel foyers.

While conditions in UK factories did improve during the 1930s, they were never safe. In 1938, T&N established a factory at Clydebank to manufacture a/c sheets. The plant operated until 1970 and at its peak it employed more than 300 workers. After delivery in hessian sacks, raw asbestos was stacked by hand in a warehouse. Bag handling was a dangerous job as the fragile bags often broke showering workers with fibre. There was no protective clothing and cleaning was done with a brush and shovel.[33] The waste was used to surface the car park.[34] One Clydebank resident remembers that Agamemnon Street, adjacent to Turner's factory, was covered in white dust which settled on cars and window sills.

The Second World War and its Aftermath

The prospect of war in 1939 saw the industry enter a boom that was to last thirty years. Asbestos was useful in shipbuilding and armaments manufacture and for that reason the Ministry of Supply in the UK declared T&N and a number of its subsidiaries Controlled Undertakings. To meet urgent contracts, dust control measures were relaxed, and old plants and the men and women who operated them were worked to exhaustion. Profits benefited. Johns-Manville did equally well, and in 1941 sales rose

[32] For a history of asbestos spray, see G. Tweedale, 'Sprayed "Limpet" Asbestos: Technical, Commercial, and Regulatory Aspects', in G. Peters and B.J. Peters (eds.), *Sourcebook on Asbestos Diseases* (1999), pp. 79–109; W.B. Reitze et al., 'Application of Sprayed Inorganic Fiber Containing Asbestos: Occupational Health Hazards', *American Industrial Hygiene Association Journal* 33 (March 1972), pp. 178–91.

[33] Johnston and McIvor, *Lethal Work*, pp. 65–66.

[34] Johnston and McIvor, *Lethal Work*, p. 201.

by 50 per cent on the strength of the demand for insulation, packings, gaskets, and refractories. Large numbers of women were employed by Johns-Manville and by 1943 one quarter of its workforce was female. In the following year to offset labour shortages, Johns-Manville recruited Jamaican and Honduran workers, and prisoners-of-war from the European theatre. They were joined by part-time labour, which included teachers, students, housewives, and servicemen on leave.[35]

In 1940, the US Congress voted funding to build a two-ocean navy, and in the following year Lend-Lease saw US dockyards made available for the repair and maintenance of British vessels. Between 1939 and 1945, almost 6,000 ships were built at US naval yards. At its peak in November 1943, there were 1.7 million workers on shipyards, and during the war a total of 4.5 million men and women were employed in the industry.[36] Shipyards are dangerous places and hectic production schedules resulted in numerous accidents. American workers were also exposed to silica dust, solvents, lead, mercury, and, of course, asbestos.

Because of its light weight, low conductivity, and imperviousness to salt water amosite was preferred to magnesia in the construction and repair of ships.[37] Following a national conference of work conditions, in 1943, the Navy Department issued a set of minimum safety requirements which covered all aspects of shipyard work from traumatic accidents to insidious injury.[38] The authors identified handling, sawing, cutting, moulding, and welding-rod salvage as the major sources of asbestos dust. Among other provisions, they called for respirators for those working with silica, asbestos, solvents, acids, gases, and metal fumes. Dusty work areas were to be segregated, ventilation installed, and workers given regular medicals. Unfortunately, the pressures of work meant that those measures were not put into effect.

Armand Stella was one of more than 600 pipe coverers who worked at the South Boston and Charlestown Naval shipyards. Stella installed covering on new destroyers and he worked on refits during which engine rooms, fire rooms, and crew quarters were refurbished.[39] Major refits

[35] Johns-Manville Corporation, *Annual Report* (1944), p. 5.

[36] Corn and Starr, 'Historical Perspective', p. 363.

[37] See W. Fleischer et al., 'A Health Survey of Pipe Covering Operations in Constructing Naval Vessels', *Journal of Industrial Hygiene and Toxicology* 28 (January 1946), pp. 9–16.

[38] *Minimum Requirements for Safety and Industrial Health in Contract Shipyards*. US Navy Department (1943).

[39] Testimony of Armand Stella, in *Edward F. Sears and Irene Sears v Pittsburgh Corning Corp.* Civ Act No 80-2647-MA United States District Court, District of Massachusetts, Boston, 13 February 1985.

involved the installation of new boilers, bathrooms, water lines, and steam pipes. Insulating those surfaces in the confined spaces of a ship produced large amounts of dust. Stella worked with a variety of Johns-Manville, Philip Carey, Keasbey & Mattison, and Eagle-Picher products. They included a/c, amosite sections, asbestos block and temperature cement, asbestos cloth, paper, and asbestos blankets. He would cut and mould sections to fit the pipes, cover them with asbestos cloth and then apply asbestos-based sealants. Armand Consalvi also worked as a pipe coverer on the Boston yards from 1940 to 1945. After the war he had a variety of jobs in which he used the same a/c, pipe covering, muslin, and asbestos rope.[40] His employment included contract work at schools in New York and at an IBM plant in Kingston.

The war encouraged the use of asbestos in a number of unlikely products. Although they were never required on the battlefield, large numbers of gas masks were manufactured, and as a preventative measure US and British troops were trained in their use. The masks made in British factories for the Ministry of Supply contained blue asbestos, which was believed to be an ideal filter.[41] The filters were assembled in layers, inserted into the canisters, and soldered in place. The 1,200 women who assembled masks at Boots factory in Nottingham worked in clouds of dust. A number of factories also operated in Canada using the same specifications as the British Army.[42] The crocidolite was initially processed at a textile plant in Asbestos, Quebec, where wool and fibre were carded in layers, cut into individual pads, and packed into boxes for dispatch to Ottawa and Montreal. There they were assembled by female workers who reduced the pads to a correct weight and volume by the manual removal of surface layers. The pads were then inserted into the canisters and tested. The finished canisters contained about 20 per cent crocidolite, by weight. The women, who assembled the masks, like the men and women who wore them in training, inhaled millions of crocidolite fibres.

The war fostered an association between asbestos, safety, and modernity which the industry was able to exploit. From the 1950s, the asbestos products used on US ships included wall and ceiling sprays, wallboards, gaskets, structural supports, boiler relining bricks, packing, caulking, gloves,

[40] Testimony of Armand Consalvi, in *Armand Consalvi and Mai Consalvi v Johns-Manville* Corp. MBL No 1 Civ Act No 81-1785-MA United States District Court, District of Massachusetts, Boston, 24 April 1984.

[41] For an account of that industry, see R. Meeran, 'The Boots Mesothelioma Cases' in G.A. Peters and B. Peters (eds.), *Current Asbestos Issues* (1994), pp. 273–84.

[42] See A. McDonald and J.C. McDonald, 'Mesothelioma After Crocidolite Exposure During Gas Mask Manufacture', *Environmental Research* 17 (1978), pp. 340–6.

bed blankets, ventilator ducts and lining, floor tiles, ceiling tiles, paint, dry walls, rope, cement and adhesives, and radiator fittings.[43] The *SS United States*, which on its maiden voyage in 1952 established a record for the fastest Atlantic crossing, was full of blue and brown fibre. In addition to bulkhead insulation and Marinite boards, the lounge chairs, cushions, sea chests, curtains, and even bedding blankets on that ship were made from asbestos.[44] In the construction industry, sprayed asbestos proved an important innovation. Before the invention of Limpet, steel girders on tall buildings had to be encased in concrete to prevent buckling. By reducing the weight of frames, building costs were also reduced. From 1950 the US National Gypsum Company marketed its own brand of spray called Thermacoustic. The first use of 'mineral fibre' for a large multi-storey building was on the Chase Manhattan Bank in New York City in 1958. The sixty-floor structure was fireproofed with over 1 million square feet of sprayed mineral fibre, mostly Limpet. A thirty-storey building would use around 200 tonnes of fibre and by 1970 half of the large multi-storey buildings in the US were insulated with sprayed asbestos.

Johns-Manville products were used in a wide variety of commercial, military, and domestic settings. They included nuclear reactors and in the Pershing, Polaris, Minutemen, and Nike Hercules rocket and missile systems. It was with some justification that Johns-Manville's management told shareholders:

> There is almost no field of human endeavour in which, at some stage or another, some J-M product does not play a part. Being one of the most diversified manufacturers in the world, the Company makes more than 400 different lines of products at its 20 plants in the United States and Canada.[45]

On the other side of the Atlantic, T&N had contracts for the London underground and the British nuclear power industry. Like Johns-Manville it enjoyed the most successful period in its history and profits rose from £500,000 in 1945 to £7 million in 1960. The company regularly paid an annual dividend in excess of 20 per cent.[46]

In the three decades after 1945, Australia was the highest per capita user of asbestos in the world.[47] Australian manufacture was dominated

[43] L.C. Jacques, 'Shipboard Asbestos Use: An Historical Perspective' in G.A. Peters and B. Peters (eds.), *Current Asbestos Issues* (1994), p. 228.

[44] Jacques, 'Shipboard Asbestos Use', p. 235.

[45] Johns-Manville Corporation, *Annual Report* (1948), p. 3.

[46] Tweedale, *Magic Mineral to Killer Dust*, p. 100.

[47] NSW Government: Cabinet Office, *Report of the Special Commission of Inquiry into the Medical Research and Compensation Foundation* (September 2004). 'Asbestos and James Hardie', Annexure J, p. 117.

by James Hardie Asbestos Pty Ltd. The company, which was founded at the end of the nineteenth century, was one of the first to realize the potential of a/c sheets. In 1916 it opened the Camellia factory near Sydney, to produce building materials. Over the next decade further plants were opened in Perth, Adelaide, and Melbourne.[48] Sheltered by tariff barriers and aided by government contracts James Hardie enjoyed great commercial success. In New South Wales (NSW) the company supplied a/c products to the Housing Commission, the Metropolitan Water, Sewage, and Draining Boards, and to numerous Shire councils.[49] Between 1945 and 1954 more than half of the new homes built in NSW were made from a/c sheets.[50] On average, James Hardie consumed 70 per cent of the fibre used annually in Australia.[51] James Hardie differed from Johns-Manville and T&N in buying most of its fibre from outside sources.[52] It did, however, have a share in the Cassiar mine in British Columbia.

From the early 1960s, Armstrong-Cork, Congoleum Industries, GAF, and Johns-Manville began manufacturing asbestos paper. Speciality products, many of which were almost pure fibre, included flame barrier papers used in aircraft and ships and facial tissues. Asbestos papers were also used to filter pharmaceuticals, wines, and beer. The market grew rapidly and by 1975 over 326,500 tonnes of asbestos had been used as an additive in paper marketed in the US.[53] At that time there were thirteen manufacturers.[54]

Johns-Manville and T&N used similar production techniques, but their market strategies were very different. T&N looked towards the British Empire and opened factories in colonial states. From the 1930s, T&N operated a/c plants at Kymore and Calcutta, in India, while in Africa it had factories in Nigeria, Kenya, Zambia, South Africa, and Southern Rhodesia. In contrast, Johns-Manville grew by opening new facilities within the vast US domestic market. When operating off-shore, it tended like Eternit to sell its expertise and license independent firms, which then returned either royalties or minority stock. During 1949, Johns-Manville entered into agreements with firms in Australia, France, and Belgium. It

[48] For a history of the company, see G. Hagan, 'James Hardie Industries 1880–1980' (Macquarie University BA thesis, 1980).

[49] Hagan, 'James Hardie', p. 32.

[50] NSW Government, *Report*, Annexure J, p. 16.

[51] NSW Government, *Report*, Annexure J, p. 121.

[52] L. Noakes, *Asbestos Supplement Department of Supply and Shipping Mineral Resources of Australia* (Summary Report No. 17, July 1945).

[53] A.D. Little, 'Characterisation of the US Asbestos Papers Market' (Final Draft Report to Sores Inc. C-79231, May 1976), p. 3.

[54] Little, 'Characterisation', pp. 17–18.

also supplied companies in Argentina, West Germany, and Chile with technical assistance.[55]

Several parallels and convergences exist between the asbestos and tobacco industries. In the decade after the Second World War cigarette consumption in the US rose dramatically. By 1952 every citizen over the age of 15 consumed, on average, nearly 200 packs of cigarettes a year. The market was dominated by six companies selling a limited number of brands which were more or less identical. The American cigarette was a cylinder of plain thin paper, three inches long, filled with a mixture of burley and flu-cured tobacco. Each manufacturer sought to impress its brand upon consumers and, to that end, by the early 1950s the industry was spending over $50 million a year on advertising. Lorillard, one of the smaller producers, had a tiny share of the market.

In the midst of its commercial success the tobacco industry was faced suddenly in 1952 with medical evidence linking smoking with lung cancer. The release of that knowledge into the public domain saw a sharp fall in the demand for tobacco products. The industry response was to introduce mentholated and filter-tipped cigarettes and in 1954 alone nineteen new brands or new versions of old brands came onto the market. Cigarette filters were made from a variety of materials, including absorbent cotton, crêpe paper, and cellulose acetate fibres. Lorillard was keen to increase its sales by introducing a brand which could be promoted for its unsurpassed safety. In 1951 its management became aware of a secret filtering material which had recently been declassified by the US Army. That material, which was so fine that it could trap particles as small as one micron, was the African blue asbestos used in gas masks.[56]

Asbestos was associated in the public imagination with safety and Lorillard was quick to recognize that it was an ideal material with which to assuage smokers' fears. Lorillard subcontracted the work to a factory near Boston, which made the new 'Micronite' filters from a mixture of crocidolite, crimp paper, and cotton acetate. The production process was entirely dry, which resulted in high dust levels within the plant. The new brand was given a clean white package to appeal to female smokers and named Kent, after Lorillard's retiring president Herbert A. Kent. The cigarette was a prestige product selling for 8 or 9 cents above the price of regular filter tips. It was promoted by Lorillard as 'the greatest health protection in cigarette history'. Despite the company's optimism, Kent captured only a

[55] Johns-Manville Corporation, *Annual Report* (1950), p. 6.
[56] For a history of the Micronite filter, see R. Kluger, *Ashes to Ashes* (1997), pp. 148–52.

small part of the market. The asbestos filters were so efficient they made it difficult to draw smoke, thereby reducing the flavour. A pack-a-day smoker was inhaling 131 million of the thinnest crocidolite fibres each year.[57]

Working with Asbestos

Work conditions in the US industry remained largely unchanged from 1930 to the late 1960s. Basil Whipple began work at the Ruberoid plant in Erie, Pennsylvania, in 1935. For twenty years he was employed on a paper-drying machine before moving into the mill board department. Ruberoid bought scrap gaskets from various suppliers, which Whipple would feed into a speed hammer mill. The mill ground the waste into a powder to which new asbestos would be added. The old gaskets came in burlaps sacks as did the asbestos, which was purchased from Johns-Manville and the Ruberoid subsidiary Vermont Asbestos. The beating room and the trimming machines were in a single building, but the grinder created so much dust that Whipple was forced to close the door when the machine was operating. There were no warning labels on the asbestos bags, no ventilation fans, and no warning signs in the mill. The union complained to the management about the dust, but Ruberoid always said it was harmless: 'Nobody was scared of asbestos then', Whipple later recalled: 'They didn't understand it.'[58]

Hugh Jackson spent his career working for Johns-Manville mostly in plant safety. From 1952 to 1960 he was manager of the Industrial Health section and he reported directly to Dr Kenneth Smith, the corporation's medical director. Jackson often visited the Waukegan factory in Illinois, where within a single complex Johns-Manville manufactured thermal insulation, textiles, paper, mill board, and sheet materials. It also made pipe coverings, shingles, and flat boards.[59] During the manufacture of shingles, asbestos, cement, and silica were combined into slurry, then pressed, dried, and cut into sections. According to Jackson, many of the jobs at Waukegan were dusty. The asbestos would arrive in 100 lb burlap bags, which along with the silica would be dumped by hand into the machines. There was no proper ventilation and the men would breathe

[57] See J.A. Talcott et al., 'Asbestos-Associated Diseases in a Cohort of Cigarette-Filter Workers', *NEJM* 321 (2 November 1989), pp. 1220–3.

[58] Deposition of Basil Whipple, in *Helen. M. Connor vs. Johns-Manville* US District Court, Western District of Pennsylvania, Case No. 80–177, Erie, PN, 30 December 1981, p. 28.

[59] Deposition of Hugh Jackson, in *Oral Forest vs. Johns-Manville Sales, Corp.* US District Court, Eastern District of Missouri Case No. 78-1346C, Denver, CO, 13 December 1979.

in fibre. That pattern was repeated at numerous factories throughout the US.

Asbestos dust scattered well beyond the factory gates. The streets around T&N's Roberts' plant in Leeds were awash with fibre and in the words of one resident: 'It was as though we were practically eating dust. I remember dust and fibre all around the streets near the factory. You could see the dust in the air. I have seen it blow around like a snowstorm.'[60] In addition, workers brought dust home on their clothes. A physician who worked for an American manufacturer described that problem in the following way: 'Once asbestos gets into the home, carried home by the workmen, it is there virtually permanently—it gets into the rugs, into the carpets, it gets suspended by movement and actually you are getting 24 hour/day exposure.'[61]

Most American, British, and French workers who came into contact with asbestos did not work in the primary industry, but were employed at oil refineries, shipyards, on building sites or in auto-repair shops. In addition to their exposure to a known hazard they shared a lack of knowledge about the dangers they faced. The actor Steve McQueen, who died from mesothelioma at the age of 50, and who is said to have worked in shipyards and for a time in a brake-repair shop, is typical of the casualties.

The US market for drum brake linings was dominated by Raybestos-Manhattan Inc. Its brake and clutch pads were marketed under the brand names Raybestos and Grey-Rock and sold to automotive warehouses, distributors and jobbers, car dealers, and repair shops. Raybestos also sold direct to 'do-it-yourself' car owners. Most drum linings were by weight between a third and three-quarters pure asbestos. Disc pads also contained chrysotile. According to the Asbestos Information Association, by 1972 worldwide over 300 million motor vehicles used asbestos-based brake linings. In that year in the US alone, more than 50,000 tonnes of asbestos was used in their manufacture.[62] As one manager of an auto-repair shop commented in 1975: 'Almost everything that moves uses some form of asbestos friction material.'[63]

All repair shops followed much the same procedures. After removing the wheel and brake drum, the brakes were usually cleaned out by compressed air or brushed by hand. Often the new linings were drilled, ground, and

60 Tweedale, *Magic Mineral to Killer Dust*, p. 168.

61 Quoted in Castleman, *Asbestos*, p. 441.

62 AIA/NA, *Asbestos and Brake Linings* (1973), p. 1.

63 Letter from Jim Lewis, Brake & Clutch Services, Dallas to the Docket Officer, US Department of Labor, 20 November 1975.

cut to fit before the wheel was reassembled. Those exposed to airborne fibre included workers in garages and shops, suppliers, rebuilders of brake shoes, and those employed in car dealer yards. The industry created further problems through the casual disposal of waste. It is possible to gauge just how dusty repair shops were by the reaction of employers in 1975, when the American health and safety body, the Occupational Safety & Health Administration (OSHA), proposed to introduce more stringent dust regulations. Jim Lewis who ran two small auto shops in suburban Dallas estimated that it would cost him $45,000 to comply with the new standard. As a result, his employees would not receive a deserved wage rise and his consumers would pay more to have their brakes repaired. In a letter to the Department of Labor Lewis wrote: 'Almost everything uses some form of asbestos friction material. Before long, with all the new safety standards coming into effect we may have the safest country in the world, but very few people will be able to afford to live here.'[64] Roy Watts, the manager of the Hastings Company, in King, North Carolina, ran a small shop rebuilding auto parts. He estimated that in order to comply with the new OSHA standards he would have to seal off the 'tear down' and relining areas used for brake shoes and clutch discs, install filter exhaust ventilation in all work areas, provide respirators and protective clothing for employees, install showers and change rooms, put up warning signs, and introduce sophisticated monitoring equipment.[65] Few repair shops ever conformed to such standards.

The situation was no better in the construction industry. In 1968 the International Association of Heat & Frost Insulators & Asbestos Workers had 18,000 members, with probably the same number working outside the union.[66] The insulation trade consisted of a large number of small businesses. By the mid-1960s, there were more than a thousand employers each on average employing less than twenty men.[67] Small employers lacked the capacity to make workplaces safe. Those few that did had little impact on the wider industry as men would move from one job to another or from one employer to the next. Insulators were part of a workforce of more than three million construction workers, which included steam fitters, electricians, carpenters, plumbers, masons, and tile

[64] Letter from Jim Lewis.

[65] Letter from Roy Watts, Plant Manager, the Hastings Company, King, North Carolina to the Docket officer, US Department of Labor, 12 January 1976.

[66] Transcript of Statement by Dr Irving Selikoff before the House Committee on Education and Labour, Select Sub-Committee on Labor Washington, 7 March 1968, p. 2.

[67] Transcript of Statement by Selikoff, p. 9.

fitters.[68] Often those tradesmen worked side by side breathing in the same dust.

Asbestos proved to be ideal for insulating power stations and oil refineries. The State Electricity Commission (SEC) operated large power stations at Yallourn, 100 miles outside of Melbourne, Australia. The Yallourn stations were fuelled with coal from a vast open-cut mine. They provided all the state's electricity as well as employment for thousands of men.[69] For many of those men and their families Yallourn was a good place to live. The SEC provided steady employment, subsidized housing and medical care for families. There was a strong community spirit and marvellous sporting facilities. But Yallourn was a company town: the SEC was the sole employer and it owned the shops, the houses, and the health clinic.

In 1945 Lyle Seear began work at Yallourn. He recalls that the power station was smoky, poorly lit, and full of uncovered machinery. Asbestos was everywhere but the workers took no notice of it. When servicing the turbines, maintenance crews would climb under the machinery where the floors were was so thick with fibre their overalls turned white.[70] The crews were given a primitive mask consisting of a piece of moulded aluminium with gauze padding. The masks made breathing difficult and so the men would not wear them. When the boilers were running, the machinery vibrated so much that the lagging would disintegrate, spilling asbestos onto the men working below. The fibre would then lie on the floor for days before being swept up with a broom. The trade unions took no interest in occupational health and asbestos was never discussed at union meetings. During Lyle Seear's years at Yallourn there was an oral tradition that men were dying of lung cancer. Seear remembers that some men would cough each night on the bus. The sickest men were the laggers and boiler makers, who had clubbed fingers and skin like parchment. At the time no one knew what was wrong with them.[71] The money was good and so the men put up with the dust. Lyle Seear only became aware that asbestos is dangerous in the late 1970s.

In 1985 the SEC began demolishing Yallourn, claiming it wanted access to the coal beneath the town. The former SEC workers have a different perspective. They believe the town was demolished to disperse the population. Many men and their wives, who were exposed to the fibre their husbands brought home on their clothes, were nearing middle age, and

[68] Transcript of Statement by Selikoff, p. 8.
[69] For a history of Yallourn, see M. Fletcher, *Digging People Up For Coal* (2002).
[70] Jock McCulloch interview with Lyle Seear, Latrobe Business Centre, Moe, 16 May 2003.
[71] McCulloch interview with Lyle Seear.

the SEC wanted to avoid the spectacle of a pandemic of ARDs in such a small community. Today the town of Yallourn no longer exists.

The Mines

Western societies could enjoy the benefits of asbestos in a vast range of products only because of the output of the industry's mines. The asbestos mines were important in another way: the major companies derived the bulk of their profits from them. The sale of fibre was crucial to Johns-Manville and as production levels rose, so too did the amount of asbestos it sold to other manufacturers. In 1951, sales of Canadian Johns-Manville were in excess of $58 million, of which just over half went to outside companies.[72] During 1957, which marked Johns-Manville's centenary, its four Canadian mines produced 568,000 tonnes of asbestos: less than 20 per cent went to Johns-Manville factories.[73] The $60 million the company invested at that time in expanding and modernizing its Canadian plants was financed largely from the sale of fibre.[74] The source of corporate profits becomes even more obvious if Johns-Manville's annual reports are examined closely. In 1972, it had 25,000 employees and total sales of almost $800 million.[75] Construction materials was the largest of the company's five divisions with sales of $347 million and a profit of $41 million. But in terms of profits the most important part of the company's operations was its thirteen mines which had sales of $88 million and profits at $20 million. Throughout the 1960s and 1970s, the profits from the mining division were often in excess of 20 per cent. By contrast the construction materials division returned barely half that figure.[76] In 1973 sales to foreign countries, including fibre, accounted for less than 20 per cent of total sales, but over one-third of the group's profits.[77] Thus, the steady, sometimes spectacular, rise in Johns-Manville's profits after 1930 (see Figure 2.1) was heavily reliant on mining.

On the other side of the Atlantic, the industry also enjoyed strong growth that was primed by consumer demand for its products and the output of the mines. T&N continued its dominance of the European industry and in 1957, for example, recorded a net profit of nearly

[72] Johns-Manville Corporation, *Annual Report* (1951), p. 5.
[73] Johns-Manville Corporation, *Annual Report* (1957), p. 9.
[74] Johns-Manville Corporation, *Annual Report* (1956), p. 9.
[75] Johns-Manville Corporation, *Annual Report* (1972), p. 1.
[76] Johns-Manville Corporation, *Annual Report* (1972), p. 3.
[77] Johns-Manville Corporation, *Annual Report* (1973), p. 5.

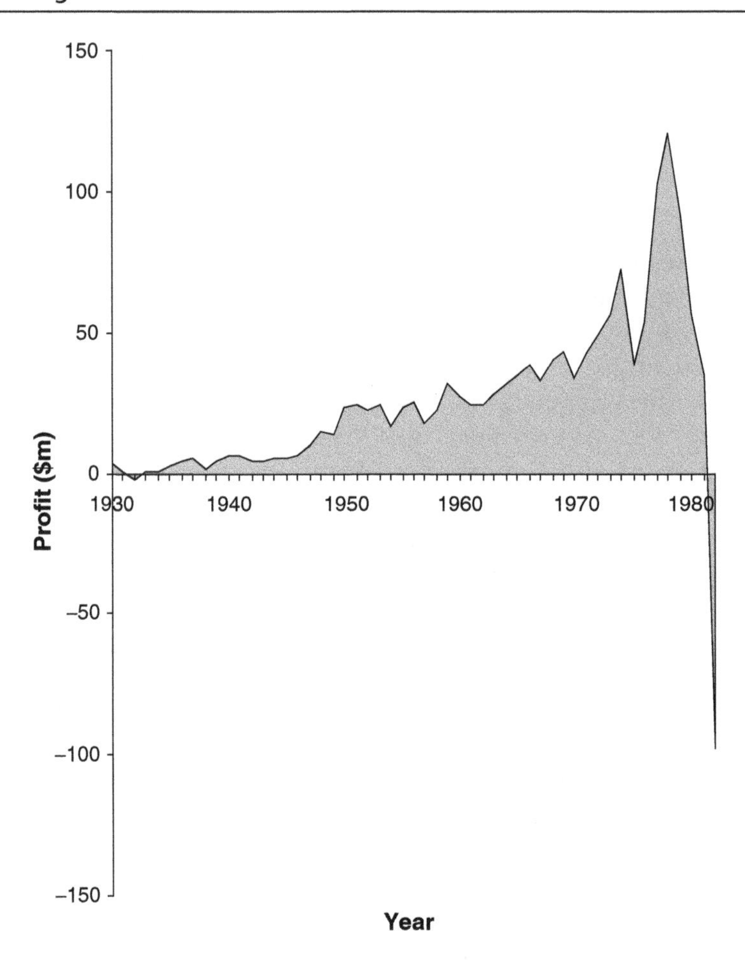

Figure 2.1. Johns-Manville Profits after Tax, 1930–82

£6 million. Over half of that amount came from its southern African mines.[78] In the post-war period, the three major Eternit groups prospered: Eternit-Belgium (Etex), Eternit-Switzerland, and Eternit-Denmark. Everite had a large share in Asbestos Investments Pty Ltd (Asbesco), which mined in both South Africa and Zimbabwe. At its peak in the 1970s, Asbesco owned Kuruman Cape Blue Asbestos Pty Ltd and Danielskuil Cape Blue Asbestos Pty Ltd, which operated mines in the Northern Cape. In 1976, Everite controlled 20 per cent of South African production.[79] Eternit also

[78] T&N Board Papers (1957).

[79] J. Myers, 'Asbestos and Asbestos Related Disease in South Africa', Southern Africa Labour and Development Research Unit, Working Papers No. 28 (June 1980), p. 10.

operated manufacturing plants in Cape Town and Johannesburg. The Belgian and Swiss companies were closely linked until 1989, when in theory at least they became competitors. In 1985, Eternit-Switzerland, owned by Stephan Schmidheiny, was the world's second largest seller of asbestos; its a/c operations in thirty-two countries produced annual sales of $2 billion. In 1997, Etex generated global sales of $2 billion.[80] The whole organism was supported by a web of uncompetitive marketing deals and price agreements that were a hangover from the cartels of the inter-war period.[81]

Far away from the corporate boardrooms in New York and London, workers toiled to produce asbestos fibre. One of the peculiarities of asbestos is that it tends to be mined in beautiful places. The Cassiar mine in British Columbia and the Libby Mine in Montana are set like 'jewels' in the North American mountain wilderness.[82] Chrysotile Arizona was established at the bottom of Ash Creek Canyon, in the shimmering desert area east of Phoenix.[83] The Shabanie mine in Zimbabwe is surrounded by conical hills, which could have been lifted from a Ming dynasty painting. The now deserted Penge mine in South Africa is adjacent to the Kruger National Park and nestles in the midst of steep hills which are cut through by the Oliphants River. The bleached landscape around Kuruman in the Northern Cape is offset by the blue Asbestos Mountains where 'eyes' of subterranean water form striking oases. Havelock in Swaziland lies in the midst of spectacular mountains and forests. The long-abandoned Wittenoom mine in the Pilbara region of Western Australia is framed by intense blue skies, bright red earth, and the hauntingly beautiful ghost gums which line the Wittenoom Gorge. Such physical beauty forms a sharp contrast to the work conditions on the mines and the impact of mining upon the health of local communities.

Asbestos miners faced many of the same problems as other miners, including low rates of pay, dangerous conditions, and intractable management. There are in addition technical factors which make asbestos mining particularly hazardous. Although the methods of extracting asbestos ore have varied from open-cut to depth-mining, the aim of processing fibre is always the same, namely to preserve the mineral's physical properties. For that reason asbestos is milled dry. After arriving at a mill, ore is fed into primary crushers, then crushed and sorted. The ore is then crushed again,

[80] *British Asbestos Newsletter*, 34 (Spring 1999).
[81] Monopolies Commission, *Asbestos and Certain Asbestos Products*.
[82] S. LeBlanc, *Cassiar: A Jewel in the Wilderness* (2003).
[83] G. Knuckey, *Chrysotile Arizona 1914 to 1945* (2007).

and the fibre lifted off by suction. The further into the milling process, the more dust is generated. The most hazardous jobs at Wittenoom, Vermont, Thetford, Penge, Havelock, and Shabanie were in the bagging sections where fibre was pressed, often by hand, into burlap sacks. As recently as the 1980s, between a third and a half of mill workers on Southern Africa mines could expect to contract asbestosis.[84] Despite industry claims to the contrary, conditions in Canada were little better.

Mining is dangerous work and although miners recognize they may have to sacrifice their own health to make a living they do not expect that the health of their families will also be put in jeopardy. In Quebec, as in Swaziland and Zimbabwe, the dry milling of fibre and the huge waste dumps which accumulated near mining communities have caused disease among the wives and children of miners.

In 1878, mining began at Thetford in Canada, but the methods of extraction were primitive and returns depended upon the richness of the ore. The Canadian mines were mostly open-cut and the covering soil, which was up to twenty feet thick, had to be removed in summer; in winter the ground froze. Quarry mining appealed to management as it required no ventilation or timbering and offered easier supervision of the workforce. There were no such advantages for labour and from November to April the men who toiled in the quarries were exposed to snow and ice. Most of the cobbing or hand processing was done on contract by teams of women who were paid between 30 and 35 cents per hundred-weight. The cost of labour was so low that a tonne of fibre could be cobbed for less than $8.[85] The Canadian mines enjoyed the advantage of proximity to the huge US market thereby avoiding the freight costs which hit Southern African and Russian producers. Initially, only long fibres, which accounted for around 2 per cent of Canadian output, were bagged and the remaining 'shorts' were discarded. The discovery that short fibres could be combined with pulp to make asbestos paper, with cement to make shingles, or magnesia to make insulation, revolutionized the market.

In the decades before the First World War, the Canadian industry was plagued by speculation and a series of price collapses led to several bankruptcies.[86] Despite the market's instability asbestos soon became an

[84] See for example Dr P. Elmes, 'Visit to Shabanie and Gaths Mines' March 1987, 14 April 1987; and 'Asbestos Dust Levels in Overseas Companies', Executive Committee Meeting, March 1981.

[85] For a history of that early period, see F. Cirkel, *Asbestos: Its Occurrence, Exploitation and Uses* (1905); and J.G. Ross, *Chrysotile Asbestos in Canada* (1931).

[86] See J. Gerin-Lajoie, 'Financial History of the Asbestos Industry' in P. Trudeau (ed.), *The Asbestos Strike* (1974), pp. 83–104.

important part of Quebec's economy and in the districts of Thetford, Black Lakes, Danville, and East Broughton the mines were the major employer. Most miners came from rural communities and they moved seasonally between asbestos and farming. The Canadian industry was the first to industrialize and by 1905 the mines were using compressed air drills. (On some South African mines hand drilling continued until the 1940s.) During the Second World War, Canadian production rose from 364,000 to 467,000 tonnes: in 1948 it reached 717,000 tonnes.[87] By 1952, the Canadian industry was totally absorbed by US companies such as Johns-Manville and Philip Carey.[88]

The conditions on the Quebec mines were harsh. In January 1949, the Montreal newspaper *Le Devoir* carried a series of articles on the village of East Broughton, a poor community in Canada's poorest province. The author, Burton LeDoux, interviewed miners and their families and described the deplorable conditions under which they lived and worked.[89] The sole source of employment for the village's 3,000 inhabitants was the Quebec Asbestos Corporation, a subsidiary of Philip Carey. The mine lay next to the Corporation's factory, which operated six days a week. The rates of pay were low and the work conditions were hard. The mill was bitterly cold in winter and there was no place for the men to eat away from the dust and noise. To keep the costs of production to a minimum, the Corporation relied upon manual labour and even the cleaning of the mill was done by hand. There was no safety equipment and workers would themselves buy goggles to keep the fibre out of their eyes. Every part of the operation, from mining to processing, to the bagging of asbestos, created clouds of dust. Le Doux found that by the end of each shift the men's clothes were encrusted with grey-green fibre, as were their faces, eyebrows, ears, and hair.[90]

The worst jobs at East Broughton were in the bagging section where *empocheurs* bagged fibre by hand. *Empocheurs* were not Corporation employees but worked for private contractors and were paid piece rates rather than wages. They would fill the burlap sacks, then shake and hand-press the fibre before sewing the sacks closed. *Pileurs* picked up the leaking sacks and piled them in the store room. The system provided the Corporation with cheap labour and sheltered the parent company

[87] Trudeau, *Asbestos Strike*, p. 100. [88] Gerin-Lajoie, 'Financial History'.

[89] B. LeDoux, 'Asbestosis at East Broughton—A Village of Three Thousand Suffocates in Dust', *Le Devoir*, 12 January 1949. See also E. Delisle and P.K. Malouf, *Le Quatuor D'Asbestos: Autour de la Grève de L'Amiante* (2004).

[90] LeDoux, 'Village of Three Thousand'.

Philip Carey from legal action by disabled workers. The men were neither warned of the dangers they faced nor did company physicians tell them when they had contracted asbestosis. The western side of the village was surrounded by tailings from the mill and there was dust in all the houses. LeDoux found whole families afflicted with respiratory illness.

The provincial government in Quebec was aware of the situation, according to LeDoux, but chose to do nothing. In February 1949, miners at Johns-Manville's Asbestos mine began a strike for better pay and conditions. The strike, which soon spread to Thetford, lasted five months and became a turning point in the history of labour relations in Quebec. It did not, however, end the threat of occupational and environmental disease. Johns-Manville's management told its shareholders that the miners wanted excessive wage increases and absurd leave entitlements. It made no mention of their demands for improved work conditions.[91]

During the 1950s there was little improvement on Canadian mines. When T&N's company physician Dr John Knox visited Thetford in December 1964, he found there was less dust on the pavements and roofs of domestic dwellings than on his previous visit twelve years earlier. But he noted that none of the new houses or the schools had been 'positioned with the idea of avoiding dust blown off the encircling (asbestos) dumps'.[92] He was particularly concerned that fibre from the 250 tonnes of waste generated each day at Bell Mine was blown over the district. A decade later T&N's senior physician, Dr Hilton Lewinsohn, found the area heavily polluted. Even the new mill at Lake Asbestos was choked with dust and he lamented that the ventilation system had been designed to enhance output rather than improve occupational safety.[93] There were ten active mining companies in the region, employing about 6,200 workers, at the mining towns of Black Lake, Thetford, and Asbestos. In places, the miners' homes were sandwiched in valleys created by the dumps.

Although there were small mines in Greece and Cyprus most of the asbestos used in the factories of Western Europe came from Canada and southern Africa. A notable exception was the Canari mine in Corsica which began producing chrysotile in 1927. In 1941, Eternit bought the mine and from then until its closure in 1965 it operated under the name of La Société Minière de l'Amiante (SMA), a fully owned Eternit subsidiary. It was an open cast operation, with tunnels blasted into the hillside

[91] Johns-Manville Corporation, *Annual Report* (1949), p. 6.

[92] J.F. Knox, 'Report on Visit to Thetford Mines, Asbestos and Montreal', 17 December 1964.

[93] 'Report from Dr Hilton Lewinsohn on Visit to Atlas Asbestos Company and the Bell Mine, Thetford Mines, Canada, 1973'.

to exploit the deposit. The mill was built on a hill above the sea and tailings were dumped down the cliff and onto the beach. In the post-war period output rose from 12,000 tonnes in 1954 to over 25,000 tonnes in 1962. The workforce also grew from 277 workers in 1952 to 350 by 1965. Although by Canadian standards output was modest, Canari put France among the world's top ten chrysotile producers.

Working conditions were always hazardous and visitors commented that in the mill, visibility was less than six feet. There was no dust extraction system, simply a rusty tube vented to the mill's exterior, and workers described conditions as being akin 'to constant snow'. There were no respirators or showers and no proper place for the men to take meals. In 1963 a journalist who visited Canari described it as 'a truly white hell'.[94] In the early 1960s, local authorities became concerned that the dumping of asbestos waste into the sea was discouraging tourism.[95] Public debate about the mine's environmental impact coincided with labour disputes over wages and working conditions. In 1965 Eternit closed the mine.

The exception to the rule of profitable mines was Wittenoom in Western Australia. The mine, which was operated by the Colonial Sugar Refining Company (CSR) from 1944 until 1966, was the only crocidolite mine outside South Africa.[96] As its name suggests, CSR was a sugar-processing company and its management never came to terms with the demands of an unfamiliar industry. Wittenoom was unable to produce fibre competitively against imports from South Africa and the mine recorded losses for most years of its operation. By 1955, it had accumulated debts of A$1,600,000. At the mine's closure that figure had grown to A$2.5 million. The failure of Wittenoom was due to a number of factors. They included the mine's isolation (it was 1,000 miles north-east of the state capital, Perth), the high cost of transport, the hardness of the host ore which destroyed ducting in a matter of hours, and CSR's lack of expertise. The major factor was the high cost of labour. To attract workers to such an inhospitable site, where summer temperatures were often over 40°C, it was necessary to pay high wages. The wages for unskilled labour were seven times higher than in Perth and many more times higher than those paid to black miners in South Africa. This meant Wittenoom fibre was too expensive for local manufacturers such as James Hardie Industries,

[94] Guy Meria, *L'Aventure Industrielle de L'Amiante en Corse* (2003), pp. 226–7.

[95] Meria, *L'Aventure Industrielle*, p. 240.

[96] For a history of Wittenoom, see J. McCulloch, 'The Mine at Wittenoom: Blue Asbestos, Labour and Occupational Disease', *Labour History* 47 (February 2006), pp. 1–19.

who would not buy CSR's crocidolite.[97] The continued failure to return a profit led the mine's closure in 1966, thirty years before the last of South Africa's crocidolite mines.

African Crocidolite Mines

The other mines which supplied British and US factories with chrysolite were in South Africa, Swaziland, and Zimbabwe. South Africa also produced almost all the world's amphiboles. The first crocidolite mines opened near Prieska in the Northern Cape in 1893 and by the 1920s mines were operating in the Limpopo Province, 200 miles north-east of Johannesburg. The asbestos deposits lay close to the surface and could be extracted by miners using simple tools. Under the so-called tributer system, which lasted until the early 1950s, the primary labour unit was the family rather than an individual male worker. Men blasted and then dug ore by hand from shallow adits and women hand-processed or cobbed the fibre using square-faced hammers. The separated fibre was then sorted by length and bagged. Children worked alongside parents in what was a largely informal system of production. It was common for owners to pay miners with chits or 'good fors' which were only reclaimable at a company store.

The asbestos mines offered one of the few local forms of employment to poverty-stricken communities. The men of the Northern Cape, who worked on the gold mines of Johannesburg, would at best see their families twice a year. Those who turned to farm labour had only seasonal employment which paid even less. Women could work as domestic servants in the white households of Kimberley or Lydenburg for R40 (US$4) a month. But under apartheid they had to leave their children behind. The pay on the asbestos mines was poor, but it did give families an opportunity to stay together.

Tributing offered mining companies many advantages.[98] Miners were paid piece rates and so the costs of 'dead mining' were borne by labour. The informality of the system meant there was only a small permanent staff of whites who checked the fibre grade, thereby further reducing production costs. Most important of all, miners and their families fell outside the provisions of the Mines Acts. As a consequence, employers avoided having to provide rations or medical care. It also enabled Cape Asbestos

[97] See Official Transcript, *Tariff Board Inquiry Re. Asbestos Fibre* (2 June 1954).
[98] See A.L. Hall, *Asbestos in the Union of South Africa* (1930).

and T&N to employ large numbers of women and juveniles, a practice forbidden under the mines legislation. In October 1940, for example, Cape's Penge mines employed 447 juveniles out of a total labour force of 1,625.[99] Young workers pressed fibre into hessian bags, and swept mill benches and floors. Those work practices were profitable for employers and Cape's management readily acknowledged the importance of the mines to its success.[100]

Under the sway of apartheid, British firms did as they liked. The mills were choked with dust and the practice of sacking disabled workers removed the most obvious cases of asbestosis from the sight of regulatory authorities.[101] Although there were no trade unions, there were frequent labour disputes.[102] In January 1945, 350 miners at Westerberg in the Northern Cape went on strike. The miners demanded higher wages, free firewood, and water, and second-hand sacking to build their huts. The men also asked Cape Asbestos to supply free planks for coffins.[103] Cape agreed to provide firewood, planks, and sacking, but rejected the demand for higher pay. Those who refused to return to work were dismissed. Following the strike, a local Methodist minister, the Reverend L.B. Tsangela, wrote in protest to the Department of Native Affairs. He pointed out that the wages were extremely low, but he was more critical of the living conditions: 'The houses into which these people live are deplorably despicable and ought not to be allowed used for human habitation.'[104]

By the early 1950s, the demand for asbestos could not be met by existing methods and Cape Asbestos, Everite, and Gefco invested in deep shafts and wage labour. Between 1950 and 1960, the value of South African output trebled. Production reached almost 200,000 tonnes in 1960, when the mines were employing in excess of 20,000 black and coloured workers. The shift to industrial mining was never simple and some of the old work practices survived. The most important was cobbing and the hand-sorting of ore by women. The host rock in South Africa is so

[99] Letter from E.W. Lowe, Acting Director of Native Labour to Secretary for Native Affairs, Pretoria, 19 October 1940. Department of Native Affairs NTS 2217 408/280.

[100] See Cape Asbestos Company Ltd, *Asbestos* (The magazine), Diamond Jubilee, 7–8 July 1953.

[101] J. McCulloch, *Asbestos Blues* (2002), p. 128.

[102] Memo from G. Mears, Secretary, Native Affairs to Director of Native Labour, Johannesburg, 13 July 1945. NTS 7684 212/332 Strike Westerberg mine: Cape Asbestos Company.

[103] Memo from Lieut-Col Howe, Department, Commissioner South Africa Police, Kimberley to the Commissioner South Africa Police, Pretoria, 21 February 1945. Native Affairs NTS 10011 188/408F, Part 1.

[104] Letter from Rev. L.B. Tsangela to Res. Native Comm. Prieska, 8 October 1945, Native Affairs NTS 10011 188/408F, Part 1.

abrasive it will tear holes in ducting, damage machinery, and fill mills with fibre.[105] Mining companies found it an advantage to mill hand-processed cobs so that little host rock entered the grinders and crushers. That reduced the length of the milling process, thereby preserving the machinery. It also produced a more valuable product as less grit was driven into the fibre.[106] The advantages of hand-processing meant that until the industry's twilight, women comprised up to half of the asbestos mine workers in South Africa.[107] It was common for young women to have their infants beside them as they worked.

The wages were poor and many of the workers had barely enough to eat. In the early 1960s there were regular outbreaks of scurvy and pellagra at Cape Asbestos mines. A large number of workers at Koegas were hospitalized in 1962 and several men and women died.[108] Cape was unwilling to prevent such easily avoidable diseases as it would have reduced profits. It was even less willing to make the investment necessary to reduce dust in the mills. Managements were well aware that to make the mines safe would have required such a massive outlay that asbestos would have lost its comparative advantage over substitutes.

There was a severe drought around Kuruman in 1975 when Veronica Thokomelo began working on the mines. Many of the women from her village of Ga-Mopedi, were employed at Riries, Merencor, or Coretsi. A bus took them to work each morning at 5 a.m. The day ended at 3 p.m. A doctor came once a month to make medical checks and prior to employment there was an X-ray. The men and women were told it was for tuberculosis. They were never told that asbestos is dangerous. At Coretsi there were three conveyor belts with different-sized stones, which the women would sort.[109] Water sprays helped them distinguish between fibre and the host rock and they also reduced the dust. But it was hard work. The heavy rocks had to be removed quickly as the ore passed into the mills. The women had no gloves and the sharp fibres got into their hands, forming asbestos corns. They were not allowed to talk while working and they had only three minutes for a tea break. There was no sick pay. Children also worked at the mines after school and in the holidays.

[105] E. Stander and J.J. La Grange, *Asbestos* (1963), p. 34.

[106] Jock McCulloch interview with Schalk Lubbe at Springbok, 14 December 1999.

[107] See J. McCulloch, 'Women Mining Asbestos in South Africa, 1893–1980', *Journal of Southern African Studies* 29 (2003), pp. 411–30.

[108] J. McCulloch, *Asbestos Blues*, pp. 105–112.

[109] McCulloch interview with Veronica Thokomelo, Olivia Kiet, Grace Tehetsho, and Nameco Toto at Ga-Mopedi Community Centre, Ga-Mopedi, Northern Cape, 16 June 2003.

The women at Riries worked longer hours and so they earned more money. They got R24 (currently US$2.40) per month, while at Coretsi the pay was R19 (US$1.90). The men stayed in compounds and they were paid R50. Over time the pay increased and when Coretsi closed in 1987, the monthly pay for a woman was R60. A family needed about R200. With two incomes from the mines supplemented by working a small parcel of land it was possible for a family to survive.[110] When the mines closed, life became very hard. The women had no income and many families were stricken with ARDs. As a young girl, Grace Tehetsho lived close to Merencor where her father and brothers worked. Grace herself began working at the mine in 1960. Many members of her family including five of her uncles have died prematurely from lung disease. The doctors at the hospital said they had tuberculosis, but Grace thinks it was the asbestos.[111]

The work was always hazardous and the miners also took fibre home on their clothes. William Nakaphala worked at a mine near Mathabatha, in the Limpopo Province during the 1970s. At the end of each shift he would be covered in fibre. As he later recalled: 'If you cannot wash [there were no showers at work] that night you go to bed just like that and the blanket will do the washing. Then you wake up in the morning looking better because all the fibre has been trapped in the blanket.'[112]

Waste from milling was dumped close to villages, and roads were surfaced with tailings. In 2006, fibre could still be seen as bright blue strips lining the roadsides around Kuruman. Women washed fibre-impregnated clothes in the river, which was the main source of drinking water for those living downstream. They also mixed tailings with mud to plaster the walls of their homes and the floors of courtyards. Children used the dumps as play areas. Tailings were preferred to river sand in making house bricks. At the now closed Havelock mine in Swaziland, huge fibre-rich dumps dwarf the primary school, which is less than 200 metres downwind from the now derelict mill. There is much fatalism in the mining villages. The people are poor and they cannot move away. They believe that their families are already affected and so it makes no difference to stay.

South African production peaked at 379,000 tonnes in 1977, but thereafter fell dramatically due to growing public awareness in the US and the

[110] McCulloch interview with Thokomelo et al.

[111] McCulloch interview with Thokomelo et al.

[112] W. Nakaphala, Northern Province, in *Final Proceedings, National Asbestos Summit*, The National Parliamentary Portfolio Committee on Environment & Tourism, 1998, 24–26 November 1998, p. 37.

European Union (EU) of the health risks. The Department of Mines in Pretoria attributed the fall to the activities of Canadian and Soviet interests who were, it believed, seeking to have their chrysotile expand into the markets vacated by amosite and crocidolite.[113] South Africa continued to produce amphiboles for customers in the Middle East and North Africa until 1996, when Merencor, the last of the mines, closed.

Conclusion

The history of asbestos is a history of contrasts. Chrysotile and crocidolite found their way into an array of innovative products and prestigious construction projects. Yet the asbestos insulation in the World Trade Center and the London underground, which was supposed to keep white collar workers and commuters safe, came from the mines of Quebec and southern Africa where men, women, and children worked in clouds of dust. There are parallels between asbestos and tropical products such as coffee, tea, cocoa, and sugar which are so important to the citizens of Western Europe and North America. Coffee and tea, which formed an early part of a globalized economy, are often produced by men and women who work for wages unimaginable to the people who frequent the cafes of London and New York. With asbestos the contrasts have been even greater. The mines paid pitiful wages and the work was so hazardous that it could ruin a man's or woman's lungs in a few months. It can *still* destroy the health of those who live in nearby communities. Perhaps the major contrast is between the vision of safety and economic efficiency used to promote asbestos products and the industry's massive investment in defending itself from litigation and state surveillance.

Seventy years ago George Orwell commented on how coal miners in Britain toiled underground so that civilized people could be warm. The asbestos story is worse in the sense that the labour it requires is so hazardous, the safety it provides to consumers so illusory, and the profits it gives investors so ephemeral. It is also worse because of its toxicity. The industry leaders have always claimed that the asbestos tragedy was caused by a lack of knowledge and the failure of medical researchers to warn corporations and governments of the risk. A careful examination of the medical literature suggests otherwise.

[113] See *Report of Department of Mines for Year Ending 31 December 1978* (Pretoria, 1979), p. 7.

3

Building the Defences

You may recall that we have written you on several occasions concerning the publishing of information, or discussion of, asbestosis.... Always you have requested that for certain obvious reasons we publish nothing, and, naturally your wishes have been respected.

A.S. Rossiter, editor *Asbestos* magazine
to Sumner Simpson, president Raybestos-Manhattan
25 September 1935

The finding [of cancer in mice] looks like dynamite.

Vandiver Brown, Johns-Manville corporate attorney
to Page Woodward, J-M executive, 21 March 1947

In February 1929, Pauline Lasin, who lived at Manville, New Jersey, became a widow. She was 37 and the mother of six small children. Pauline's husband, John, who for a decade had worked for Johns-Manville, had died of asbestosis. The family was Ukrainian and Pauline Lasin was left with no way of supporting her children. Soon after her husband's death, she was approached by a lawyer who encouraged her to file a claim against Johns-Manville for $40,000. During the preliminary hearings Johns-Manville mounted its standard defence.[1] It did not deny that asbestos had caused John Lasin's death, but denied culpability. The company argued that his illness was due to the risks of employment and those risks were part of the terms of the contract made by John Lasin. He knew or should have known that asbestos is dangerous. It also claimed that Lasin contributed to his own death by his negligence. He had failed to take reasonable care by wearing a face mask. Like so many plaintiffs, Pauline Lasin was persuaded by Johns-Manville's lawyers to drop the case.

[1] *Pauline Lasin plaintiff vs. Johns Manville Corporation* US District Court District of New Jersey, November 1929.

Subsequently, she may have received some charitable payment from the company.

The Lasin case was just one of tens of thousands filed in US courts from the late 1920s. All dusty industries including asbestos manufacture and those glass, steel, and iron foundries which used sand faced the same problems of lung disease and litigation. A turning point in public awareness of the dangers of silica came with the Hawk's Nest incident. Hawk's Nest was a West Virginian hydroelectric project undertaken from 1927 by Union Carbide. Three thousand workers, many of them black migrants from the South, toiled in clouds of almost pure silica dust. Hundreds of men died from silicosis, creating so much bad publicity for Union Carbide that their deaths led eventually to a Congressional inquiry.[2] According to Dr Anthony Lanza, the assistant medical director of the Metropolitan Life Insurance Company (MetLife): 'Silicosis and asbestosis burst upon the amazed consciousness of American Industry during the period 1929–1930. Arising out of the period of economic depression, the situation with respect to silicosis and asbestosis became manifest as a medico-legal phenomenon of a scope and intensity that was at once preposterous and almost unbelievable.'[3] In New York State alone in 1933 over $30 million in silicosis lawsuits were filed against the foundry industry.[4] The hazard posed by free silica dust and asbestos fibres was very different to the heavy machinery or hot piping which threatened factory workers with a range of traumatic injuries. The effects of dust are insidious and the damage to health often takes years to become visible. The longer the period between injury and impairment, the less incentive there is for an employer to improve work safety.

The crisis of the 1930s was the first of three crises faced by the asbestos industry. The second was lung cancer in 1940 and the third and deepest crisis began with Dr J.C. Wagner's discovery of the link between asbestos and mesothelioma in 1959. In each instance, the industry leaders developed techniques which enabled them to calm public fears, frustrate legitimate claims for compensation, and placate regulatory authorities. Those techniques ranged from suppressing evidence of risk, intimidating medical researchers, and generating favourable publicity. The industry has always emphasized the slowness of medical discovery pointing out that evidence about disease took decades to emerge. While that contains some truth, it was the industry's interference in the

[2] M. Cherniack, *The Hawk's Nest Incident* (1986).

[3] A. Lanza (ed.), *Silicosis and Asbestosis* (1938), pp. 64, 405–6.

[4] D. Rosner and G. Markowitz, *Deadly Dust* (2006), p. 81

research process rather than the complexities of the asbestos diseases which slowed down discovery. A number of key studies which identified high levels of asbestosis among miners, for example, never reached the public domain thereby delaying for decades the recognition of risk.

The boundaries to knowledge of asbestos disease were geographical and political. States are limited spatially, but the larger companies were vertically integrated and so they knew what was happening on mines in Quebec and Southern Africa. They also knew about disease among factory workers in Western Europe and North America. There were further boundaries between management and regulatory authorities which found expression in two bodies of medical evidence. On the one side was the literature in the public domain often published in scientific journals and sometimes commissioned by state authorities; on the other side was the industry-funded research which usually remained secret. Over time the gap between those two literatures grew wider and wider. The industry also exerted influence through the work of experts who mediated the public perception of risk. In the US the most prominent of those experts was Dr Anthony Lanza (1884–1964) who for more than thirty years dominated debates in the US about pneumoconiosis.

The emergence of knowledge about asbestosis should have been relatively simple. It is an occupational disease largely confined to mines and factories and which has a single, obvious cause. The asbestos mines were full of dust and the work was arduous, which made them the ideal place to carry out research. Yet with notable exceptions the early literature focused upon factory workers in Western Europe and North America rather than the men, women, and children who toiled on Canadian and southern African mines. The US parent companies always claimed there was little, if any, asbestosis in Canada. Disease in the textiles industry was due, they argued, to work conditions in particular factories, rather than the inherent dangers of the fibre. That strategy moved discussion away from the material itself to the specific contexts in which it was used. The industry also drew a sharp distinction between asbestosis and other forms of pneumoconiosis. Silicosis and asbestosis, which are both life-threatening diseases, differ in one important respect. Men or women with silicosis are highly likely to contract pulmonary tuberculosis. Those with asbestosis are not. Johns-Manville and its competitors used that synergy to play down the dangers of asbestos. Asbestosis is such a mild ailment, they argued, that a worker will suffer virtually no impairment.

Over the past twenty years, litigation in US courts has spilled vast quantities of documents into the public domain. That material includes unpublished medical studies and correspondence between senior management, medical researchers, and lawyers. Raybestos-Manhattan was the major US manufacturer of brake linings. The company was founded in 1929, and from that date until 1948 Sumner Simpson was its president. After his death in 1953 his secretary gathered together his papers. They consisted of letters and memos between himself and Vandiver Brown, legal counsel for Johns-Manville. There is also correspondence with scientific bodies including the Mellon Institute of Industrial Research and the Saranac Laboratory, and there are unpublished medical studies by the Industrial Hygiene Foundation of America. There are letters from A.S. Rossiter, the editor of the trade journal *Asbestos*, and George Hobart, senior partner with the New Jersey law firm which acted for Johns-Manville. As we shall see, the Sumner Simpson papers show that from the early 1930s Raybestos-Manhattan and Johns-Manville monitored the medical literature and circulated copies of the latest research to senior management. The correspondence also reveals a determination to prevent knowledge of risk becoming public. That in turn involved controlling medical discovery.

The Saranac Laboratory in upstate New York was one of the premier institutes researching occupational disease. Saranac had been founded in the 1880s by Dr Edward Trudeau to treat tuberculosis, but soon branched out into the study of associated lung diseases.[5] Under the leadership of Dr Leroy Gardner and later Dr Arthur Vorwald, it conducted some of the most important research into silicosis and asbestosis, invariably with industry funding. MetLife along with the asbestos conglomerates was among Saranac's sponsors, and during the 1930s it commissioned a number of studies. Although Gardner and his successors never testified in court on behalf of industry, the Laboratory's financial dependence on external funding did influence research outcomes.

In November 1936, Raybestos-Manhattan and Johns-Manville funded experiments into asbestosis at Saranac. Having made that decision, Vandiver Brown wrote to the then director Dr Gardner:

> It is our further understanding that the results obtained will be considered the property of those who are advancing the required funds, who will determine whether, to what extent, and in what manner they shall be made public. In the

[5] L. Blinn, 'The History of the Saranac Laboratory 1885–1958', unpublished typescript (1959).

event it is deemed desirable that the results be made public, the manuscript of your study will be submitted to us for approval prior to publication.[6]

Under that code favourable results were used to defend existing work conditions. When litigation threatened, those same results became evidence that employers could not have known of the dangers. If the results were unfavourable they were suppressed. From the early 1930s, while the industry leaders such as Johns-Manville and T&N invested in research, they also required company doctors to keep abreast of the latest data and to attend scientific conferences.[7] At a general meeting of the Asbestos Textile Institute at Thetford in June 1965 Karl Lindell, chairman of Canadian Johns-Manville, noted with some pride: 'The knowledge possessed by the industry's doctors on the biological effects of chrysotile asbestos was not surpassed elsewhere in the world.'[8] Lindell's observation had been true for more than thirty years. Johns-Manville, Raybestos Manhattan, Eternit, and T&N knew everything about asbestos.

Asbestosis: Hiding the Evidence

Until the end of the nineteenth century, the industry was so small that individual physicians rarely saw a sufficient number of asbestos workers to be alerted to the dangers of airborne fibre. Consequently, the scattered citations in the medical literature from that period are more significant when read with hindsight.[9] A breakthrough occurred in Britain in 1906 with the observations of Montague Murray, who in evidence before a commission on industrial disease described the case of a man who had worked with asbestos for ten years. On autopsy, the man was found to have extensive fibrosis of the lungs.[10] A survey of British industry in 1910, which included a number of asbestos factories, failed to discover any specific danger, and it was not until 1924 that a second death from asbestosis was recorded.[11] Over the next six years, the

[6] Letter from Vandiver Brown, Johns-Manville, to Dr L. Gardner, Saranac Laboratory, New York, 20 November 1936. See also B.I. Castleman, *Asbestos: Medical and Legal Aspects* (2005), p. 52.

[7] Minutes General Meeting Asbestos Textile Institute (ATI), 4 June 1965.

[8] Minutes General Meeting ATI, 4 June 1965.

[9] For a history of the development of knowledge about asbestosis, see Castleman, *Asbestos*, pp. 1–40, 191–4, 601–4.

[10] See M. Greenberg, 'Classical Syndromes in Occupational Medicine: The Montague Murray Case', *AJIM* 3 (1982), pp. 351–6.

[11] W.E. Cooke, 'Fibrosis of the Lungs Due to the Inhalation of Asbestos Dust', *BMJ* (26 July 1924), ii, p. 147.

British Medical Journal published several articles on asbestos and by the late 1920s evidence of asbestosis was becoming obvious in a number of British asbestos textile factories.

The most detailed of the early English studies was completed in 1927. It was the work of Ian Grieve, a local doctor in Armley, Leeds, whose practice brought him into contact with workers from J.W. Roberts's asbestos-textile plant. The factory, which employed about 100 men and women, used crocidolite to manufacture asbestos mattresses for locomotives. Grieve's medical thesis, 'Asbestosis', provided harrowing details of occupational disease.[12] The production techniques, such as hand-beating mattresses to remove lumps, meant that everything in the small factory was covered in blue dust and workers could develop asbestosis within five years. In a series of case studies, using X-rays to confirm his diagnoses, Grieve found that the disease was debilitating, progressive, and almost inevitable. A working life could last as little as fifteen years, with some workers dead at 37. Roberts's response was to install dust extractor fans, but the factory remained one of the worst asbestosis black spots in the UK.

Grieve paid tribute to the 'courtesy and enthusiasm' of the Roberts' family and he seems to have been given a free hand in examining his patients. Grieve's account shows that the factory only functioned with the industrial lubricant of higher than average wages and acts of paternalism. Respirators were also still needed; indeed, Grieve felt that they should be worn continuously and was critical of workers, especially 'empty-headed' girls, who failed to comply. Grieve's thinking on occupational illnesses was conventional. He felt that the complacency of workers was largely responsible for what he termed the 'indifference of medicine' towards asbestosis. He believed that asbestosis should be made a compensable disease, but he never questioned the production of such a lethal material, or the cramped factory layout and its proximity to workers' housing.

While Grieve was conducting his research Roberts' factory was bought by T&N. That company took no notice of Grieve's work, and the Roberts' factory remained one of the most dangerous in the UK. It was not until the 1990s that T&N sought out a copy of Grieve's thesis, because of its relevance in litigation.

The rising concern about the British industry led eventually to a government enquiry in 1928 headed by E.R.A. Merewether and C.W. Price. Merewether examined 363 workers, the majority of whom were

[12] I.M.D. Grieve, 'Asbestosis' MD thesis (September 1927). See also G. Tweedale, 'Management Strategies for Health: J.W. Roberts and the Armley Asbestos Tragedy, 1920–58', *Journal of Industrial History* 2 (1999), pp. 72–95, and Grieve Obituary, *BMJ* (24 February 1973), p. 495.

employed in textile factories and had more than five years' exposure. Full medical examinations supplemented by work histories showed that a quarter of employees had fibrosis, and that the incidence and severity of the disease increased with the length and degree of exposure. After ten years employment, industry workers had a 50 per cent chance of contracting the disease.[13] Asbestosis destroys the body's reserve capacity to resist other infections, which can make even moderate disability life-threatening.[14] Merewether and Price concluded that asbestosis can be fatal, it has an insidious onset, and can occur with less exposure than silicosis. In that sense, it is a more serious hazard. In a follow-up article published in 1933 Merewether noted that of those who work in dusty conditions, 'continually from leaving school, few will survive ten year's exposure and none will reach the age of 30'.[15] The best means for prevention are dust suppression, regular medical checks, worker education, and as a last resort the use of respirators.[16] Since 80 per cent of the asbestos imported into the UK came from Canada, any major variation in the incidence or nature of the disease between UK and US factories was unlikely. As a result of the Merewether and Price study, from 1931 the British industry was subject to a regulatory framework of dust control, medical surveillance, and compensation. Germany passed parallel legislation for Aryan workers in 1936, and Italy, in 1943. In the US there was no national dust legislation until 1970.

Merewether and Price cited a number of reasons why it had taken physicians so long to identify the asbestos hazard. The industry was relatively new, and the clinical resemblance between asbestosis and tuberculosis had obscured many cases. So too had the composition of the workforce.[17] In the US, the UK, France, and Italy, around a quarter of employees were women who prior to marriage would spend brief periods in asbestos textile factories.[18] Such employment was sufficient to produce asbestosis, but not sufficient for the symptoms to become visible. The labour process was another factor hiding the disease rate. The asbestos-textile industry required less arduous work than did the foundries, mines, and construction sites where silicosis was so common. The relatively light

[13] E.R.A. Merewether and C.W. Price, *Report on the Effects of Asbestos Dust on the Lungs and Dust Suppression in the Asbestos Industry* (1930), p. 10.

[14] E.R.A. Merewether 'A Memorandum on Asbestosis', *Tubercle* 15 (1933–4), pp. 69–81, 109–18, 152–9.

[15] Merewether, 'Memorandum on Asbestosis', p. 152.

[16] Merewether and Price, *Report on Effects of Asbestos*, p. 17.

[17] Merewether and Price, *Report on Effects of Asbestos Dust*, p. 17.

[18] Lanza, *Silicosis and Asbestosis*, p. 419.

work in asbestos-textile plants could often be done by a man or woman with diseased lungs.[19]

In their report, Merewether and Price established two important principles. They found that a careful physical examination offered a better method of diagnosis than did X-rays alone. Their other insight concerned the disease's insidious nature. The lung has a reserve capacity, which fibrosis or scarring gradually destroys. A worker will feel healthy until that reserve capacity is erased. From that point, the disease would progress rapidly and death would often result from secondary ailments, such as bronchitis or heart disease.

In June 1930, abstracts of Merewether and Price's study appeared in the *Journal of the American Medical Association,* a publication which reached an estimated 80 per cent of US physicians.[20] Two months later the ILO's silicosis conference was held in Johannesburg. It attracted delegates from Canada, the UK, Australia, Italy, the Netherlands, and Germany. Dr Leroy Gardner from Saranac and Dr Albert Russell from the US Public Health Service also attended. While the conference focused on the prevention and compensation of silicosis, Dr Gardner gave a presentation on asbestos dust.[21] So too did three Italian physicians.

In 1933, Philip Ellman, a physician employed at a tuberculosis clinic in London, published a study based primarily on asbestos textile workers. Ellman noted that asbestosis developed so slowly that symptoms often appeared years after a worker had left the industry. He listed the symptoms as dyspnoea (shortness of breath), a dry cough, anorexia, cyanosis (a blueness of the skin), emaciation, finger clubbing, and asbestos corns. In fatal cases the onset is more rapid than silicosis and dyspnoea is usually disproportionate to the clinical evidence from X-rays.[22] Ellman agreed with Merewether about the disease's severity. 'The average length of employment in fatal cases,' he wrote, 'is only one half that of silicosis.'[23] Throughout the 1930s, the European literature continued to emphasize the seriousness of asbestosis.

During the 1920s, the US industry was still in its infancy, and there were none of the dramatic pressures on production which were to come

[19] Merewether 'Memorandum on Asbestosis', p. 113.

[20] Castleman, *Asbestos*, p. 14.

[21] L. Gardner, 'General Technique of Dust Exposure', in *Silicosis: Records of the International Conference held at Johannesburg 13–27th August 1930* (1930), pp. 53–4.

[22] P. Ellman, 'Pulmonary Asbestosis: Its Clinical, Radiological and Pathological Features, and Associated Risk of Tuberculous Infection', *Journal of Industrial Hygiene* 15 (1933), pp. 165–83.

[23] Ellman, 'Pulmonary Asbestosis', p. 182.

with the Second World War. Even so, US factories were full of dust and conditions were particularly bad in asbestos-textile mills. In 1929, the industry faced both the Great Depression and a flood of litigation. Most accident insurance for US workers was held by private companies such as MetLife. To protect themselves, employers and insurers fought to have silicosis placed under state workers' compensation schemes. These were no-fault systems in which claims usually had to be made within one or two years of leaving employment. That in turn greatly reduced the number of successful claims. Some states used the principle of 'limited and graduated liability', which meant that the law only applied to those working in the industry at the time legislation was passed. Rosner and Markowitz have shown how, through a series of complex manoeuvres, the US industry was able to erase silicosis from public discourse.[24] In particular, it removed the controversy from the political arena and into the hands of experts where the crisis could be mediated to its advantage. The concept of Threshold Limit Values (TLVs) for dust was one of the keys to that achievement (see below).[25]

In early 1929, the industry leaders approached MetLife, America's largest insurer, to conduct a hygiene survey. Johns-Manville and Raybestos were aware of the UK research and wanted to know if asbestos was an occupational hazard in US factories. To answer that question, between 1929 and 1931 MetLife conducted a survey of five textile mills on the Atlantic seaboard. It involved an assessment of dust levels, a physical examination of workers including X-rays of 126 employees, and a study of dust-exhaust systems. The project was the first of its kind and it could well have been used to build upon the British research. Unfortunately, the results were never published. Among those who assisted with the project was Dr Leroy Gardner from Saranac, who was part-funded by MetLife. The study began with a description of the labour process, with each factory identified by a letter rather than by name. Asbestos arrived at the factories in burlap sacks, which were emptied onto the floor, shovelled into pug mills, and processed. The semi-processed fibre was then moved into skips from one machine to another, each time by hand. After processing, the asbestos was mixed with cotton, again by hand.[26] Apart from Plant E, MetLife found the dust levels were not excessive except in the preparation

[24] Rosner and Markowitz, *Deadly Dust*, pp. 75–104.

[25] G. Markowitz and D. Rosner, 'The Limits of Thresholds: Silica and the Politics of Science', *American Journal of Public Health* 85 (1995), pp. 253–62.

[26] Metropolitan Life Insurance Company (MetLife), 'Effects of the Inhalation of Asbestos Dust Upon the Lungs of Asbestos Workers' (New York: Industrial Health Service, Policyholders Service Bureau, 1931), pp. 9–10.

rooms where readings up to 82 million particles per cubic foot (mppcf) of air were recorded.

In contrast to Merewether and Price, the MetLife researchers relied upon X-rays for classifying employees between positive (asbestosis), doubtful (may have the disease), and negative (free of disease). The physical examinations were cursory and featured a chest expansion test and some general questions about health. The MetLife researchers noted that in its early stages the symptoms of asbestosis are usually inconclusive, yet they were confident of making a reliable diagnosis. Of those with less than five years' employment twenty-eight were positive or doubtful and of the total of 108 men in the cohort only seventeen were free of disease.[27] There were eighteen women of whom only three were negative.[28] A total of sixty-four employees complained of shortness of breath while sixty-nine complained of a persistent cough, both classic symptoms of asbestosis. The authors commented: 'Several of those classed as negative (on the X-rays) stated that they were "short-winded" and were so recorded, but too much emphasis should not be placed on these statements of subjective symptoms.'[29] An unusual enlargement of the heart, a feature consistent with severe pulmonary disability, was found in forty-eight subjects and is cited as the survey's most notable finding: 'It is probably that this is a compensatory enlargement due to the additional work put upon the heart in efforts to pump blood through the fibrosed lungs.'[30] The subjects in the MetLife report faced the same grim future as the asbestos textile workers studied by Grieve, Merewether, and Price.

Those parts of the survey devoted to dust suppression gave priority to economic rather than medical factors. In Plant C, for example, experimental dust-extraction equipment had reduced exposure levels by half but the authors were unimpressed: 'It is neither practicable nor economically desirable to install such equipment as will make the air entirely free of dust, even if this were possible.'[31] Air controls disturbed the carding and weaving operations which in turn increased production costs.

MetLife found that the injury caused by asbestos is of a milder type than silicosis, a conclusion which ran against its own findings that asbestosis is associated with cardiac enlargement and breathlessness. Equally disturbing were the low dusts levels in the sample, which led MetLife to conclude: 'It is possible that asbestos may cause pneumoconiosis more readily than free silica.'[32] The study presents few details about the workforce. It

[27] MetLife, 'Effects of Inhalation', p. 14.
[28] MetLife, 'Effects of Inhalation', p. 16.
[29] MetLife, 'Effects of Inhalation', p. 14.
[30] MetLife, 'Effects of Inhalation', p. 17.
[31] MetLife, 'Effects of Inhalation', p. 11.
[32] MetLife, 'Effects of Inhalation', p. 20.

appears, however, that as in the Merewether and Price sample there was a high turnover with few workers being above the age of 50. Of the 106 who were positive or doubtful, 85 had less than ten years' employment. MetLife recommended that the industry address the dust problem, new employees be physically examined, employees X-rayed annually, and the industry sponsor studies of the effects of asbestosis on the heart.[33] The survey found that it would be difficult and costly to reduce dust to a point at which there would be no further disease. That insight in part explains MetLife's resistance to the safety innovations in Plant C. Such innovations were expensive and the data suggested that in any case they would not prevent disease. The Merewether and Price study was released in the UK during the study period, but is not cited in the body of the report. That omission is surprising since the British literature is mentioned as having inspired the survey.

In 1930, MetLife also surveyed the Manheim plant operated by Raybestos-Manhattan in Pennsylvania. That survey differed from the other studies in that it involved a more thorough physical examination of the workforce. The researchers found high dust levels in the carding, spinning, and broad loom operations with counts up to 43 mppcf.[34] MetLife examined forty-five employees all of whom had at least three years' service. Of the twenty who were X-rayed, ten had asbestosis. That included one man with just five years' exposure in the carding room.[35] The report concluded that while asbestosis was common, the disease was mild. Once again no evidence was presented to justify that conclusion.

In combination with the UK research, the MetLife studies presented the industry with a choice: either to make workplaces safer and thereby reduce the disease burden, or to close the plants. It did neither, but instead employed a series of legal and technical initiatives to defuse what had become a political problem. The global industry began to establish a number of professional and semi-autonomous associations to defend work conditions in mines and factories, conduct medical research, and assuage public fears. As shown in Table 4.1, those associations would eventually include the Asbestos Information Association of North America (AIA/NA), the Asbestos Textile Institute (ATI), the Industrial Hygiene Foundation (IHF), the South Pacific Asbestos Association (SPAA), and the Asbestosis

[33] MetLife, 'Effects of Inhalation', p. 20.

[34] MetLife, 'A Study of Dust Conditions in the Manheim Plant of the United States Asbestos & Rubber Division of the Raybestos-Manhattan Inc' (New York: Industrial Health Service, 1930).

[35] MetLife, 'Study of Dust Conditions', p. 15.

Research Council (ARC). The trade association of the Quebec industry was particularly powerful, because by the 1950s its members accounted for around 65 per cent of the world's chrysotile production.[36] A Quebec Asbestos Producers' Association (QAPA) was formed in 1931, to be succeeded by the Quebec Asbestos Mining Association (QAMA) in 1948, and eventually by an Asbestos Institute in 1984. Individual companies also participated in groupings of allied industries, such as thermal insulation, asbestos brake linings, asbestos cement, asbestos paper, and floor coverings. Overall, a hundred such associations were created.

The IHF was one of the most important associations in the 1930s shaping scientific and public debate about occupational disease. In response to continuing litigation a group of executives from the foundry, iron, steel, and asbestos industries attended a symposium on dust at the University Club in Pittsburgh in January 1935. Vandiver Brown was there and made a record of the delegates' concerns.[37] He noted that all agreed on the need to counter the menace of ambulance-chasing lawyers and unscrupulous doctors, arrive at a diagnosis of dust disease which would preclude unjustifiable claims, make various diseases compensable under workers' compensation laws, and resolve the problem of dust control and establish standards for measuring risk. The delegates agreed to set up an organization to present their case to state authorities and the public. Initially titled the Air Hygiene Foundation (AHF) and established in association with the Mellon Institute in Pittsburgh, it had close contacts with Saranac and with the Universities of Pennsylvania and Harvard.[38] Anthony Lanza was involved in the formation of the AHF, which initially attracted twenty members: two years later that had grown to 168. By 1940, the AHF's 225 members included Johns-Manville, US Steel, and Union Carbide. In 1941 it changed its name to the Industrial Hygiene Foundation and later still to the Industrial Health Foundation. The role of the IHF included providing consultancy services for members, holding training courses, and conducting surveys of workplaces. The Foundation had a reference library and from 1937 its staff prepared abstracts for members from a list of seventy medical and technical journals. One of its first achievements was in promoting the concept of TLVs.

[36] Minutes of ATI Air Hygiene Committee Meeting, 7 September 1955.

[37] Vandiver Brown Memo re: Mellon Institute of Industrial Research Symposium on Dust Problems–Pittsburgh, 15 January 1935.

[38] The Mellon Institute of Industrial Research in Pittsburgh housed a number of laboratories. It was one of the first organizations wedding scientific research, university education, and commercial technology. See Rosner and Markowitz, *Deadly Dust*, pp. 105–7.

In 1927 the American Chemical Society had published a list of noxious gases with exposure levels below which no acute effects were supposed to occur. These TLVs were based on the assumption that if exposure to toxins could be kept below a certain level then workplaces would be safe. The idea of a TLV was particularly attractive to industry. Specific hazards could be negotiated between experts drawn from industry and government agencies, thereby shifting debate about risk away from the press and trade unions. Labour had no scientific expertise and therefore no role to play in such discussions. In the case of asbestos and silica, the favoured instrument for measuring dust was the midget impinger. This device drew a sample of air through a liquid medium to trap particles, which were then examined microscopically. The instrumentation was simple to operate and remained in use in America for forty years. In less than a decade, TLVs became the method for setting workplace standards in the US.[39] In Western Europe the emphasis was upon general engineering controls, plant design, and work practices (and as a last resort the payment of compensation) rather than compliance with an abstraction like TLVs. Merewether and Price's study did not refer to dust measurements and neither did the UK regulations of 1931.

At the National Silicosis Conference held in Washington in April 1936, the chairman of the AHF Alfred Hirth endorsed the setting of a dust standard for free silica. Dr Lanza suggested 5 mmpcf as a threshold. The foundry industry soon agreed, as 5 mmpcf could be achieved with existing technologies.[40] Johns-Manville management was keen to adopt the same threshold for asbestos dust, since that level had already been achieved in many plants.[41] Its proposal appeared generous given the industry's claims that asbestosis was a less serious disease than silicosis.

The TLV for asbestos of 5 mppcf was adopted by the Public Health Service in 1938 and for the next thirty years it was used as a voluntary code. The American Conference of Governmental Industrial Hygienists (ACGIH), an informal club of city, state, and federal health officers (which actively solicited industry participation), adopted the standard in 1946 as part of an annual list of TLVs for toxic vapours and dusts. Whether or not the 5 mppcf threshold is seen as a victory or defeat for asbestos manufacturers depends upon one's perspective. TLVs were more ambiguous than industry or regulatory authorities were usually willing to admit. Thresholds were engineering rather than medical artefacts, and 'good

[39] On the development of the US standards, see Castleman, *Asbestos*, pp. 215–35.

[40] Markowitz and Rosner, 'The Limits of Thresholds', pp. 253–62.

[41] Castleman, *Asbestos*, p. 235.

practice' was shorthand for what was commercially acceptable. Professor Philip Drinker, a Harvard engineer, spoke on the question of risk at the Second Saranac Symposium on Silicosis in June 1935. Drinker commented that the key issue was not whether a standard had been shown to be 'hygienically perfect, but [whether] it was attainable and represented what all then agreed would be "good practice"'.[42] For those managers who had read Merewether and the MetLife studies it was a resounding victory.

The IHF helped the industry to weather its first crisis, but there was little if any improvement in conditions for factory workers. One of the few US studies which entered the public domain during that decade was completed in 1938. Dr Waldemar C. Dreessen of the Department of Public Health conducted a survey of three asbestos-textile factories in North Carolina. Its purpose was to study the effects of long-term exposure, identify the processes that created dust, and hopefully identify a level below which disease would not occur.

Dreessen found that in the carding rooms the readings were all above 73 mppcf and he noted that British factories used more dust-control equipment than their US counterparts. The conditions were so bad that after a minimal period 8 per cent of employees had asbestosis. Most of those complaining of breathlessness or chest pain also had a persistent cough.[43] Three quarters of that small group with more than fifteen years, employment had asbestosis. Of the sixty-nine former employees who were traced and examined, forty-three were diseased.[44] Dreessen concluded that in comparison with anthracosis (black lung) and silicosis, the dust levels required to produce asbestosis were very low. Although his findings were damning, Dreessen underestimated the risk. As two of his colleagues from the Public Health Service later wrote: 'The incidence of rank disease [in the textile factories] was so great that prior to the beginning of the study out of a total of less than 600 employees the plants discharged 150 workers suspected of having asbestosis.'[45] In that way the industry wiped out much of the data that Dreessen was seeking to collect.

[42] P. Drinker, 'Engineering Methods in the Control of Silicosis', in B. Kuechle (ed.), *Second Symposium on Silicosis . . . at Saranac Lake, 3–7 June 1935* (1935), pp. 176–81.

[43] W.C. Dreessen et al., A *Study of Asbestosis in the Asbestos Textile Industry. Public Health Bulletin No 241, August 1938* (1938), p. 51.

[44] Dreessen et al., Study *of Asbestosis*, p. 68.

[45] J.R. Lynch and H.E. Ayer, 'Measurement of Asbestos Exposure', *Journal of Occupational Medicine* 10 (1968), pp. 21–4.

The Mines: A Parallel Universe

The mines where T&N and Johns-Manville made a large share of their profits were even more hazardous than their textile factories. In the absence of effective trade unions and little, if any, state regulation, employers could do as they liked in company towns like Thetford (Quebec), Penge (North Eastern Transvaal), or Shabanie (Southern Rhodesia). The absence of state regulation was reflected in the paucity of published research. For the public and the regulatory authorities there was a single reference to a South African study of three miners in Merewether and Price, and there was also a body of company science emphasizing the absence of disease in Canada. After that came silence.

The most important study of asbestosis in Southern Africa was completed in 1930 by Dr George Slade, a physician employed at a chrysotile mine in the Eastern Transvaal. The New Amianthus mine was operated by T&N and Slade, who had recently been appointed resident physician, was concerned about dust in the mill. Management agreed that Slade could conduct a survey of occupational disease as the basis for an MD thesis. He enrolled at the University of the Witwatersrand in Johannesburg in 1928 and began his research at about the same time as Merewether and Price.[46] The New Amianthus mine was isolated, and because he had no X-ray equipment Slade relied upon a physical examination of the workforce. He examined the men's chests and skin tone and he asked them questions about their health and work practices. Those examinations revealed that 70 per cent of employees had shortness of breath and almost all suffered from weight loss. Slade concluded that over half the mill and underground workers had asbestosis.[47] By matching work histories against medical profiles, he was able to establish an association between exposure levels and disability. Working independently of Merewether and Price, Slade reached much the same conclusions about the seriousness of the disease and the best methods for diagnosis and prevention. The results at New Amianthus were even worse than those of MetLife, and suggested that T&N should either have closed the mine or invested in dust-suppression equipment. It did neither. Instead, it fired George Slade.

[46] See E.R.A. Merewether and C.W. Price, *Report on the Effects of Asbestos Dust on the Lungs and Dust Suppression in the Asbestos Industry* (1930).

[47] G.F. Slade, 'The Incidence of Respiratory Disability in Workers Employed in Asbestos Mining with Special Reference to the Type of Disability Caused by the Inhalation of Asbestos Dust' (University of Witwatersrand MD thesis, 1930).

Having completed the first South African thesis on asbestos disease and the first MD at the University of the Witwatersrand, Slade left the industry. He never published his results. Slade spent the rest of his career as a general practitioner in Johannesburg before retiring with his wife to Jersey, in the Channel Islands. He died in 1978. The South African Medical Register has no record of Slade having practised medicine and the vast T&N archive contains no reference to Slade having worked for the company.[48] The copy of his thesis held at the University library was destroyed in a fire in the early 1930s. The only remaining copy eventually found its way into the hands of J.C. Wagner in 1954. Wagner cites the work in his own thesis and from there it entered the literature, thirty years after it was completed.[49] Slade's thesis remains the only substantial study of asbestosis among South African chrysotile miners.

Canada was the world's leading producer of white asbestos. It had a more sophisticated economy than South Africa and in theory at least Canadian workers were protected at law from injury in the workplace. In 1930 Dr Frank Pedley, then director of the Department of Industrial Hygiene at McGill University in Montreal, published two articles on asbestosis. MetLife, which held the insurance for Canadian Johns-Manville, had set up the department three years earlier. The scientists may have regarded the funding as a route to independent research. If they did, it was not a view shared by MetLife, who spoke of securing a 'mortgage on McGill' for certain services that would render a 'tangible quid pro quo'.[50] Writing in the *Canadian Medical Association Journal*, Pedley claimed that mining and milling in Canada might not present the same risk associated with manufacture in the UK.[51] In his second article, Pedley remarked that there had been no cases of asbestosis in Canada, and that there was no evidence of a hazard.[52]

In May 1930 one of Pedley's colleagues Dr J.H. Stevenson, gave a paper on occupational disease to members of the Quebec Asbestos Producers' Association. Stevenson, a physician with Canadian Johns-Manville, explained that asbestosis occurred rarely in Canadian mine workers. While there may have been cases in the distant past Johns-Manville was such a fastidious employer it routinely transferred mill workers to other

[48] G.F. Slade, 'Incidence of Respiratory Disability'.

[49] J.C. Wagner, 'The Pathology of Asbestosis in South Africa' (University of Witwatersrand PhD, 1962).

[50] G. Knight to Dr Knight, 19 January 1926.

[51] F. Pedley, 'Asbestosis', *Canadian Medical Association Journal* 22 (1930), pp. 253–4.

[52] F. Pedley 'Asbestosis', *Canadian Public Health Journal* 21 (1930), pp. 576–7.

jobs after ten years' service.[53] Stevenson wished he had access to the X-rays of South African miners so he could become more familiar with the disease. When provincial authorities began reviewing the Workmen's Compensation Act in 1940, Johns-Manville wrote in protest to the Quebec government. It claimed that there had been no cases of asbestosis at its mines and any reference to that disease in the legislation would only encourage fraudulent appeals for compensation.[54] Over the next half century, claims about the absence of asbestosis in Canada were repeated *ad nauseam* by the industry.

Despite their protestations, Pedley and Stevenson were as aware as Johns-Manville's board of occupational disease at Thetford. In conjunction with the mining companies in 1930, MetLife commissioned Drs Pedley and Stevenson to study asbestosis. The project involved 141 Canadian Johns-Manville employees at Asbestos and fifty-four men at Thetford. Eighty-five per cent of the Thetford miners had ten years' service compared with only 22 per cent for those at Asbestos. Pedley and Stevenson began with a review of the miners' general health.[55] Unlike their UK colleagues, Pedley and Stevenson did not take work histories.

The general health of both groups was very poor, especially for men who were fully employed. Half of those in each group needed dental care; 10 per cent of the men from Asbestos had hearing loss, while over half of the Thetford subjects had marked industrial deafness; 27 per cent from Asbestos and 13 per cent from Thetford had infected tonsils; and 10 per cent of both groups had hernias, a serious complaint in men required to do heavy work. Around 10 per cent of the Thetford cohort had asbestos corns or raised areas of skin, similar to a small wart, caused by imbedded fibres. The clinical examination of the men's lungs proved even more disturbing. Basal rales (chest crackles) were common, but the most frequent complaint was shortness of breath. Of the forty men examined by Dr Stevenson, fourteen (35 per cent) had stage one asbestosis. Of the fifty-six men X-rayed at Thetford, almost half had asbestosis as did seventeen of the 141 men at Asbestos.[56] The high incidence at Thetford was consistent with the longer average period of service and in his report Pedley wrote: 'From this table one would conclude that asbestosis is not likely to develop with an exposure of less than five years, but that

[53] 'Asbestosis' paper presented by Dr J.H. Stevenson to the Quebec Asbestos Producers Association, Sherbrooke, 23 May 1938.

[54] Vandiver Brown Memo re. Quebec Occupational Diseases Legislation, 15 April 1940.

[55] Frank Pedley, 'Report of the Physical Examinations and X-Ray Examination of Asbestos Workers in Asbestos and Thetford Mines', Quebec, November 1930.

[56] Pedley, 'Report of Physical Examinations', pp. 9–13.

with exposure of more than ten years there is a great possibility of it developing.'[57] Pedley concluded that while the disease was common it was of little importance since there was no accompanying incapacity. In reaching that conclusion, he ignored his own finding that shortness of breath was common. He also ignored the medical literature from Britain.

The study by Pedley and Stevenson was never published, but they did send a copy of their results to Dr Lanza. Henceforth, the views expressed by Lanza about asbestosis varied according to the audience he was addressing. In private, he acknowledged that claims about the absence of disease in Quebec were motivated by the self-interest of the asbestos companies.[58] In public, Lanza endorsed the fallacy that the Canadian mines were safe.[59] In 1939 Lanza wrote that there was disease in asbestos-textile plants, but not on the mines.[60] Again in 1940, he repeated the lie that the mines were free of asbestosis.[61] It is significant that from the early 1930s, Lanza's employer, MetLife, assessed asbestos workers as such a high risk group that it added a 40 per cent loading to their health insurance premiums.[62] Lanza had little sympathy for labouring men and women and even less if they were Afro-American. In April 1933, he wrote to a colleague Dr Emery Hayhurst about the Hawk's Nest disaster: 'This business in West Virginia [Hawk's Nest] must be a terrible mess. Somewhere I got the impression that many of the workers were Negroes. I am beginning to believe that the Negro succumbs more rapidly to silicosis than the white man, and the thought occurs to me that this may possibly be due to the greater frequency of syphilis among the Negro labouring classes.'[63]

The outbreak of the Second World War brought an immediate increase in production and a worsening in work conditions, yet industry maintained its deception about occupational safety. In April 1940, Dr J. H. Stevenson gave a paper at Sherbrooke, Quebec, on the safety of Canadian mines. He explained to his audience (the QAPA) that Johns-Manville employees were X-rayed annually, and of the 507 men at Asbestos with more than ten years' service, only 4 per cent showed early signs of fibrosis.

[57] Pedley, 'Report of Physical Examinations', p. 13.

[58] Letter from A.J. Lanza, MetLife, to Manfred Bowditch, Saranac, 13 December 1937.

[59] A. Lanza, W.J. McConnell, and J.W. Fehnel, 'Effects of the Inhalation of Asbestos Dust on the Lungs of Asbestos Workers', *Public Health Reports* 50 (January 1935), pp. 1–12.

[60] A.J. Lanza and R.J. Vane, 'Industrial Dusts and the Mortality from Pulmonary Disease', *American Review of Tuberculosis* 39 (1939), pp. 419–38.

[61] Castleman, *Asbestos*, p. 19.

[62] Letter from Buell to Dr J.C. McConnell, Metropolitan Life, 22 August 1939. See also Castleman, *Asbestos*, pp. 4–5.

[63] Lanza, MetLife, to Dr Emery Hayhurst, 17 April 1933.

In fact, there was so little disease in Canada it was almost impossible to conduct research ('we have never had reason to have autopsies on our men'). The company had a thriving Quarter Century Club, but none of them ever died from lung trouble, because of the local 'immunity from asbestosis'. Stevenson did admit that the Thetford mill was full of dust and he wondered: 'If it would be possible to have small ventilated chambers erected in certain locations in the mills where operators of machines could be kept, a part of the time, while still watching their machines, so that they could get clean air.'[64] There was dust and disease on South African mines, but in Canada there was supposedly only dust.

State authorities may have believed Stevenson, but Johns-Manville's competitors were less gullible.[65] Stevenson's paper was distributed within the industry and copies were sent to T&N in the UK. T&N operated mines in Southern Africa, including New Amianthus where Slade had worked a decade earlier. It also had mines in Quebec. It was not surprising that T&N's chairman Sir Samuel Turner was irritated by Stevenson's claims. He commented that asbestosis was difficult to diagnose and that it was common for more cases to show up at *post mortem* than during clinical examination.[66] Sir Samuel was certain that if the UK Medical Board had examined Johns-Manville's employees they would have found plenty of disease. J.A. Cann, a senior executive with T&N, agreed.[67] Cann noted that X-rays were not reliable and for that reason British physicians always took detailed work histories, something Stevenson and Pedley had failed to do. Cann was familiar with the Canadian mines and during a visit to the King Mine, in Quebec, he had found conditions to be very poor. There was not enough draught to remove even palpable dust and pollution was particularly bad in the bagging areas. Disease was inevitable, he believed.

The lack of effective state regulations makes it difficult to assess the industry's efforts in complying with the voluntary TLV of 5 mppcf. It is certain, however, that with the outbreak of the Second World War work conditions deteriorated. An indication as to how bad is found in the Sumner Simpson Papers. Following complaints about asbestos factories, in early 1944 the War Production Board ordered the state of New Jersey to conduct a survey. Without bothering to conduct their own survey, the state authorities accepted MetLife's assurances that the Manville plant

[64] Stevenson, 'Asbestosis' paper, 15 April 1940.

[65] For a review of Stevenson's clinical methods see M. Greenberg, 'A Report on the Health of Asbestos, Quebec Miners 1940', *AJIM* 38 (2005), pp. 230–7.

[66] Sir Samuel Turner Memo, 25 May 1940.

[67] Memo from J.A. Cann to C.S. Bell, London, 3 June 1940.

was safe. However, the industry's private survey by the ATI told a different story. The head engineer of the IHF, W.C.L. Hemeon, was commissioned to assess asbestos-textile plants, including Johns-Manville. His report was so damning that like the rest of the Sumner Simpson Papers it remained confidential until it was produced during litigation in 1977.

Hemeon found a serious health problem and a lack of information about occupational disease across the industry. Of the ten plants he visited only two conducted regular medical checks. At those plants—the Raybestos-Manhattan plant at Manheim and Johns-Manville's factory at Manville, New Jersey—20 per cent of the workforce had asbestosis.[68] Even more disturbing was Hemeon's finding that disease was occurring in plants adhering to the voluntary TLV of 5 mppcf. 'The information available', Hemeon wrote, '*does not* permit complete assurance that five million [particles] is thoroughly safe nor has information been developed permitting a better estimate of safe dustiness'.[69] There was also in his opinion an urgent need to develop improved techniques for diagnosis.[70] While none of the factories had systematic health records Hemeon believed that Dr Lanza at MetLife had X-rays dating from the 1930s which could provide the data necessary to make workplaces safer.[71] Hemeon was probably unaware of the secret literatures held by the US and British industries and it is interesting to reflect upon the kind of report he would have written if he had read Pedley and Slade.

The suppression of evidence also occurred in Western Europe. In 1939, Dr Enrico Vigliani, acting chief physician for the Turin section of the National Bureau of Propaganda for the Prevention of Accidents (ENPI), conducted a cross-sectional study of asbestos workers at four Italian factories. The ENPI was an occupational health institute set up under the Fascists to promote the ideology of corporatism. Naturally it was dominated by industry. The four factories used in the study produced asbestos textiles and most of their employees were women. The factories included the Cape Asbestos subsidiary S.A. Capamianto, in Turin, which used South African crocidolite.[72]

[68] W.C.L. Hemeon, 'Report of Preliminary Dust Investigation for Asbestos Textile Institute, June 1947' (1947), p. 15.

[69] Hemeon, 'Preliminary Dust Investigation', p. 22.

[70] Hemeon, 'Preliminary Dust Investigation', p. 16.

[71] Hemeon, 'Preliminary Dust Investigation', p. 19.

[72] Professor Enrico C. Vigliani, *Studio Sull' Asbestosi Nelle Manifatture Di Amianto* (Cirie': Stabilmento Tipografico Giovanni Capella, 1940). For a critique of Vigliani's study, see E. Merler, 'A Cross-Sectional Study on Asbestos Workers Carried Out in Italy in 1940: A Forgotten Study', *AJIM* 33 (1998), pp. 90–3.

The 439 workers in the sample were given a physical evaluation and a chest X-ray. A medical history was also taken. Dust levels measured with a thermal precipitator were very high with one sample showing a reading of 5,300 ppcc. Workers were supplied with personal masks made from cotton, but they were rarely used. Instead, they wrapped handkerchiefs over their mouths to keep out the dust. Using conservative methods of evaluation, Vigliani found that in some work categories 47 per cent of subjects had asbestosis. Among those groups doing the dustiest jobs the figure was as high as 90 per cent. Vigliani recommended that work conditions be improved through ventilation, a system of pre-employment evaluation, and repeated medicals. None of those recommendations were implemented, nor did Cape attempt to apply Vigliani's findings to its South African mines.

After the war the suppression of evidence of asbestosis in Canada continued. In June 1946, Dr George Wheatley from MetLife visited Thetford where he met the new medical director Dr Paul Cartier. Wheatley thought the survey of workers which Cartier had just begun would uncover a lot of asbestosis as well as tuberculosis. He wrote to his colleague Anthony Lanza: 'The more thorough the job the more problems they are going to discover. Since asbestosis is compensatable, this is going to cost them [the industry] much money.'[73] Cartier's survey was never published, but less than two years later the Thetford mines were shut by a protracted strike over pay and occupational safety.

T&N physician John Knox gave a truer picture of ARDs in Thetford, when he visited Cartier in 1952 at the Thetford Mines Industrial Clinic (a modest facility operated by the industry). Knox found, when looking at Cartier's X-rays of workers, 'a considerable amount of asbestosis . . . of sub-clinical type'.[74] In a memo to his bosses, Knox continued: '[Cartier] agrees with me that it is undesirable to reveal to a worker evidence of an early reaction shown only on a radiograph.' Knox noted that the 'conception of asbestosis as a progressive disease is not popular here'. At T&N's Bell Mine, Knox described 'snowstorm conditions' in most areas and he expressed considerable doubt when told that 'old-timers' were fit and well, because dust control in the mill was ineffective, hand shovels were used, and dust exposure was heavy. Knox concluded that there were about 400 people 'at risk in the mine and mill'. Johns-Manville's mill was as bad.

[73] Memo from George Wheatley to A.J. Lanza, 15 June 1945.

[74] Dr John Knox, visit memo, 17 October 1952.

Knox would get well used to observing such conditions, both in North America and Europe. In 1960, on a visit to a T&N Ferodo subsidiary in Normandy, he toured the asbestos textile factory at Condé-sur-Noireau. He noted that since 1954 there had been forty-seven asbestosis cases and thought that there would be many more. He also noted the reason: '[N]o ventilation . . . fluff and fibre seemed to abound everywhere. . . . No men wore respirators.'[75] He added: '[U]ncontrolled inhalation of asbestos dust (chrysotile) has led to a high incidence of fibrous lung complications.' Yet the local doctor in Caen, Dr Porin, said that he had never seen asbestosis. Knox felt this was 'strange', yet in 1964 Porin was appointed by the government as the resident compensation expert to examine workers in the region! Knox mentioned a local medical tradition that blamed the popularity of cider (calvados) for any chest diseases. The Chambre Syndicale D'Amiante believed that, like Canadians, French workers had 'immunity' to asbestosis and blamed British cases on urban pollution. Later Knox would attend an industry conference in Caen, which made even him wax cynical. He believed it was a delaying tactic: 'A further breathing space would then be created of 10–15 years when all trouble could be blamed on the "old days".'[76]

Lung Cancer

The link between lung cancer and asbestos was slow to emerge. Lung cancer is a more complex disease than asbestosis. There are in addition other causes of lung cancer besides asbestos, most notably cigarette smoking, which slowed down medical discovery. Workers at T&N factories were dying from lung cancer as early as 1932 but the connection between those deaths and asbestos was not easily proven.[77] A series of articles in American and British medical journals in the 1930s drew attention to the coexistence of cancer and asbestosis, but the studies were scattered and no firm conclusions were drawn by their authors. In 1938, a German physician, Martin Nordmann, published a study of two asbestos textile workers. Nordmann also reviewed the literature and concluded that asbestos causes lung cancer.[78] His findings found ready acceptance in the German academy and during the Second World War lung cancer

[75] Report of Knox visit to French Ferodo, 22 June 1960.
[76] J. Knox, Caen Congress report, 29–30 May 1964.
[77] G. Tweedale, *Magic Mineral to Killer Dust* (2001), p. 142.
[78] For a brief discussion of Nordmann, see Castleman, *Asbestos*, pp. 42–3.

in combination with asbestosis became a compensable disease for Aryan workers.[79] Eternit had subsidiaries in Germany, but chose not to apply that knowledge to its factories elsewhere.[80] Czechoslovakia introduced a parallel scheme after the war.[81]

Probably the most important asbestos research carried out at Saranac was on lung cancer. In 1940 Dr Leroy Gardner tested a cohort of white mice for asbestosis.[82] After inhaling fibre, over 80 per cent of the mice developed pulmonary cancer. Gardner was aware of the existing European literature and he immediately notified Johns-Manville attorney Vandiver Brown.[83] He in turn sent copies of Gardner's findings to 'several interested companies' including T&N. Gardner's results, which would have lent weight to studies from Italy, France, and Norway linking asbestos with cancer, were not published. When Gardner died suddenly in December 1946, Lanza and the QAPA took possession of the study. Saranac was dependent upon the industry's patronage and Lanza sat on Gardner's results for three years. During that period there were strikes on the Canadian mines over work conditions which made the Saranac results more sensitive. In August 1946, Thomas Gatke, president of the Gatke Corporation wrote to Raybestos-Manhattan about the failure of Dr Gardner's work at Saranac to 'live up to expectations'. Gatke's conceded that experts had knowledge, but he wanted data to counter what he called 'country doctors who set themselves up as experts but who actually have had no experience beyond what they read in books'.[84] It was as if having claimed publicly that there was no risk for those who worked with asbestos, the industry leaders were determined to believe their own fictions. A report produced by Saranac in September 1948 saw Gardner's original findings corrupted and the phrase pulmonary cancer replaced by the anodyne term 'adenomas'.[85]

In 1950, Dr Arthur Vorwald, who had replaced Gardner as director of Saranac, conducted a number of animal inhalation tests to prove that

[79] For an intriguing account of occupational health and safety legislation in Nazi Germany, see R. Proctor, *The Nazi War on Cancer* (1999), pp. 107–13.

[80] R.F. Ruers and N. Schouten, *The Tragedy of Asbestos* (2005), p. 22.

[81] Castleman, *Asbestos*, p. 61.

[82] For an account of that history, see 'Early Saranac Cancer Tests with Asbestos', Castleman, *Asbestos*, pp. 49–59. See also D.E. Lilienfeld, 'The Silence: The Asbestos Industry and Early Occupational Cancer Research—A Case Study', *American Journal of Public Health* 81 (June 1991), pp. 791–800; G.W.H. Schepers, 'Chronology of Asbestos Cancer Discoveries: Experimental Studies of the Saranac Laboratory', *AJIM* 27 (1995), pp. 593–606.

[83] Castleman, *Asbestos*, p. 50.

[84] Letter from Thomas Gatke, President Gatke Corporation to Mr J. Rohrbach, Vice-president of Raybestos-Manhattan, 12 August 1946.

[85] Castleman, *Asbestos*, p. 56.

asbestos did not cause cancer. The research was again funded by the Quebec industry and Vorwald agreed that the results would only be available to the 'underwriting companies'. The results were never published, but it is reasonable to assume that if he was competent, Vorwald would have reached the same conclusions as had Gardner ten years earlier.[86] The Seventh Saranac Symposium held in September 1952 was attended by Dr Lanza and representatives from Johns-Manville and T&N. It also attracted the leading British authority Dr Edward Merewether, who gave a paper showing that the lung-cancer rate among asbestos workers in the UK was rising. Merewether drew on the knowledge he had accumulated as head of the Factory Inspectorate. His published report in 1949 had emphasized the incidence of lung cancer in asbestos workers.[87] He noted that the cases of massive fibrosis after less than ten years' exposure were a thing of the past, but as workers lived longer, more were developing cancer. Merewether suggested that the same was probably happening in North America and he dismissed outright the idea that Canadian chrysotile was less harmful than other fibres.[88]

Industry-based researchers in the UK were reaching the same conclusions. Dr Hubert Wyers, the medical officer for Cape Asbestos in the UK, published a study in 1949 of 115 fatal cases of asbestosis among which he found seventeen cancers of the lung and pleura.[89] In addition to confirming the association between asbestosis and lung cancer, Wyers suggested that many workers were dying of asbestosis before they were being diagnosed with cancer. Over the next five years Merewether's findings were corroborated by a number of researchers culminating in a study of T&N's Rochdale factory by Richard Doll.[90]

In the early 1950s, Richard Doll (1912–2005) was at the start of an illustrious career in epidemiology that would help confirm the link between smoking and lung cancer. Alongside his smoking studies, Doll was also interested in exploring occupational cancers. The growing literature on asbestos and lung cancer caught his interest and, as a start, Doll contacted Wyers at Cape in 1948. Doll was, of course, unaware of

[86] On Saranac Laboratory Cancer Studies for QAMA, see Castleman, *Asbestos*, pp. 66–9.

[87] Ministry of Labour and Factory Inspectorate, *Annual Report... for 1947* (London, 1949), pp. 66–81.

[88] Castleman, *Asbestos*, p. 77.

[89] H. Wyers, 'Asbestosis', *Postgraduate Medical Journal* 25 (December 1949), pp. 631–8.

[90] S.R. Gloyne, 'Pneumoconiosis: A Histological Survey of Necropsy Material in 1205 Cases', *The Lancet* 260 (14 April 1951), i, pp. 810–14: L. Breslow et al., 'Occupations and Cigarette Smoking as Factors in Lung Cancer', *American Journal of Public Health* 44 (1954), pp. 171–81.

industry knowledge and the Saranac experiments and he was sceptical about asbestos as a cancer agent, citing as authorities Cartier, Lanza, and Vorwald! He wanted to use Cape's data, so that any association of asbestos with lung cancer could be 'logically proved'.[91] This was the language of epidemiology, a statistical science that was now staking its claim to superiority over older approaches that had relied on observations by physicians and pathologists to recognize occupational diseases. Initially, however, the statistics were not forthcoming. Doll's correspondence with Wyers extended in a leisurely fashion for several years, but by the mid-1950s had produced nothing. By then, the UK Factory Inspectorate had reached its own conclusions. In 1953, Dr Arthur McLaughlin, a prominent Factory Inspector, listed asbestos as a carcinogen in the leading medical journal *The Lancet*, without even feeling the need to comment on it.[92]

Soon after, Doll was contacted by T&N. Dr John Knox, the company physician, had attended the Seventh Saranac Symposium in 1952 and was sufficiently concerned about pulmonary cancer to raise the issue with his directors. In the following year, they granted him permission to approach a statistician (Doll), who could analyse the prevalence of lung cancer cases at the company's Rochdale plant. Knox then supplied Doll with the required data. For Doll, it was relatively straightforward analysis to compare deaths at the plant with those in the general population. Given the sensitivity of the issue, it is surprising that an outsider was entrusted with this task. Knox seems to have expected that the analysis would show that the apparent high incidence of lung cancer at Rochdale was an artifact (caused by the circumstances of asbestosis autopsies, where pathologists examine the lungs so minutely for other diseases). Possibly T&N expected the findings to remain the property of the company, like Gardner's experiments. If so, both Knox and T&N got a shock. Doll's analysis showed an elevated risk of lung cancer among asbestos textile workers at Rochdale—results that Doll felt should be published. T&N tried to block him, with one of the company directors—almost certainly Norton Morling—even attempting to intimidate the editor of the medical journal in which Doll intended to publish the paper. The company failed to derail publication and Doll's article (shorn of Knox's name) was published in 1955.[93]

Doll's paper offered the scientific 'proof' that asbestos was a carcinogen. It would subsequently generate enormous interest amongst attorneys and

[91] Doll to Wyers, 29 June 1948.
[92] A.I.G. McLaughlin, 'The Prevention of Dust Diseases', *The Lancet* (1953), ii, p. 49.
[93] R. Doll, 'Mortality from Lung Cancer in Asbestos Workers', *BJIM* 12 (1955), pp. 81–6.

historians.[94] But at the time, its impact was difficult to discern. The usage of asbestos continued to expand and T&N's profits mounted. This may have been due to the optimistic tenor of Doll's paper. He flagged a cancer hazard, but at the same time hinted that dust control measures demanded by the government in 1931 had solved the problem. This conclusion, which suggests that the pressure exerted by T&N had not been without some effect, was the weakest part of the paper. Aside from the incompleteness of the data supplied by T&N, the follow-up period under the 'new' dust-controlled conditions was only twenty-one years (for a disease that typically took twenty-five years or more to develop). Doll had not had enough time to assess accurately whether the risk had been reduced or eliminated. However, the T&N directors were sufficiently encouraged by Doll's findings (and the lack of bad publicity) to invite him to continue to track the Rochdale workforce.

For the asbestos labour force, Doll's epidemiology proved a mixed blessing. As Doll's reputation grew (he was knighted in 1971), he appeared to epitomize the detached, curiosity-driven scientist. However, as an Oxford University-based scientist, Doll was remote from the dirty realities of asbestos use in the construction and shipbuilding trades, and although he had been trained as a physician he never treated sick workers. He was content to refine his data on Rochdale factory workers—a very select group in an atypical factory—while never feeling impelled to look at ARDs among workers in dustier sectors of the industry. He never criticized T&N or the asbestos industry: indeed, he believed that industry was 'beginning to behave responsibly' and that governments should only attribute risk after scrupulous research.[95] It was a philosophy that had obvious attractions for commercial interests, because scrupulous research easily translated into inaction, delays, and hair-splitting over results.

For the next thirty years, Doll was a trusted adviser of T&N. The Knox-Doll cohort, based on asbestos textile workers and not insulators, formed an integral part of the industry's defence of its product and working conditions in its factories. By the early 1960s, Doll and T&N were still confident that the data suggested that in Rochdale 'the specific occupational

[94] Castleman, *Asbestos*, pp. 41–132; B. Castleman, 'Doll's 1955 Study on Cancer from Asbestos', *AJIM* 39 (2001), pp. 237–40; C. Beckett, 'An Epidemiologist at Work: The Personal Papers of Sir Richard Doll', *Medical History* 46 (2002), pp. 403–21; M. Greenberg, 'A Study of Lung Cancer Mortality in Asbestos Workers: Doll, 1955', *AJIM* 36 (1999), pp. 331–47.

[95] R. Doll, 'Occupational Cancer: A Hazard for Epidemiologists', *International Journal of Epidemiology* 14 (March 1985), pp. 22–31.

hazards to life have been completely eliminated'.[96] It was a persuasive idea, but not one that was supported by Dr Bill Kerns, the physician who replaced John Knox at T&N: he resigned in the early 1960s because he had been appalled at dust levels at Rochdale and at T&N's attempt to censor his work. However, with the Factory Inspectorate taking little interest in ARDs, Doll's (and T&N's) voice was the dominant one in any debate on the asbestos hazard in the UK. It was always a reassuring voice. T&N and Doll consistently presented a picture of a concerned, hygienic industry that was shedding its hazardous past.

To underline that message, in 1957 the leading UK asbestos companies founded the Asbestosis Research Council (ARC). This was largely the initiative of T&N, who in 1956 suggested that the leading companies (T&N, Cape Asbestos, and British Belting & Asbestos) should start an organization to coordinate research. The ARC appeared to be distinct from the asbestos companies, a fiction fostered by its carefully chosen and official sounding name. It allowed company physicians and industry scientists to present themselves to the public as members of a scientific organization, not a commercial company.[97]

The coincidence of the ARC's founding with the publication of Doll's cancer paper is obvious, even though the ARC was ostensibly set up to prevent, diagnose, and treat *asbestosis*. By then, of course, neither complex scientific research nor laboratory work was necessary to understand the dangers of asbestos. The hazards could be appreciated simply by looking at the impact of the dust on the workforce and by examining pathological data. But the ARC was specifically excluded from conducting epidemiological studies and most of its research effort would be spent on dust counting and animal experiments. This work was not entirely without merit, but it was only distantly related to workers who continued to inhale dust. The ARC's main job was to delay government regulation and to defend the commercial interests of the industry. As we shall see in Chapter 4, the ARC would interlock with several other industry-financed organizations, where the main thrust was towards public relations. The ARC's research was useful in this context, because it suggested that various 'problems' needed to be resolved by time-consuming research rather than doing what was obvious.

[96] J.F. Knox, R.S. Doll, and I.D. Hill, 'Cohort Analysis of Changes in Incidence of Bronchial Carcinoma in a Textile Asbestos Factory', *ANYAS* 132 (December 1965), i, pp. 527–35. See also J.F. Knox, S. Holmes, R.S. Doll, and I.D. Hill, 'Mortality from Lung Cancer and Other Causes among Workers in an Asbestos Textile Factory', *BJIM* 25 (1968), pp. 293–303.

[97] G. Tweedale, 'Science or Public Relations? The Inside Story of the Asbestosis Research Council', *AJIM* 38 (December 2000), pp. 723–34.

In Canada, the cancer problem was also being researched. In 1956, QAMA decided to finance a study of cancer in the mining district of Quebec—thereby pre-empting government plans to conduct a similar survey. The task was entrusted to two IHF researchers, Daniel C. Braun and David Truan, who reported in 1957.[98] The study looked at 6,000 workers, among whom nine lung cancer cases were found, with three 'suspected'. It was concluded that this would be roughly what could be expected in Quebec province (which was used as the control). However, this method diluted the significance of the cancer cases, because a truer comparison with the mines would have been a rural area, not Quebec province, which included Montreal and Quebec cities. On the other hand, it was not all good news: the original report had noted a significant incidence of lung cancer amongst those with asbestosis (12.5 per cent), which was similar to the percentage reported in other cohorts. One section in the confidential report was therefore entitled 'Asbestosis and Lung Cancer'.

In 1958, a condensed version of the results was published in an academic journal of the American Medical Association.[99] The article repeated the favourable results from the unpublished report, but when it came to the cancer findings some important changes were made. Specifically, the following statement in the unpublished draft did not appear in the final article: '[T]he results suggest that a miner who develops the disease asbestosis does have a greater likelihood of developing cancer of the lung than a person without this disease. We suspect, however, that under-reporting of asbestosis cases has led to a fallacious finding in this connection.'[100]

The cancer discussion had proved unpalatable to QAMA and its counsel Ivan Sabourin. Johns-Manville had no liking for it either. Company physician Kenneth Smith had written to Sabourin, while the manuscript was being revised:

> We have noted deletion of all references to the association of asbestosis and lung cancer.... While we believe that this information is of great scientific value, we can understand the desire of the QAMA to emphasize the exposure of the asbestos miner and not the cases of asbestosis. We are also in agreement with the deletion of the reference to smoking and lung cancer.[101]

[98] D.C. Braun, 'An Epidemiological Study of Lung Cancer in Asbestos Miners for Québec Asbestos Mining Association, September 1957'. Marked confidential, restricted copy No. 27.

[99] D.C. Braun and T.D. Truan, 'An Epidemiological Study of Lung Cancer in Asbestos Miners', *AMA Archives of Industrial Health* 17 (June 1958), pp. 634–53.

[100] Braun, 'An Epidemiological Study', p. 76.

[101] Kenneth Smith to Sabourin, 30 December 1957. T&N director John Waddell was later told by the QAMA that the findings 'would have offended the tobacco interests, since it was

The editor of the *Archives of Industrial Health*, Herbert Stokinger, was delighted to accept the paper, telling Braun that it was 'a model of epidemiologic method'.[102] Stokinger was a key member of the ACGIH, and as such had influenced the TLV debate. He had theorized that cancer in asbestos workers in other countries was not paralleled in Canada, so Braun and Truan had confirmed his suspicions.[103] Elsewhere the paper was soon denounced for its methodological flaws, inter alia, by Wilhelm Hueper and Thomas Mancuso. But it was nevertheless cited as an authentic piece of work by some of the scientific community and it was to be useful in litigation. The episode has been described as a 'cynical circle of events', in which companies first changed the Braun and Truan paper to suit their immediate needs and then attempted to rely on the corrupted study to escape liability in court.[104] It was a striking example of how trade associations interfered in the scientific process to erect defences around their product.

Mesothelioma

Mesothelioma is the most frightening of the ARDs. It is always fatal and its latency period of up to sixty years makes its destruction of the human body seem inexorable. Worse still, the disease can result from relatively trivial exposure which puts at risk anyone who comes into contact with asbestos. Nothing in the early literature suggested that chrysotile was any safer than the amphiboles. Given that chrysotile accounted for 90 per cent of global production, if it caused mesothelioma then the asbestos industry was doomed. By the time evidence about mesothelioma emerged in 1960, asbestos products were so widely used in Western economies that tens of millions of people were potentially at risk.

Mesothelioma of the pleura was known to physicians in the nineteenth century and from 1870 there is a small literature on the subject. By 1921, the term 'mesothelioma of the pleura' was being used by UK researchers, but until the 1950s the tumour was so rare that some physicians doubted

much more critical of the effects of smoking than of asbestos'. Waddell's record of discussions with J. Beattie and K.V. Lindell, 15 December 1965.

[102] Stokinger to Braun, 20 January 1958. [103] Castleman, *Asbestos*, pp. 95–8.

[104] D.S. Egilman and A. Reinert, Letter re. 'Corruption of Previously Published Asbestos Research', *Archives of Environmental Health* 55 (January/February 2000), pp. 75–6.

its existence. Apart from an autopsy there was no certain means to distinguish mesothelioma from primary cancers of the lung.[105]

In 1954 the South African Government Mining Engineer asked the Pneumoconiosis Research Unit (PRU) in Johannesburg to examine the problem of occupational disease on the asbestos fields. Nothing was known about disease rates and the Department of Mines was keen to see if blue fibre caused the same diseases which had been identified in North America and the UK. The Department's request led to the appointment of a young scientist named J.C. Wagner to the PRU. Wagner's work was ostensibly on asbestosis. No research had been done on that subject in South Africa since George Slade in 1930, and Wagner hoped to produce significant results. At medical school he had been told nothing of mesothelioma, which was believed to be extremely rare.[106] As a returned soldier, Wagner was older, more experienced, and more independent than most postgraduates. That independence proved vital to the research he was to do over the next six years.

Wagner began working on asbestosis, then shifted in midstream to study the unique tumours that were appearing in the Northern Cape. It was a brave decision, made more difficult by the growing reluctance of the mining industry to tolerate his research.[107] With the help of Dr Kit Sleggs, who was the director of the West End Tuberculosis Hospital in Kimberley, Wagner collected cases of mesothelioma among men and women who had lived close to the mines. The publication in 1960 of Wagner's paper is generally credited with being the first to emphasize the association with asbestos. The study was based on an analysis of thirty-three cases of pleural mesothelioma, with all but one patient having a proven exposure to asbestos.[108] Only eight of the thirty-three had occupational exposure, but twenty of the remaining twenty-five had as infants lived near the mines. It is ironic that what was to become one of the great occupational health discoveries of the twentieth century was based principally on cases drawn from outside the workplace. Wagner's article changed the understanding of the dangers of asbestos and suggested a nexus between

[105] See I. Webster, 'Methods by which Mesothelioma can be Diagnosed', in H.W. Glen (ed.), *Asbestos Symposium* (1978), pp. 3–8.

[106] Jock McCulloch interview with Dr Chris Wagner, Weymouth, Dorset, 22 March 1998.

[107] For an account of that discovery, see J. McCulloch, 'The Discovery of Mesothelioma on South Africa's Asbestos Fields' *Social History of Medicine* 16 (2003), pp. 419–36.

[108] The paper was submitted for publication in April 1960. See J.C.Wagner, C.A. Sleggs, and P. Marchand, 'Diffuse Pleural Mesothelioma and Asbestos Exposure in the North West Cape Province', *BJIM* 17 (1960), pp. 260–5.

work, the environment, and cancer. His research could not have come at a worse time for the South African industry, which had invested heavily in new plants to meet rising global demand. Wagner's article caused a storm in Johannesburg, and it was partly for that reason that in 1962 he accepted a position as a pathologist with the Medical Research Council in South Wales. He worked in the UK until his retirement twenty-five years later.[109]

J.C. Wagner enjoyed a brilliant career in the UK, but as we will show in Chapter 5 it is one which illustrates the hazards of working on asbestos research. Scientists like Wagner and his brother-in-law Dr Ian Webster, who supervised his thesis at the PRU, were interested in abstract questions, which should have been far removed from commercial influence or state interference. They soon discovered otherwise. Because mesothelioma threatens bystanders as well as labour, Wagner's discovery had immediate implications for the mining, manufacture, and use of asbestos products. Working at a research centre far removed from the major universities of Europe and North America, Chris Wagner reconfigured the perception of asbestos risk. For both him and Ian Webster that discovery was to become a burden.

Conclusion

The period from 1929 to 1935 was crucial for the industry. It was during those years that independent researchers identified the symptoms and causes of asbestosis. In addition, studies commissioned by Johns-Manville, MetLife, T&N, and the IHF showed that around half of those employed in Canadian, US, and Southern African mines and asbestos textile factories would eventually develop a serious disease for which there was and remains no cure. By 1935, the industry leaders knew there was no cheap technology to remove dust from the workplace. Therefore the best method for reducing injury was to inform workers about the hazard. That was not done. Instead the theses by Ian Grieve and George Slade disappeared and important research commissioned by MetLife and the IHF was suppressed.

One of the major challenges facing management was how to respond to evidence from British asbestos textile factories showing asbestos dust to be

[109] McCulloch interview with Wagner.

even more hazardous than free silica. The US manufacturers soon arrived at an answer. They cited the supposed absence of disease on Canadian mines to argue that the studies by Merewether and Price did not apply to North America. Surprisingly, no one seemed to notice that the same fibre from the same mines in Quebec, which was causing so much asbestosis in the UK, was harmless when used on its own side of the Atlantic. By 1935 the industry had created two articles of faith. The first, the supposed absence of asbestosis on the mines proved that disease further along the chain of production was due to the conditions in specific factories. It was not due to the high toxicity of asbestos. Over time the evidence used to support that claim became more and more tenuous, yet the myth of disease-free mines survived into the 1960s and beyond when the mesothelioma crisis erupted. The second article, favoured by US manufacturers, was that asbestosis is a less serious illness than silicosis. Workers with asbestosis could supposedly continue in their jobs and live productive lives without discomfort. Therefore even if asbestosis was common that was no reason for regulation.

A third article of faith, which first appeared after 1964, concerned the heavy exposure levels which the industry suddenly acknowledged as having been common in the 1930s and 1940s. That admission, which contradicted long-held industry claims about work safety, was used to explain the high rates of disease among older workers. In effect the industry rewrote its own history. In 1938, Anthony Lanza had commented: 'As soon as the hazard was realized, industrial firms fabricating asbestos took energetic steps to control the dust so that it is probable the cases of asbestosis will become uncommon.'[110] Nothing could have been further from the truth. In 1951, Owens-Illinois commissioned Saranac to survey its Kaylo insulation factory at Sayreville, New Jersey. The plant used various raw materials including chrysotile and amosite.[111] The survey revealed asbestos and silica dust levels more than ten times above the threshold of 5 mppcc. Saranac recommended improved housekeeping and a pre-employment medical program, the same recommendations made twenty years earlier in the UK by Merewether and Price.

One of the important lessons the industry learned during the first crisis was how to use professional associations. The IHF was influential in debates about dust levels from the mid-1930s and it continued for

[110] Lanza, *Silicosis and Asbestosis*, p. 420.

[111] Industrial Hygiene Survey by Saranac Laboratory, 'Report: Kaylo Division Plant Owens-Illinois Glass Company, Sayreville, NJ, May 29, 1951', pp. 6–7.

decades to carry out research on the industry's behalf. General Motors (GM) was typical of the IHF membership. The automobile industry was a major user of asbestos and GM manufactured its own asbestos-based brakes, brake linings, and clutch facings. It also used asbestos in gaskets, adhesives, and electrical components. The company took an active interest in medical research and in 1938 GM President William S. Knudsen established an annual award for the 'most outstanding contribution to industrial medicine'. The recipients included Leroy Gardner and Tony Lanza. General Motors joined the IHF in 1942. By 1949, the GM vice-president H. W. Anderson was on the IHF Board of Trustees and GM's industrial hygienist Frank Patty was on the Chemical and Toxicological Committee. Like their Union Carbide and Johns-Manville counterparts, GM physicians had close ties with Saranac. What is notable about the IHF was its methods. The Foundation claimed to be acting on the basis of science rather than commercial interest. Its research was supposedly independent and as a result the IHF presumed to speak with authority on health and safety. In practice, as Vandiver Brown put it, the IHF was 'the creature of industry'.[112] Where IHF research identified an occupational hazard, as did Hemeon in 1947, the data were neither passed on to regulatory authorities nor released to independent scientists. By adhering to the industry's code of silence, the IHF violated what is arguably the most basic tenet of scientific inquiry, the free exchange of knowledge.

The US system of industrial hygiene worked on the principle of 'harmonious relations'. Until the passing of the Occupational Safety and Health Act in 1970, federal government inspectors had no right of entry to workplaces. In addition, the centrepiece of occupational safety in US factories, the TLV, was itself flawed. As early as 1940, Lanza acknowledged the inadequacy of dose-response information as a guide to risk. British technologists were highly critical of the US standard. During a discussion at a conference in New York in 1964, Gordon Addingley from the British Belting & Asbestos Company was blunt: 'We [the British industry] also have our ideas about the American limit of 5 mppcf as a safe limit. We know that there is no scientific basis for that limit whatever.'[113] The midget impinger was an imprecise instrument, which counted dust particles rather than fibres. Even where measurements were accurate, each

[112] Brown to C.J. Stover, publisher of *Asbestos*, 4 December 1936.

[113] H.E. Whipple (ed.), 'Biological Effects of Asbestos', *ANYAS* 132 (31 December 1965), i, p. 335.

part of a factory would have different dust concentrations and different ratios of benign dust to asbestos. Beneath those technicalities lay the fact that TLVs reflected economic imperatives rather than the interests of those on the shopfloor.[114] Most of the TLVs were set to a level at which visible injury would not occur, but were inappropriate with the insidious cancer hazards posed by asbestos and chemicals. According to Castleman: 'Whatever guidelines and standards existed in theory, there was *in practice* a greater degree of worker protection afforded in the United Kingdom than in the United States.'[115]

There is no better way to gauge the industry's success in shrouding the asbestos hazard in silence than the US Public Health Service's 1962 review of manufacturing. The Service found there was no reliable information on the number of asbestos workers in the US. Its best estimate was 10,000 in the primary industry and perhaps twenty or thirty times that number that were exposed in downstream manufacture.[116] Even less was known about disease rates. The diagnostic criteria for asbestosis were often poorly defined and there were no studies of workers who had left the industry. According to the Public Health Service in 1962, no reliable data were available on the incidence of asbestosis.

As the result of litigation, a number of key studies including the MetLife results from the 1930s, and Saranac's research on lung cancer from 1940, have come to light. Those documents suggest how decisions about occupational health were made. We know that suppression of knowledge had an impact upon employees, the scientific community, regulatory authorities, trade unions, and even the consumers of asbestos products. It is also clear that the purpose of suppression was to frustrate regulation and save money on workplace safety. Suppression always involved keeping knowledge of an occupational hazard from the people with the most to lose, namely the men and women who worked with asbestos. The industry's overall strategy, which remained unchanged during the three crises, was to shift the hidden costs of production onto labour or the state.

The industry learned a number of lessons during the 1930s. It learned how to control information and how to suppress evidence of disease; it

[114] See D. Egilman, 'The Asbestos TLV: Early Evidence of Inadequacy', *AJIM* 30 (1996), pp. 369–70.

[115] Castleman, *Asbestos*, p. 260.

[116] US Public Health Service, *Objectives and General Plan for Occupational Health Study of the Asbestos Products* (Public Health Service: Industry Division of Public Health, 21 August 1962), p. 2.

learned how to a reconfigure a problem about work conditions into a scientific challenge to be mediated by experts. The industry learned how to transform the systematic doubt characteristic of good science into a political weapon. It learned how to corrupt company doctors and how to use professional associations to assuage public fears. And it learned how to behave ruthlessly towards employees like John Lasin. All of those lessons proved valuable in the lingering crisis over mesothelioma.

4

The Challenge of Mesothelioma and Irving J. Selikoff

> [He] is short, he is round, he is old. Normally these attributes combine unprepossessingly in one man, but not in Irving Selikoff. He has the aura of a savant of the previous century, someone who might have been sculptured by Rodin. There is the cane, for one thing, and the fluent French, when that's necessary, and, at the core, the sheer dignifying weight of accomplishment. Selikoff is America's most prominent researcher in the field of asbestos disease.
>
> J. Hooper
> 'The Asbestos Mess', *New York Times*
> 25 November 1990

The discovery of mesothelioma and the publication of the South African work in the British medical press in 1960 had little discernible impact on the asbestos industry—at least, as regards world production of the mineral. In the 1960s, the leading companies like Johns-Manville (whose profits as shown in Figure 2.1 continued to rise through the 1960s) were still 'rich, sassy, and riding high'.[1] Between 1960 and about 1980, world production of asbestos was untrammelled and it entered its fastest period of growth.

In order to explain this paradox, it is necessary to explore events in the aftermath of the discovery of mesothelioma. This period, covering about twenty years after 1960, was characterized largely by battles over government regulation. These battles largely concerned three countries: Britain (the leading European consumer), Canada (the leading Western producer), and the US (a consumer of over a quarter of the world's asbestos

[1] M.M. Swetonic, 'Death of the Asbestos Industry' in J. Gottschalk (ed.), *Crisis Response; Inside Stories of Managing Image under Siege* (1993), p. 289.

production). South Africa, perhaps surprisingly, played little part in those battles. The political and scientific situation in that country—as we will explain more fully in Chapter 5—ensured that knowledge about mesothelioma caused no public controversy. But dissemination of knowledge was less circumscribed in America. It was here that the international alarm that asbestos was a major twentieth-century public health hazard was first raised. The key event was a conference organized in New York City in 1964.

New York Academy of Sciences Conference

A conference on the 'Biological Effects of Asbestos' was the idea of Dr Irving Selikoff (1915–1992) and Dr Jacob Churg (1910–2005), who worked at Mount Sinai Hospital in New York. They successfully canvassed the New York Academy of Sciences, and thereafter Selikoff took the lead in organizing the gathering, which was to be international in scope. Selikoff was then a relatively unknown 49-year-old New York physician, but the conference was to establish his international reputation in asbestos.[2] Brooklyn-born, from an immigrant Jewish background, Selikoff's early medical education was blighted by anti-Semitism and the Second World War, so that his medical diplomas had to be sought in Scotland and Australia. But once he began practising medicine, he soon made his mark. He became a chest physician in New Jersey and in the early 1950s was involved in the pioneering use of the drug isoniazid for treating tuberculosis. Isoniazid brought Selikoff the first of his many honours—the Lasker Award (among whose recipients featured many future Nobel Prize winners)—and his first exposure in the national media. A lucrative practice on Fifth Avenue might have beckoned, but instead Selikoff chose to establish a clinic in a New Jersey working-class community. It was there that he became aware of asbestos and its health consequences.

Some of the workers that Selikoff examined in his clinic were from the local UNARCO asbestos plant.[3] In 1961, he wrote to the company asking for access to medical records so that he could study these workers, but he was rebuffed.[4] In 1962, therefore, Selikoff contacted the locals

[2] M. Greenberg, 'Biological Effects of Asbestos: New York Academy of Sciences 1964', *AJIM* 43 (2003), pp. 543–52.

[3] Union Asbestos Rubber Company, founded in 1926, had plants in Paterson, NJ (later moved to Tyler, TX), and Bloomington, IL. Its specialty was Navy insulation products, using amosite supplied by Cape Asbestos.

[4] P. Brodeur, *Expendable Americans* (1974), pp. 7–9, 12–14.

of the International Association of Heat & Frost Insulators & Asbestos Workers (the main insulating or lagging trades union). Initially, they were suspicious. In 1955, they had naively contacted the Industrial Hygiene Foundation about their members' lung illnesses but had received only reassurances.[5] However, Selikoff's tact and commitment to discovering the truth won over the union. With access to the insulators' work and medical profiles, Selikoff assembled a team for a scientific study. He was in an excellent position to do so, having by then become director of a new Environmental Sciences Laboratory at Mount Sinai Hospital in New York City. He recruited Dr E. Cuyler Hammond, the American Cancer Society's director of statistics and epidemiology, who had already published a landmark study confirming the link between smoking and lung cancer. Like Selikoff, he was interested in applying the latest epidemiological techniques to occupational cohorts—an idea that was still relatively new. They were joined by a research associate of Selikoff's, Janet Kaffenburgh, who worked with Hammond in preparing a list of the men in the study. The list profiled their ages and dates of employment, with data on causes of death, derived from union headquarters records. These data were subsequently transferred to file cards that were sorted on the living-room floor of Dr Hammond's house by him and Mrs Kaffenburgh. The pathologist Jacob Churg verified the cause of death. What these individuals discovered became headline news at the New York conference.

It took place over three days in October 1964 at the Waldorf Astoria Hotel, with 300–400 present. It was a veritable *Who's Who* of leading scientists involved with asbestos, though there were some notable absentees. Of the fifty-three papers presented, only two were by representatives of the American or Canadian asbestos industry. The British industry (with seven papers) was better represented, but generally asbestos industrialists shunned the gathering. This was partly because they were unwilling to see the agenda seized by Selikoff—whose motives they already distrusted—and partly because they knew that the conference would be commercially damaging.

Selikoff emerged as the leading light at the conference. Besides an opening address, he was also involved with the presentation of four papers. Although Selikoff's work had already been published in the medical press, the conference showcased his work on insulators. Epidemiology is usually a thankless task, involving the analysis of thousands of individuals and

[5] John W. Kane, Secretary of IAHFIAW, to IHF, 13 August 1955; and n.d. reply from Dr C. Richard Walmer.

variables, and often producing only the most ambiguous results. Yet Selikoff's conclusions from a dataset which involved a relatively small number of men (632) were strikingly clear.[6] Asbestos insulation work could be lethal. Selikoff's first study published in 1964 covered workers who were on the union rolls in 1943.[7] When these men were tracked to 1962, they showed that insulators had an excess death rate of 25 per cent, with a heavier mortality than normal from not only asbestosis, but also lung cancer, mesothelioma, and stomach/colon/rectal cancer. This was alarming news for insulators. It was also highly damaging for the asbestos industry. Selikoff, by launching independent research, had unwittingly embarrassed the American industry, which for decades had shown little interest in studying its workers even in the major factories, let alone in the building trades and shipyards. Moreover, the objectivity of Selikoff's studies meant that they could never be seriously challenged. Even the flintiest asbestos industrialist had to admit that insulation was not a long-lived profession.

That would not be the end of it, because Selikoff's identification of a cohort of asbestos workers was a tool for further explorations. In 1968, Selikoff and his team produced another classic study which illustrated the dangerous synergy between tobacco and asbestos, by showing that asbestos workers who smoked had ninety times the risk of developing an asbestos-related cancer than individuals who were non-smokers and were not exposed to asbestos.[8] A decade later, with the help of Herbert Seidman (another epidemiologist from the American Cancer Society), Selikoff and Hammond were able to provide further data on the smoking/asbestos risk, demonstrate that even short-term exposure to asbestos resulted in significant cancer risk up to thirty years later, and extend the cohort to include every man (17,800) in the insulators' union.[9] It was the largest study of asbestos workers worldwide, which Selikoff was to continue until his death.

[6] S.D. Stellman, 'Issues of Causality in the History of Occupational Epidemiology', *Sozial- und Praventivmedizin* 48 (2003), pp. 151–61; L. Garfinkel, 'Asbestos: Historical Perspective', *CA: A Cancer Journal for Clinicians* 34 (1984), pp. 44–7.

[7] I.J. Selikoff, J. Churg, and E.C. Hammond, 'Asbestos Exposure and Neoplasia', *JAMA* 188 (1964), pp. 22–6.

[8] I.J. Selikoff, E.C. Hammond, and J. Churg, 'Asbestos Exposure, Smoking, and Neoplasia', *JAMA* 204 (1968), pp. 106–12.

[9] Hammond, Selikoff, and Seidman, 'Asbestos Exposure, Cigarette Smoking, and Death Rates', *ANYAS* 330 (1979), pp. 473–90; H. Seidman, I.J. Selikoff, and E.C. Hammond, 'Short-Term Asbestos Work Exposure and Long-Term Observation', *ANYAS* 330 (1979), pp. 61–89; I.J. Selikoff, E.C. Hammond, and H. Seidman, 'Mortality Experience of Insulation Workers in the United States and Canada, 1943–1976', *ANYAS* 330 (1979), pp. 91–116.

A key feature of Selikoff's research was the rising incidence of that dreaded disease *cancer*, especially mesothelioma. The latter hovered over the New York conference. It caught the attention of the delegates, the press, and over forty years later still jumps from the pages of the conference proceedings. Both Wagner and Ian Webster gave presentations at the conference, providing further evidence from South Africa and Britain of the link between mesothelioma and asbestos. They emphasized that victims often only had slight exposure, did not necessarily work in the asbestos mines, and often did not have asbestosis. Besides Selikoff's demonstration of mesothelioma amongst American insulators, physicians in Northern Ireland confirmed the same disease amongst Belfast shipyard workers. London physician Molly Newhouse highlighted the occurrence of mesothelioma amongst individuals who had merely lived near a London asbestos factory. A Liverpool pathologist found mesotheliomas in a lorry driver, a typist, and in workers recycling old asbestos sacks. In other words, the risk was not simply confined to individuals who worked in the asbestos industry. Wilhelm Hueper, an American cancer expert who had long warned about the cancer risk of asbestos, emphasized the new at-risk groups: repair men, maintenance men, engineers, mechanics, laboratory technicians, office workers, medical personnel, truckers, railroad workers, yardmen, construction workers, shipyard workers, automobile plant and garage employees.[10] In short, asbestos had now become an environmental problem (an aspect that is more fully considered in Chapter 7).

The industry delegates at the conference concentrated on less controversial topics, such as dust counting and geology. Only two industry papers were related to epidemiology: one by Dr Walter Smither on asbestosis at Cape's factory in London, and the other (part authored by Richard Doll) on lung cancer amongst T&N's Rochdale workers, which offered the reassuring possibility that asbestos-related lung cancer had been eliminated. Generally, the industry took refuge in obscurity, disingenuousness, and hypocrisy. For example, Cape director Dr Richard Gaze cast doubt on the linkage between asbestos and mesothelioma, arguing that other minerals and elements could be to blame—conveniently forgetting that his own company had only two years previously commissioned a report that demonstrated the opposite. Dr Kenneth Smith, at Johns-Manville, in a high-flown speech that cited Thomas Jefferson, urged the conference

[10] H.E. Whipple (ed.), 'Biological Effects of Asbestos', *ANYAS* 132 (31 December 1965), i, p. 189.

to examine 'our problem as honest scientists with no ulterior motive, in a scientific manner'.[11] Presumably, Smith did not include Selikoff in his group of honest scientists, as only a few days before the conference he had denounced Selikoff to his bosses as 'ambitious and unscrupulous, and... out to make a name for himself at the expense of the asbestos industry'.[12]

The industry was particularly vexed by the conference press bulletins which were issued daily, some of which highlighted cancer and environmental problems. Although the bulletins were no more than statements of fact, the Cape and T&N contingent concocted their own press release (with the backing of English physicians, such as Dr John Gilson) which deplored the 'sensationalist' tone of the conference, because it could cause 'unjustified fears' amongst the public. Selikoff ignored them. The proceedings of the conference were published and thousands of copies of the fat 700-page buff-coloured volume were sent out in 1966, thus alerting occupational health physicians and government doctors that asbestos was now a potential public health hazard. Forty years later, the proceedings remain in print.

The conference greatly increased Selikoff's profile, so much so that in the following year he was in Washington explaining asbestos hazards to President Lyndon B. Johnson. Apart from his scientific work, the mere presence of Selikoff became a problem for the asbestos industry, since he emerged as a powerful spokesperson for asbestos victims. This was a logical outgrowth of Selikoff's activities at Mount Sinai Hospital. What had begun as a fashionable urban hospital on Upper East Side Manhattan was by the 1960s flanked by poor working-class areas. It was this changing social and physical environment that prompted Mount Sinai to form America's first urban department of community medicine, where Selikoff's Environmental Sciences Laboratory was located. Besides connecting directly with urban problems, Selikoff could enjoy the resources of a major clinical facility and a busy medical school. He soon assembled a large team that included contemporaries, such as Bill Nicholson and Herb Seidman (both statisticians), and others (such as Philip Landrigan) who would carry his work forward. This team would eventually undertake additional statistical studies, including important projections by Nicholson on morbidity and mortality from ARDs into the

[11] Whipple, 'Biological Effects', p. 685.

[12] *'Johns-Manville v. Home Insurance, et al.'*: Judicial Council Co-ordination Proceeding No. 1072. Memo, 22 May 1986, p. 141.

twenty-first century,[13] and clinical field studies of asbestos-exposed workers in various trades, including the shipyards, Quebec asbestos mining, sheet-metal workers, and of household members of asbestos-exposed workers. Selikoff also helped launch the Collegium Ramazzini as an international forum for occupational health issues.

Selikoff was a bear of a man, with a natural inclination to dominate any gathering. He loved organizing conferences, in which he played the role of master of ceremonies to the full, often by sitting centre stage and directing the proceedings and discussions. His relationship with his professional colleagues was gregarious and cosmopolitan, occasionally tinged by the tensions created by Selikoff's complex personality. A colleague recalled: 'Like many Russian Jewish men, [Selikoff] had gargantuan visions, appetites and moods... [with] an enormous capacity for seeing the big picture.'[14] He would often rise in the early hours of the morning to start work in his New Jersey home, before driving into Manhattan across the Washington Bridge to start at Mount Sinai at 7.00 a.m. or earlier—the beginning of an eighteen-hour day. This legendary energy, however, was not directed towards the accumulation of wealth or the customary trappings of success. What Selikoff wanted was to use community medicine as an effective vehicle for social change.

His preferred method was to work from the inside of the system, which (as he well knew) was marked by a grossly unequal power relationship between American industry and labour organizations. However, he was acutely aware of the human consequences of occupational injury and was prepared to publicize ARDs as widely as possible. As a natural communicator, who knew how to work almost any kind of audience and who certainly enjoyed the limelight, Selikoff and the media were natural bedfellows. Indeed, Selikoff has been described as the first media-savvy medical man. This was to bring Selikoff brickbats from his medical critics who regarded fraternizing with the media as unprofessional and political, yet saw nothing political about their own work with asbestos companies.

For the industrialists, Selikoff was the physician from hell: they had no scientific studies to counter him and lacked his flair for publicity. Johns-Manville's public relations man Matthew Swetonic, no slouch himself at the media arts, conceded that Selikoff was 'one of the most talented

[13] W.J. Nicholson et al., 'Occupational Exposure to Asbestos: Population at Risk and Projected Mortality 1980–2030', *AJIM* 3 (1982), pp. 259–311.

[14] Personal communication to G. Tweedale from Dr Elihu Richter, 17 December 2006.

medical publicists of the age'—or as he put it, a Walter Cronkite of the scientific community.[15] Selikoff always provided good copy and was always accessible:

> There was never any: 'Let me think that over and I'll get back to you'.... He understood deadlines and what constituted a 'good' story, and was always patient, never condescending... [and]... he perfected almost to an art form... the 'white coat approach.' Whenever he appeared on television or was entertaining reporters at Mount Sinai, he never failed to appear in a white lab coat, as if he had just emerged from examining a patient or performing some experiment. While a standard technique today for conveying scientific credibility, in the late 1960s it was new and fresh and never failed to impress.[16]

It was the sudden public prominence of Selikoff and his findings that launched the industry's major defence of its product. David Egilman and Robert Proctor have identified a number of tactics used by manufacturers of toxic materials to mount a defence. These have included organizing front and public relations companies, influencing the regulatory process, discrediting critics, and manufacturing an alternative science (or history).[17] In the following sections of this chapter, these strategies are explored; in Chapter 5, a detailed study is presented of the industry's alternative science.

A Dangerous Man

Shooting the messenger is a well-recognized syndrome in science, especially in the field of occupational health, where scientists who threaten commercial interests are—to say the least—not applauded.[18] The American industry had its sights trained on Selikoff immediately after the New York conference. The ATI drafted a lawyer's letter to Selikoff and the New York Academy of Sciences that warned Selikoff and Hammond about the dangers of talking to the press and triggering 'damaging and misleading news stories'.[19] When this 'frightener' did not work,

[15] M.M. Swetonic, 'Why Asbestos?'. Paper presented to ATI, Arlington, Virginia, 7 June 1973.

[16] Swetonic, 'Death of Industry', p. 296.

[17] R. Proctor, *Cancer Wars* (1995), pp. 100–32. S.R. Bohme, J. Zorabadeian, and D.S. Egilman, 'Maximising Profit and Endangering Health: Corporate Strategies to Avoid Litigation and Regulation', *IJOEH* 11 (2005), pp. 338–48.

[18] R.R. Kuehn, 'The Suppression of Environmental Science', *American Journal of Law and Medicine* 30 (2004), pp. 333–69.

[19] Cadwalder, Wickersham & Taft letter, 26 October 1964.

smear tactics followed. Selikoff's Jewishness and foreign qualifications offered scope for questioning his medical credentials and the quality of his research. While that line of attack was being evaluated, the industry carefully monitored Selikoff's public appearances and scavenged for anything that could be used against him. As early as 1965, one industry correspondent wrote: 'Our present problem is to find some way of preventing Dr. Selikoff from creating problems and affecting sales. A direct approach (attacking his character) might be more damaging than helpful and I am only suggesting that we explore, at this time, all avenues open to us.'[20]

The industry was keen to confine any debate to specialist forums, though even here Selikoff could do damage. As one industry executive noted, having attended one of his seminars: 'For whatever reason they attended the seminar, they [the attending physicians] were nearly all scared to death of asbestos when they left.'[21] After hearing Selikoff speak, Johns-Manville vice-president Jack Solon commented to a colleague: 'The man Selikoff, is an excellent presenter; he is in constant command of the situation and he is convincing. If you'd like to hear how convincing—I'll gladly let you have a set of the tapes to listen to.'[22] In June 1970, Selikoff gave a seminar at the University of Toronto. As usual, it was attended by industry representatives who recorded the talk for 'error analysis'.[23] Four months later a general meeting of the ATI, held at Charlotte, North Carolina, discussed at length the best way to counter Selikoff. Dr J. Goodman, a consultant with Raybestos-Manhattan, remarked that Selikoff was a 'dangerous man and that the asbestos industry was going to have to learn to combat his tactics'.[24] Goodman rejected the suggestion that the ATI protest to the grievance committee of the American Medical Association (because it rarely acted on complaints) and instead advised pressuring the Medical School at Mount Sinai. At an international conference of asbestos industrialists in London in 1971, the British industry reiterated the demand for Selikoff to be countered at medical forums. Solon's colleague Bill Raines replied that this was not so easy: 'I restrained my

[20] F.H. Edwards to J.M. Briley, 29 November 1965. Confidential 'Report: Irving J. Selikoff, M.D.'

[21] F.H. Edwards to J.M. Briley, 29 November 1965. Confidential 'Report: Irving J. Selikoff, M.D.'

[22] Report on the Recent Presentation in Toronto by Dr. Selikoff. F.J. Solon, Denver, 18 May 1973.

[23] Swetonic to W. Raines and Carl Thompson, Hill & Knowlton, 30 June 1970. Postgraduate Seminar on Asbestos Disease, Toronto.

[24] Minutes General Meeting of ATI, Charlotte, NC, 4 February 1971.

reply to the understatement that we at Johns-Manville and, subsequently the AIA [Asbestos Information Association], have recognised this need for at least six years but have thus far been unable to find the qualified individual who is willing to do the job.'[25]

Solon had an extended correspondence with Selikoff in which he invariably addressed his adversary as 'Dear Irving'. In a letter dated 16 March 1973, Solon ended with the comment: 'I am pleased with our recent exchanges—and I hope you are, too.'[26] A week later Solon contributed to a confidential memo titled 'Discredit Selikoff'. It featured two columns: one containing statements by Selikoff and the other a list of point-by-point rebuttals, so that every statement by Selikoff could be undermined.[27] In the hope of discrediting Selikoff, Johns-Manville routinely copied and circulated recordings of his public statements to company executives and allied medical researchers. Solon sent copies of one such presentation to Dr Marcus Key, the US Assistant Surgeon General and Director of National Institute of Occupational Safety & Health (NIOSH), to the heads of the AIHA (American Industrial Hygeine Association) and the ACGIH, the Secretary of Labor, and the Assistant Secretary at OSHA. Solon explained to his colleagues: 'All this is by way of letting important people in government and the professions know the kind of misleading acts Dr Selikoff continues to commit. This is the kind of positive aggressive approach we intend to take on our environmental affairs problems.'[28]

The suspicion of Selikoff was transatlantic. Robert Murray, the Trades Union Congress' (TUC) medical adviser, felt unable—thirty years after the event—to forgive Selikoff for inviting the press to the New York conference, because 'the media started off from there the campaign against asbestos', which Murray thought was 'misguided'.[29] Even those better placed to recognize Selikoff's medical contribution, such as T&N physician Dr Hilton Lewinsohn, on occasion demeaned him. Lewinsohn told T&N directors, after a conference in Lyon in 1973 (where Selikoff had disassociated himself from various pro-industry public announcements):

Selikoff was obviously under stress and behaved in an emotionally disturbed manner... [though he] behaved in his usual charming manner towards me and

[25] Jack Solon, Minutes International Conference of Asbestos Information Bodies, London, 24–25 November 1971.

[26] Confidential letter from Solon, Denver, to Selikoff, New York, 16 March 1973.

[27] Memo: Discredit Selikoff. W.L. VanDerbeek: Environmental Policy Committee, Johns-Manville, 30 March 1973, to F.J. Solon.

[28] Memo: Subject—Important letter Dr Selikoff from F.J. Solon to Environmental Policy Committee, Johns-Manville, 18 June 1973.

[29] Murray testimony, *Milton Jenkins v. Fibreboard*, County Court Los Angeles, 2 July 1993.

was very friendly indeed. I think that he is jealous of the UK's success in handling the asbestos and health problem and because of his emotional instability readily becomes involved in the political problems in the USA, allowing himself to do so in order to boost his alter ego.[30]

In January 1974, representatives from US and UK industries met in Montreal to review the future of asbestos. On the agenda again was neutralizing Selikoff with 'a medical man capable of standing [him] off . . . possibly, one man in North America and one in Europe'.[31] Unfortunately for the industry, no one of sufficient calibre emerged to stand off the emotionally unstable New Yorker. In 1976, the industry's lobby group, the AIA/NA (see below) recruited Hans Weill, a pulmonary disease specialist at Tulane University in New Orleans, to advise them on the state-of-the art defence and to identify medical experts for use in civil defence suits. But he was not really a counter to Selikoff, apparently agreeing that an early performance in the witness stand had been 'terrible'.[32] Another consultant hired in 1976 was Philip Enterline, a professor of biostatics at the University of Pittsburgh; however, he mostly confined his activities to a low-profile literature review that made the controversial claim that America suffered from an 'international lag' in recognizing the lung cancer hazard.[33] The companies could rely on their old ally Dr George Wright of St Luke's Hospital in Cleveland, Ohio, but he was no headline-grabber.

Tellingly, the industry was not unanimous in condemning Selikoff. In its more candid moments, even the industry accepted that Selikoff was 'a healthy nemesis'.[34] One asbestos-cement (a/c) plant manager attended a three-hour marathon talk that Selikoff presented to the AIA/NA in New York in 1974 and commented:

I was very impressed by him and his sincerity and dedication to his studies. I had a misconception of his motives and activities before this meeting. I talked to him, personally, during lunch and I believe his motives are aboveboard. We all know about his stand and the things he has published, which seem to be substantiated.[35]

[30] Lewinsohn to David Hills, 6 March 1973.

[31] Minutes of the Joint Meeting of the AIC/UK, IAI/NA and QAMA, 24 January 1974, Montreal, Canada. AIA/QAMA.

[32] 'Observations and Impressions of Dr Hans Weill', n.d.

[33] P.E. Enterline, 'Asbestos and Cancer: The International Lag', *American Review of Respiratory Disease* 118 (1978), pp. 975–8.

[34] QAMA, 10 August 1967, in Murray Bay, Quebec.

[35] Horace Beasley memo re. AIA meeting 21 February 1974 to Capco (Birmingham, Alabama), 25 February 1974.

This was not a surprising comment, given that Selikoff's pronouncements on asbestos during the 1960s and 1970s were strikingly temperate. He deliberately avoided confrontation with the industry—partly because, as a born optimist, he believed that faced with irrefutable evidence of a hazard, the leaders of the big corporations would change their behaviour (a view that his colleagues at Mount Sinai did not always share). On occasion, Selikoff actually avoided the limelight. He gave testimony in two of the early landmark legal cases, but thereafter avoided the drama of the courtroom and the role of the expert witness, not only because it would have been a drain on his time and made his confidential trade union medical files open to legal scrutiny, but also because he felt that antagonizing industry would not help his broader agenda. A colleague recalled: 'He always sought to find a means to cooperate with industry... even if it did not achieve much', because he believed long-term goals would be better served.[36]

In 1968, Selikoff stated that it was 'idle and largely inaccurate to "blame the industry" for the present situation'—though he did see public health authorities as the key to solving the crisis.[37] In that year, a joint health programme was launched by Selikoff's Mount Sinai team in association with Johns-Manville. In an article publicizing the partnership, Selikoff stated: 'It is much nearer the truth to say that both industry and labor shared in ignorance and neglect of the problem. Science and medicine are also at fault here for inadequate attention to environmental and occupational health.'[38] On the national stage, Selikoff's comments were no less conciliatory. When he gave evidence in 1970 to the Senate hearings into the draft proposals for the OSHA, he asked rhetorically: 'Who killed Cock Robin?', then answered: 'No one.... We have all failed.'[39] In 1977, Selikoff and his Mount Sinai scientists published a major textbook, *Asbestos and Disease*. It began with a foreword by Johns-Manville medical director Paul Kotin (a friend of Selikoff), contained no calls for banning asbestos, but did have sections on dust control—illustrated with photos of workers mixing a/c in enclosed bags and spraying asbestos within sealed areas. The book ended with a section titled 'Cooperative Effort', which

[36] Jock McCulloch interview with Stephen Levin, Mt Sinai Hospital, New York, 27 October 2006.

[37] Selikoff statement, 7 March 1968, before the House Committee on Education and Labor, Select Sub-Committee on Labor, HR 14816, the Occupational Safety & Health Act of 1968.

[38] I.J. Selikoff, 'Partnership for Prevention—The Insulation Industry Hygiene Research Program', *Industrial Medicine* 39 (April 1970), pp. 21–5.

[39] Quoted in J. MacLaury, 'The Job Safety Law of 1970: Its Passage was Perilous', *Monthly Labor Review* (March 1981), pp. 18–24.

had an upbeat view of the relationship between industry, labour, and government.

Only by 1980, it seems, did Selikoff begin talking in terms of a public health catastrophe—though this was still sooner than most of his contemporaries. On his deathbed, he apparently regretted he had waited so many years to get asbestos banned and wondered whether the effort he had directed at research was misspent: 'I could have saved a lot of those boys instead of waiting.'[40] By then, Selikoff's 'partnership' with industry was at an end. In that year, Johns-Manville complained directly to him about the adverse effect that his articles were having on sales and that if such 'nonsense' continued 'the industry would be "regulated" out of existence by sensationalism.'[41] Selikoff was also warned that the industry might be forced to destroy his credibility through litigation. In 1982, another review was requested by Johns-Manville of the 'serious question of whether Selikoff actually has a medical degree'.[42]

The dubious claims about Selikoff not having a medical qualification surfaced in *The Washington Times* on 21 April 1989—coincidentally when Selikoff was facing a barrage of criticism for his view that asbestos in buildings might be a problem (see Chapter 7). The report concluded that 'if the epidemiological evidence against Dr Selikoff's work isn't enough to persuade bureaucrats to rethink our asbestos policy, perhaps the dispute about his credentials is'. This attack was renewed at greater length by another journalist, Michael J. Bennett, who wrote a book in 1991 lambasting Selikoff and sneering at his explanation that his education had been affected badly by the war.[43] This smear campaign was never likely to succeed, though bizarrely attempts to discredit Selikoff have continued to the present day. In 2004, asbestos industry consultant Peter Bartrip tried something even the industry never dared while he was alive: an academic rationale for the hoary claim that Selikoff had no medical degree.[44] The accusation is false, but the fact that it was made at all is a backhanded compliment to Selikoff's enduring influence.

[40] Quoted in E.D. Richter and R. Laster, 'The Precautionary Principle, Epidemiology and the Ethics of Delay', *Human and Ecological Risk Assessment* 11 (February 2005), pp. 17–27. See also J. McCulloch and G. Tweedale, 'Science is Not Sufficient: Irving J. Selikoff and the Asbestos Tragedy', *New Solutions* 17 (2007), pp. 293–310.

[41] *'Johns-Manville v. Home Insurance'*, p. 204.

[42] *'Johns-Manville v. Home Insurance'*, p. 207.

[43] M.J. Bennett, *The Asbestos Racket: An Environmental Parable* (1991), pp. 94–7.

[44] J. McCulloch and G. Tweedale, 'Shooting the Messenger: The Vilification of Irving Selikoff', *IJHS* 37 (2007), pp. 619–34. See also D. Egilman et al., 'P.J.W. Bartrip's Attack on Irving J. Selikoff', *AJIM* 46 (2004), pp. 151–5.

Welcome to Asbestos Country

While attempting to tarnish Selikoff's reputation, the industry moved quickly to implement another key strategy—the organization of front institutions. Within a year of Selikoff's New York conference, the ATI met in Thetford Mines, Quebec.[45] The minutes of that meeting have survived and make fascinating reading. It involved all the heads of the Canadian and American asbestos industry (with their legal counsel) besides those from the leading overseas companies such as T&N. Invited guests included Dr Lewis J. Cralley, an official from the US Public Health Service (a key industry ally who helpfully remained unconvinced that asbestos caused mesothelioma). The hundred attendees were warmly welcomed to 'asbestos country with its man-made mountains'. Michael Messel, the president of Lake Asbestos of Quebec, reviewed the Canadian industry, which employed over 6,000 people and exported annually nearly 1.5 million tonnes of chrysotile. The health discussion of the workers was led by QAMA attorney Ivan Sabourin, who argued that cancer was not necessarily linked with asbestos. This was elaborated by Dr Kenneth W. Smith, who again stressed the necessity of the industry 'telling the truth'. Only the year before, Smith had argued that he did not believe that there was a serious health hazard and denied he had ever seen any asbestos-related lung cancers at Johns-Manville.[46] He reiterated to his Thetford audience the long-standing mantra of the Quebec industry: the overall health of Johns-Manville long-time employees compared favourably with industry in general, and pointed to the steadily expanding membership of the Quarter-Century Club.

Sabourin outlined the action that the industry would take in dealing with its non-existent problems. In particular, QAMA intended to rekindle the links of the 1920s between industry and academe, by seeking an 'alliance with some university, such as McGill, so that authoritative background for publicity can be had'. By 1966, the Quebec asbestos mining industry had funded and helped establish the Institute of Occupational & Environmental Health (IOEH) at McGill University. The credit line ran directly across the international border, as it was Johns-Manville who provided over half of QAMA's funds. This Institute had as its lead researcher, the English epidemiologist Dr Corbett McDonald, whose scientific work is described in Chapter 5. The Quebec industry had followed two models in

[45] Minutes ATI General Meeting, Hotel Le Provence, Thetford Mines, 4 June 1965.
[46] *'Johns-Manville v. Home Insurance'*, pp. 139–40.

Table 4.1. Asbestos Industry Trade Associations

Date	Trade Association	Location
1931	Quebec Asbestos Producers' Association—renamed Quebec Asbestos Mining Association (QAMA) in 1948	Canada
1944	Asbestos Textile Institute (ATI)	US
1957	Asbestosis Research Council (ARC)	UK
1967	Asbestos Information Committee (AIC)—renamed Asbestos Centre in 1979	UK
1970	Asbestos Information Association/North America (AIA/NA)	US
1976	Asbestos International Association (AIA)	UK
1982	Comité Permanent Amiante (CPA)	France
1984	Asbestos Institute (renamed Chrysotile Institute in 2004)	Canada

setting up the Institute. First was the Asbestos Research Council (ARC) in the UK, whose leading members liaised with the Canadians. Apparently many of the proposals put forward by the Canadians were 'elaborated largely from the *modus operandi* of the ARC'.[47] Second was the Tobacco Research Council (TRC), which had pioneered the blending of research and public relations. The Canadians felt that the TRC had been over-publicized, but believed that they could keep a lower profile by controlling the information flow from the IOEH.

Across the Atlantic, the ARC was also shifting through the gears. Its annual funding was boosted from £7,000 at the start of the 1960s to about £30,000 in 1970 (the budget had reached over £80,000 a year by the mid-1970s). The ARC also found a new home under the umbrella of a privately financed Institute of Occupational Medicine in Edinburgh. The ARC, which was the creature of the leading companies such as T&N and Cape Asbestos, began sprouting offshoots. In 1967, the ARC sponsors (with another firm, Central Asbestos) established the Asbestos Information Committee (AIC) to handle publicity and lobbying (see Table 4.1). To provide expert advice to the AIC, the industry turned for help to American public relations advisers Hill & Knowlton, the largest PR firm in the world that had honed its skills in defending the tobacco companies. The AIC was ostensibly separate from the ARC with an office in London (actually it was in Hill Knowlton's premises), but in reality they were only different sides of the same coin. Its annual budget in the mid-1970s was similar to the ARC: by the end of the 1970s it had reached nearly £200,000 a year.

[47] J. Waddell to R.M. Bateman, 12 April 1966.

With the IOEH (Canada) and ARC (UK) established, the triangle was completed in 1970 in New York with the creation of the AIA/NA. This was designed to bring 'the communications side of the [asbestos] problem under control'.[48] Again, it was Hill & Knowlton who advised Johns-Manville (and seven other US companies) to set up a similar organization to the ARC. Initially, Johns-Manville planned to fund and staff the AIA/NA, which would deal with the media, work with the government to regulate workplace conditions, and eventually set up a legal arm to assist the whole industry in defending itself against liability. The latter activity would soon swallow the entire trade association.

By the end of the 1960s, the key pillars of the industry's defence were in place in the US, Canada, and the UK, with a number of other information bodies at various stages of evolution in other asbestos-using countries. Inevitably, international cooperation followed. In 1971, M.F. Howe of the ARC organized an International Conference of Asbestos Information Bodies in London. When Howe addressed the 35 delegates, he noted that until very recently the AIC was the only asbestos information body in the world: now eleven such bodies were being formed. The keynote of the conference was inevitably public relations and how the industry was using a positive spin to stress the merits of asbestos to the public, while defending the product from what he believed were unjustified attacks, especially from the media. Nowhere in the conference was there any discussion of the physical suffering caused by ARDs, or medical treatment or financial help for victims. The words of the delegates were comforting. According to Dr Walter Smither, asbestosis was compatible with a 'reasonably full life' and chrysotile—even amosite—could be used safely.

A Johns-Manville's executive, Bill Raines, attended the conference as a representative of the AIA/NA and noted that overall in Europe the threat of government regulation was muted. For example, according to Raines, the Italian government paid little attention to the asbestos hazard and 'industry is in no hurry to call it to their attention'. But Howe warned that sooner or later the tempo would increase everywhere: 'You will find that, as we and our American colleagues have found, that time is not on your side... [because] sleeping dogs wake up suddenly and use their voices and their teeth.' He characterized the problem as one of 'attacks and defences' and urged the delegates 'to look to your defences now'.[49]

[48] Swetonic, 'Death of Industry', p. 296.

[49] 'International Conference of Asbestos Information Bodies', London, 24 and 25 November 1971, pp. 65–6.

Galvanized by this meeting, by 1973 discussions were underway regarding the forging of even closer international links between the ARC, the AIA/NA, and QAMA. As Howe put it: 'The problems of asbestos and health move freely across national boundaries, particularly where there is a common language. It is suggested that in the defence of asbestos, the asbestos industries in USA, Canada and the UK are particularly interdependent, and very close collaboration seems necessary.'[50] This was realized in 1974, when the hoped-for conference between the American and British information bodies and QAMA took place in Montreal. By now, QAMA's support for the IOEH was bearing fruit and McDonald's studies were described as of 'great value to the mining industry in the defence of its position'.[51] Reassuringly, it was said that attacks on asbestos were generally neither strong nor vociferous among the French-speaking Canadian media in Quebec—something one delegate attributed to the trouble of translating Selikoff's statements into French. Even in his absence, Selikoff stalked the Montreal proceedings. On the basis that it was best to 'know your enemy', the delegates agreed that engagement with Selikoff was not be avoided, but on the contrary was to be encouraged at international conferences, so that 'Selikoff and his people are properly *marked*'.[52]

By the mid-1970s, plans were well advanced to launch a fully fledged international body. It was a logical development from the London conference in 1971 that had been made easier by the economic linkages (such as cartels, technical exchanges, and commercial agreements over fibre) that had operated globally for decades between the leading firms. In 1976, the AIA was incorporated, at a registered office in London with Sir Neville (Jimmy) Stack as its first director-general. As a former Air Chief Marshall and Gentleman Usher to the Queen, Stack was evidently hired more for his high-level social contacts than for his technical knowledge. The AIA spanned the globe and involved companies such as Johns-Manville in America, James Hardie in Australia, Eternit in Belgium, QAMA, and the South African mining interests. During the late 1970s, the AIA staged annual 'summits', mostly in Europe, involving countries such as France, Canada, Italy, the UK, the US, Germany, and the Benelux countries. By 1984, the AIA had twenty-eight members (in thirty-four countries), and

[50] M.F. Howe, 'Notes on Closer Collaboration between AIC/ARC, AIA/NA and QAMA', 4 September 1973.

[51] Minutes of Joint Meeting of AIC/UK, AIA/NA and QAMA, Montreal, 24 January 1974.

[52] T&N memo copied to David Hills, re. Meeting of AIC/ARC, AIA/NA and QAMA, Montréal, 24 January 1974.

when it ran its biennial conference in the Park Lane Hotel, London, Sir Richard Doll was amongst the speakers.

The launch of the AIA in London reflected the importance and influence of the European asbestos industry. Asbestos consumption in the leading European countries (UK, France, Germany, and Italy) was well over 700,000 tonnes a year in the early 1970s, surpassing the US and even the US and Canada combined. The AIA, based in Europe and with most of its meetings taking place in London, became heavily involved in the defence of the European a/c industry at a time when the European Economic Community (EEC) suddenly began to take an interest in regulating the industry. The AIA reacted to forthcoming EEC directives (the first time the EEC had legislated for asbestos) by forming a special Advisory Council to pressure legislators. The AIA opposed a ban on crocidolite (still used in a/c in countries such as Belgium), the phasing-out of asbestos, and the listing or compulsory use of substitutes. Opposing a ban on crocidolite was neither politically nor scientifically respectable for most British manufacturers, but the strategy suited Eternit interests in Europe and was also received favourably by UK crocidolite users such as Cape Asbestos and by South African interests (AIA minutes were routinely copied to the South African industry). The pressure produced results; the EEC's directive in 1983 was a bland affair that only banned spraying, the use of asbestos in toys, and plugging compounds. Crocidolite was allowed in some uses and was not completely banned by the EEC until 1991.

The AIA may have been Eurocentric, but it acted as a useful bridge between the Old World and the New. By the beginning of the 1980s, the AIA and its Canadian contacts had begun to channel their efforts into masterminding a World Symposium in Montreal, which the AIA believed would provide 'an unbiased atmosphere and that hopefully... much of the distorted publicity which had been doing asbestos so much harm would be corrected'.[53] In this the French played a part. In 1982, the Comité Permanent Amiante (CPA) was formed in France to operate at the interface between industry, government, medicine, and trade unions. It was funded and controlled by the French asbestos industry (Eternit and St Gobain), but it was legitimized by the collusion of the French government and medical authorities. In effect, French public policy on asbestos was manipulated from the CPA's Paris address: 10 Avenue de Messine.[54] A key individual in the foundation of the CPA was Marcel Valtat, a political

[53] AIA Memo and Notes to Members, 31 July 1980.

[54] F. Malye, *Amiante: Le Dossier de L'Air Contaminé* (1996); C. Faesch, *Salariés de L'Amiante, Employés de L'Indifférence* (2002).

fixer, who in the 1960s had set up a high-level public relations agency for industrial interests. The CPA chairman was Jean-Pierre Hulot, who was linked through his own lobbying consultancy (Europraxis) with key industrial and government communications agencies. Europraxis received F700,000 a year from the asbestos industry, which was spent on public relations. Hulot liaised with the Canadians and he is reported to have contributed to the Montreal Symposium through subsidies paid by the French government.

The Montreal conference was a grand affair, with 660 delegates from forty-eight countries and over 150 journalists. It fully achieved the industry's aims. The AIA noted with satisfaction that the press, such as the *British Medical Journal*, had largely followed the Canadian industry's line that asbestos was something with which society was going to have to live for some time yet.[55] The World Symposium was soon followed by an Ontario Royal Commission on asbestos, which to no one's surprise gave a green light to Canadian chrysotile exports. This was the prelude to the formation in 1984 of the last major lobby group, the Asbestos Institute, based in Quebec. As asbestos was forced into retreat in the US and Europe, the Asbestos Institute took over the role of the AIA in acting as a nexus of the worldwide web of asbestos organizations. Its job was considerably eased by the generous support of a range of donors that included most notably the Federal and Quebec Government.

In October 1983, the industry's organizational strengths were apparent at a meeting of the International Labour Office in Geneva. Industry representatives from Canada, South Africa, Belgium, Australia—to name only a few asbestos-producing countries—were so numerous that they outmanoeuvred workers' groups. Of the twenty-one delegates involved, sixteen had industry connections (some were from asbestos information bodies) and the chairman was nominated by the Canadian government. Incredibly, rank-and-file trades unionists were barred from any involvement in discussions and the employers were adamant that substitutes for asbestos should not be mentioned. Nancy Tait, who represented a UK victims' group, walked out in disgust. The proceedings were dominated by employers' attempts to introduce its codes for the continued use of asbestos under 'safe' and 'acceptable' thresholds.

[55] AIA Notes for Associations, 31 August 1982. See also Canadian Asbestos Information Centre, *Asbestos, Health and Society Proceedings of the World Symposium on Asbestos* (1982), J. Dunnigan, 'Closing Remarks', pp. 539–42; D. Gloag, 'Can Society Live with Asbestos?', *BMJ* 284 (12 June 1982), pp. 1728–9.

Shoot-Out over Thresholds

Despite the threat of mesothelioma and the rising trend in ARDs, and despite Selikoff, in the 1960s and the 1970s no one was going to close down a multinational industry that employed thousands of workers. In the 1960s, the American asbestos industry alone employed 200,000 and that number does not include the many thousands involved in the downstream use of the material. Governments, companies, and many trade unions accepted that reality and began to look for alternatives to banning the product. One was to cease using asbestos in the most hazardous occupations, such as lagging. The other was to undertake intensive research into asbestos fibres, so that a more sophisticated understanding of the mechanisms of carcinogenesis might save lives and the industry. Ending crocidolite use had obvious attractions for some countries, if they did not mine the fibre or use it extensively. Another option was to use substitutes for some applications and begin to phase out asbestos. The most attractive option, however, was to control exposure to asbestos dust. On the basis of the 'dose makes the poison' (one of the industry's favourite maxims), it was argued that safe levels could be found, which would not expose workers to fatal illnesses.

Beginning in the 1960s, dust control became a major battleground between government, industry, and the trade unions. Until that time, the global industry had been unregulated, in the sense that no country had set a legally enforceable dust threshold. Dust controls on the South Africa mines, for example, were virtually non-existent. In the US and Europe, dust control in the asbestos-using industries—especially lagging and building work—was also lacking (essentially because the nature of the work meant that dust could not be controlled within the confines of ships, power stations, and high-rise blocks). In most of the asbestos factories in the US and Europe, dust control was scarcely more impressive. In a few factories in the UK (which was probably the world leader in dust control), dust extraction limited some of the worst effects, so that workers found they could see their mates through the haze of the workshops. But even in these factories, some areas were unregulated and workers were offered little protection, apart from inadequate masks. Anyone who worked with asbestos was breathing fibres; so too were asbestos users outside the factories.

Improving dust control meant some method of measuring dust exposure and also linking such data with the incidence of ARDs, so that a safe threshold could be found. By the 1960s, the midget impinger had been superseded by the membrane filter. This involved passing sampled air through a porous cellulose ester membrane and then counting the trapped fibres with an optical microscope. The result was usually expressed as so many fibres per cubic centimetre (f/cc) or fibres per millilitre (f/ml). But it was difficult to monitor workers on every job, especially outside the factories, and there were also other problems. Should workers be individually sampled or only the general level of dust in the factory? What about unusual peaks in dust exposure or the effects of multiple exposures? How were environmental exposures to be monitored? Counting fibres under an optical microscope was also enormously difficult, when those fibres were often splitting and mixed with contaminants. Crucially, most asbestos fibres would not be seen anyway, because only an electron microscope was capable of resolving the tiniest fibres. Even if fibres could be counted, devising a safe level was hindered by the fact that no reliable dust data existed before the 1960s to correlate with contemporary ARDs. It was for good reason that some critics described the results as 'guesswork' rather than scientific fact. Most workers found dust counts incomprehensible anyway. In 1972, Selikoff received this blast of common sense from asbestos insulation worker J.H. Mackley:

> Could somebody please tell me what the hell 5f/ml of dust is? Can you spot that amount of dust? Is it equivalent of sawing a piece of six-inch pipe-covering in half? Or is it similar to having a room full of dust that looks like a snowstorm? If the dust is too much for the working area—what can I do to regulate it?...please let me understand what is hazardous to my health. I think it would be a lot better to eliminate the dust and try to clean up the trade.[56]

Political and economic imperatives, however, demanded an asbestos threshold. In America, crude attempts to set a scientific threshold had involved the ACGIH, the non-governmental body of engineers and toxicologists that offered an open door to industry influence. Its TLVs for workplace exposures to toxic substances such as asbestos had been adopted by the federal government. But the TLVs were non-binding, never enforced, and have been criticized for being 'speed limits without tickets'.[57]

The first important attempt to set a threshold occurred in Britain in the 1960s in the aftermath of the publicity about mesothelioma.[58] It involved

[56] J.H. Mackley (insulation worker) to Selikoff, 31 December 1972.

[57] B.I. Castleman, *Asbestos: Medical and Legal Aspects* (2005), p. 223.

[58] M. Greenberg, 'The British Approach to Standard Setting: 1898–2000', *AJIM* 46 (2004), pp. 534–41; Greenberg, 'Revising the British Occupational Hygiene Society Asbestos Standard: 968–1982', *AJIM* 49 (2006), pp. 577–604.

the British Occupational Hygiene Society (BOHS), which was an unofficial and self-appointed body of industry physicians and scientists. Amongst its members were representatives of the leading asbestos companies, who, as government regulation loomed in the 1960s, collected data upon which a threshold could be based. The BOHS concluded that two fibres of *chrysotile* asbestos per cubic centimetre (2 f/cc) would provide a workable threshold. In lay terms, this was two fibres per thimbleful of air. It sounds very little—yet it could mean a worker inhaling hundreds of millions of asbestos fibres every day. This threshold was supposed to cut the risk of asbestosis to about one per cent over a working life, but the threshold took no account of cancer because the BOHS had not factored this into their calculation (even though it was obvious that mesothelioma had triggered the search for the threshold). Moreover, even the data were highly questionable, based as they were on a selective survey from a single factory—T&N's factory in Rochdale—conducted by its physician Dr John Knox. Even the BOHS admitted that the data were 'scanty' and deficient. However, in 1969 the government quickly adopted this shoddy piece of pragmatism as a legal threshold. In addition, the industry and the government informally agreed to phase out crocidolite, though the latter still found its way into the country in imported a/c sheets. The asbestos industry reluctantly acquiesced in the two-fibre threshold, not least because the government made it plain that initially it would not be enforced.

The political and scientific wrangling over this threshold demonstrated the value to the industry of its lobby organizations, particularly the ARC. The latter provided a perfect mediator between government and industry interests, with the ARC infiltrating the BOHS and also pressuring the government to ensure that the threshold was not too stringent. This setup was also useful when it came to defending the two-fibre standard. Within only a year or so of the standard's ratification, Knox's successor at T&N, Dr Hilton Lewinsohn, unintentionally destroyed its credibility. Lewinsohn had re-examined Knox's data and concluded that both Knox and the BOHS had woefully underestimated the incidence of asbestosis. Lewinsohn published his conclusions in an article in an obscure English medical journal and the results went unnoticed in the UK. However, the article did not escape the attention of Selikoff, who was appalled to discover that the British threshold that he had been recommending appeared to be unsafe. He challenged Lewinsohn and his colleagues to explain the inconsistencies in their data and argued that the standard should be verified independently and possibly overturned.

It was a challenge that had relevance on both sides of the Atlantic. In the early 1970s, several regulatory agencies—notably OSHA, NIOSH, and the Environmental Protection Agency (EPA)—were becoming increasingly concerned about asbestos, so raising the spectre of regulation for the industry. Since the British had beaten the path, the American government was minded (after laborious consideration) to follow Britain in establishing an identical threshold. In 1971, OSHA began regulating occupational asbestos exposure.[59] First, it adopted a five-fibre threshold as an emergency standard. Then in the following year, OSHA/NIOSH opened a lengthy debate between government, the industry, trade unions, and independent medical experts. A series of hearings in Washington provided the setting for a shoot-out between Selikoff's Mount Sinai team and the asbestos industry, as represented by its Canadian scientists.

Neither the American government nor the US asbestos industry had any liking for a threshold that was more stringent than two fibres. Selikoff's questioning of the UK standard, therefore, faced an immediate backlash, led by T&N, members of the Medical Research Council (namely Dr John Gilson), and trades union medical advisers (such as Dr Robert Murray).[60] Documents that have only recently come to light, demonstrate quite clearly that the opposition to Selikoff was international and operated at a high-level. In the summer of 1973, the American industry sent a delegation to the UK to help orchestrate a plan to silence Selikoff's 'misleading' statements. The delegation included QAMA's counsel, Ivan Sabourin, John Marsh (Raybestos-Manhattan and president of the AIA/NA), and Matthew Swetonic. The group met with British industry personnel and also UK government representatives. The Americans noted that the British 'intend to take this matter, right down to the wire', and were supported by Dr Corbett McDonald, who was on a sabbatical leave from Montreal and apparently thought that the idea of sending a strong letter to Selikoff was 'excellent'.[61] The group also met Stuart Luxon, a member of the UK Factory Inspectorate, who in March 1972 had appeared as an expert witness in support of the two-fibre threshold at the OSHA hearings. Apparently, Luxon suggested ways in which they could 'set Selikoff back somewhat... to take the pressure off the asbestos industry'.[62]

[59] J.F. Martonik, E. Nash, and E. Grossman, 'The History of OSHA's Asbestos Rulemakings and Some Distinctive Approaches that They Introduced for Regulating Occupational Exposure to Toxic Substances', *American Industrial Hygiene Association Journal* 62 (March/April 2001), pp. 208–17.

[60] Castleman, *Asbestos*, pp. 275–81.

[61] AIA/NA Memo of Trip to UK, 23 June—1 July 1973.

[62] AIA/NA Memo of Trip to UK, 23 June—1 July 1973.

Within this arena of vested interests, the industry and particularly the AIA/NA were able to counter the medical evidence of independent experts such as Selikoff with arguments that in David Michael's words, 'manufactured uncertainty'.[63] The industry argued that the health effects of asbestos were ill-defined, that other products might be responsible, and that more research was needed (an ironic argument from an industry that had sponsored few medical studies before the 1960s and, when it had, ignored the results). On the whole, the industry was well pleased with the outcome. In 1973, Swetonic observed:

> The [AIA/NA] expended tremendous efforts during the six-month period leading to the promulgation of last June's OSHA standards on asbestos. I think it is a gauge of the effectiveness of the total industry involvement in this most crucial matter that of the eleven main requirements in the standards, the industry position was accepted totally by OSHA on nine of the eleven, about 50 per cent on a tenth, and totally rejected on only one.'[64]

By 1974, industry insiders reported that the AIA/NA had moved so 'very close to the [OSHA] people in Washington' that it had apparently been requested by OSHA to write up its own workplace regulations: 'a tremendous opportunity for the industry'.[65]

Five fibres per cubic centimetre was adopted in America until 1976 (when the two-fibre threshold was introduced), thus giving the industry time to clean up. Despite its deficiencies, 2 f/cc operated as the threshold in America until 1983 and in the UK until 1982 (when belatedly the UK government implemented the recommendation of its own advisers that the threshold be lowered to 1 f/cc for chrysotile).[66] Introducing asbestos standards had proved a very slow process, mainly due to the industry's opposition. However, it should be noted that regulation was even slower to appear elsewhere. In the late 1970s, asbestos exposures in many countries were still unregulated. South Africa had no effective standard. France was unique among asbestos producers in the industrialized world for having no asbestos TLV until in 1977 it banned sprayed asbestos and set a limit of 2 f/cc.

[63] D. Michaels and C. Monforton, 'Manufacturing Uncertainty: Contested Science and the Protection of the Public's Health and Environment', *American Journal of Public Health* 95 (2005), Supp 1, pp. 39–48. See also T.H. Murray, 'Regulating Asbestos: Ethics, Politics, and the Values of Science', in R. Bayer (ed.), *The Health and Safety of Workers* (1988), pp. 271–92.

[64] Swetonic, 'Why Asbestos?'.

[65] T&N memo copied to David Hills, re. Meeting of AIC/ARC, AIA/NA and QAMA, Montréal (24 January 1974).

[66] By the 1990s in America, a threshold of 0.1 f/cc had been introduced.

Thresholds had two problems. First, the discussion was conducted entirely in terms of asbestosis, when the major concern was cancer. As Selikoff made clear, the only safe threshold for the prevention of cancer was zero exposure. Second, TLVs distracted attention from improving basic work practices, which as the insulator Mackley had pointed out demanded common sense rather than peering down a microscope. As the industry entered the era of regulation, those work practices were shown to be woefully deficient. Dust counting and thresholds did have one merit: they forced scientists and the government to look hard at working conditions and provide data that threw a glaring light on dust in the industry. It was soon established that insulation workers had often suffered levels of exposure up to 100 f/cc (with spraying asbestos producing fibre levels that could swamp the counting apparatus). The shipyards were particuarly dusty and were a seedbed worldwide for 'hotspots' of mesothelioma.[67] South African mines had dust levels that compared unfavourably with factories in the 1920s.

American factory dust levels were not much better. Perhaps the most shocking factory—though it was not unique—was the Tyler plant of UNARCO in Texas, which manufactured asbestos pipe insulation. In 1962, it had been bought by Pittsburgh Corning. A decade later, the company dismantled and buried its production facilities after a government survey had highlighted 'major industrial hygiene deficiencies' and described the factory as an 'unholy mess'.[68] These deficiencies, hair-raisingly and tragically described by Paul Brodeur, included dust levels over 200 f/cc. Even shaking disused burlap bags from the Tyler plant—which were recycled and sold to gardening centres—raised dust clouds of nearly 500 f/cc.[69] The company had been fined a couple of hundred dollars for such health violations, even though nearly all the workers employed more than ten years had asbestosis. Lung cancer and mesothelioma were rife.

When Johns-Manville reviewed its dust counts in 1969 it noted that a significant number of the readings were above even the modest levels of the American dust threshold. As one Johns-Manville director (whose wife was later to die of mesothelioma) put it: 'we have a dirty house and now we have to pay for it'.[70] In the country that was the fount of American fibre, Canadian asbestos factories produced record levels of dust and

[67] See B. Burke, 'Shipbuilding's Deadly Legacy', *The Virginian-Pilot* (6–10 May 2001).

[68] NIOSH, 'Asbestos Survey: Pittsburgh-Corning Corporation, Tyler, Texas'. Project No: 71–45, 7 December 1971.

[69] Brodeur, *Expendable Americans*, pp. 48, 50, 54, 68.

[70] Dr Kent Wise noting the comments of Cliff Sheckler, 1969. *Johns-Manville v. Home Insurance*, p. 160.

disease. The Ontario Royal Commission pronounced Johns-Manville's plant in Scarborough as 'a world-class industrial disaster' (with past levels at 40 f/cc). However, it was not the worst example. At the Holmes factory in Sarnia, Ontario, and its associated Caposite Insulation plant (which processed amosite), the government found in the early 1970s that the dust levels were the highest it had ever recorded—over 800 f/cc! Yet the factory was not immediately shut down: when it was, it left behind an epidemic of ARDs that surfaced in the late 1990s.[71]

In Europe, dust levels above the 2 f/cc threshold appear to have been universal in the 1960s and 1970s, even at the better factories such as T&N in Rochdale. French factories in Normandy and in Clermont-Ferrand were well over accepted limits.[72] French government inspector's reports of an Eternit factory at Thiant emphasized the deficient dust control as late as 1992, confirming workers' testimonies of the high dust levels.[73] In Australia in the early 1970s, none of James Hardie's plants met dust-control standards, even though the company physician favoured a 4 f/cc threshold. Australian state officials described some of the processes as 'mind-boggling' because of their hazards. One of Hardie's bag-emptying facilities was a 'corrugated iron structure, ten or twelve feet square, with a bunch of guys lifting bags, tipping it out, and shovelling raw asbestos down a hole. They couldn't see each other.'[74]

Live with Asbestos!

Until the media began logging the body count, these dusty working conditions remained largely off-stage. The public was more likely to see the industry's public relations efforts, which formed a key part in the defence of the product.

In some countries, the media was less essential to the industry's survival. In South Africa, the stifling of free speech under apartheid helped enormously in stilling both scientific and public criticism of the industry. In Canada, local pride in a major industry and wide-ranging political and scientific support made media campaigns largely superfluous. It was

[71] J. Brophy and M. Parent, 'Documenting the Asbestos Story in Sarnia', *New Solutions* 9 (1999), pp. 297–316. See also D. Smith, *Consulted to Death* (2000).

[72] Collectif Intersyndical Securite des Universites-Jussieu CFDT, CGT, FEN, *Danger! Amiante* (1977), pp. 26–7, 55–113.

[73] Odette Hardy-Hemery, *Eternit et L'Amiante 1922*–2000 (2005), pp. 199–200.

[74] G. Haigh, *Asbestos House* (2006), p. 112.

elsewhere that the industry's publicity machine was needed and was at its most sophisticated, particularly in the US and the UK. The budgets of the AIA/NA and ARC/AIA were directed largely at public relations. Hundred of thousands of dollars were earmarked by the American industry; while in the UK by the mid-1970s asbestos firms spent £500,000 on advertising campaigns that were designed to ensure the survival of asbestos. These campaigns resurrected many of the selling points that had been used for asbestos when it first became popular in the early twentieth century. In those days asbestos was referred to as the 'world's most wonderful mineral', and portrayed as a bulwark against the fiery elements. In the 1960s and 1970s, the industry desperately wanted that to remain the public's perception.

Internal memoranda from the leading companies and their public relations consultants have survived. So, too, have a large number of information leaflets and press statements from bodies such as the AIC. These show how the industry's campaign was orchestrated by Hill & Knowlton, who, in an interesting use of language, described their work as 'fire fighting/fire prevention'. One strategy was to blame someone else, either the media or the public or Congress. 'Asbestos-related disease does exist', admitted Johns-Manville, 'thus it is perhaps understandable that people would cast about for an "asbestos scapegoat". What is inexcusable is the manner in which lawyers, the media and even some in the "public interest" arena have sought to exploit the tragedy of asbestos-related disease through the repetition of inaccuracies, half-truths and exaggerations'.[75] Another strategy was to argue that the problem should be put into 'perspective'. This was to be done by media monitoring, position papers, the use of spokespersons, a media hotline, and the seeding of favourable news stories. Briefings were drawn up for executives and other company personnel, so that awkward questions could be deflected. Hill & Knowlton arranged for company executives and physicians to be given special confrontational TV training. The message to be communicated, both at press conferences and through television, was that disease associated with asbestos is rare, the public is not at risk, asbestos saves lives, and proper controls had been introduced. Responses drafted by Hill & Knowlton were not always convincing; one model reply was worded as follows: '[M]esothelioma is considered a very rare disease and apparently nobody knows much about it, including me.'[76]

[75] J-M Corporation, *Annual Report* (1978), p. 23.
[76] Interim Position Statements: Asbestos and Health, 3 March 1967.

The industry's most difficult task was to explain the rising incidence and mortality from ARDs. This was attributed to 'old' conditions and the result of exposure up to forty years ago, 'when medical knowledge concerning asbestos was only a fraction of what it is today'.[77] Johns-Manville told its shareholders in the 1970s that its knowledge of lung cancer in asbestos workers was of 'rather recent vintage'.[78] It was also argued that Selikoff's insulators were not truly 'asbestos workers'. This enabled the industry to admit that the insulators' experience was 'terrible', while denying any responsibility for it, even though the same insulators had been supplied with fibre from the leading companies and had often been employed under contract with them. According to the industry, Selikoff's reports related to a 'relatively small group of workers who removes and/or install a *variety* of insulation materials, including some which contain asbestos'.[79] These implausible contentions—that the industry was ignorant before Selikoff appeared, that it was not liable for insulators' ARDs, and that other products were to blame—would soon be tested in the courts and found wanting.

As regards the contemporary situation, especially the threat of mesothelioma, asbestos users and the public were reassured that asbestos was benign. In 1975, the AIC published a leaflet that argued that fibre was 'locked in' and rendered innocuous by another material: in a/c this was cement, in brake- and clutch-linings baked resin, in gaskets rubber or synthetic rubber.[80] The idea that asbestos fibre was safely encased in cement was a favourite line with Eternit companies, who also argued that a chemical reaction with the cement removed the fibres' toxicity.[81] Max Schmidheiny of Swiss Eternit, while dismissing Selikoff as 'a crank who did research to make money', argued that his company's products were 'not dangerous because the fibres are embedded in cement. Completely safe'.[82] In fact, asbestos was never locked-in and workers who sawed a/c sheets or worked on brakes and clutches were to develop mesothelioma just like factory workers. Tellingly, the idea that fibres underwent some kind of chemical change was described as a 'notorious theory' by Eternit's fellow British and American industrialists, who felt that it could be disastrous to their interests if they were associated with it.[83]

[77] J-M Corporation, *Annual Report* (1978), p. 24.
[78] J-M Corporation, *Annual Report* (1978), p. 25.
[79] J-M Interim Position Paper for Use in Reply to Possible Queries, 31 May 1966.
[80] *Asbestos and Your Health* (1975).
[81] R.F. Ruers and N. Schouten, *The Tragedy of Asbestos* (May 2006), pp. 23–4.
[82] W. Catrina, *Der Eternit-Report* (1989), p. 79.
[83] M.F. Howe to W.P. Howard, 18 September 1973.

Countering anxieties about mesothelioma was to be accomplished in a number of ways. First came the reassurance—repeated innumerable times—that ARDs could only be caused by inhaling lots of dust over a long period of time. Thus, the AIC told the public: 'only if asbestos dust is inhaled in substantial concentrations over a considerable period of time is it considered to constitute a possible danger'.[84] Second came the argument that mesothelioma was caused by blue asbestos, with other varieties of the fibre—especially chrysotile—being innocuous. The so-called 'chrysotile defence' (as we shall see in the following Chapter 5) became a major plank in the industry's armoury. The industry also tried to cast doubt on the mineral's link with mesothelioma. Even in the 1970s, the industry described the linkage as uncertain. One industry leaflet on mesothelioma pointed out that in twenty-eight of twenty-nine cases in Zurich no asbestos contact was detected and no asbestos was demonstrated in the tissue of the lungs and the tumours. The AIC agreed that there was now an 'association' with asbestos, but suggested that other co-carcinogens were necessary before mesothelioma could develop.[85]

The AIC and the AIA/NA knew that many of these 'facts' were scientifically unsupportable and that their evidence was selective, but the propaganda enabled the industry to advance the lie that there was no risk to the public. One widely circulated AIC publication stated: 'Half a century of world research had revealed absolutely no credible evidence of risk to the general public from environmental exposure.'[86] Yet the industry was well aware, since the early 1960s, of cases of mesothelioma caused by environmental contamination (see Chapter 6), but whenever these were broached, the industry as ever cast doubt on them or blamed imperfect knowledge. 'Nowadays, there should be no risk.... Cases of disease, alleged to be associated with exposure to dust from living near an asbestos factory in this country, are isolated examples, dating back to a period of many years ago when the need for precautions was not as well understood, as it is today.'[87] Risks in upstream and downstream activities were dismissed. One information leaflet dismissed the hazard from asbestos brake linings, arguing that when these were dismantled 'the amount of asbestos found is minute'. However, it added: 'The dust is very fine and it is good practice to remove it by vacuum, rather than by air blast'.[88] This was sensible advice, as brake repairers would eventually die of mesothelioma, too. *Asbestos and the Docker* (AIC, 1975) blamed the few

[84] *Asbestos—Safety and Control* (1976). [85] *Mesothelioma* (1976).
[86] *Asbestos—Killer Dust or Miracle Fibre?* (1976). [87] *Asbestos—Safety and Control* (1976).
[88] *Asbestos—Safety and Control* (1976).

reported dockers' mesothelioma cases on their supposed involvement in installing or stripping asbestos insulation. Thus, it was said, hardly any dockers had suffered from ARDs.

As we shall demonstrate in Chapter 5, the industry was able to influence and even manipulate the scientific evidence about ARDs. The industry's propaganda machine was then able to recycle this science, feeding it into pronouncements for public consumption. In 1972, for example, the AIC published a leaflet on the findings of a key advisory committee of the International Agency for Research on Cancer (IARC) in Geneva—a body that the asbestos industry had infiltrated. Selikoff, who also attended, disassociated himself from the IARC report because it was so laden with propaganda.[89] Many of the industry's mantras found their way into the proceedings,[90] especially the view that there was no risk to the general public and that environmental cases in South Africa only related to 'conditions many years ago'. It was also said that many cases of mesothelioma had no connection with asbestos and that pleural plaques were not necessarily associated with asbestos. Hill & Knowlton mingled doubt about the asbestos hazard ('there are experts on both sides of the issue'[91]), with pleas for a 'proper perspective' that described the mineral as simply another of life's hazards:

> Literally hundreds of substances—as common and everyday as fuel oil, charcoal broiled steaks, iron rust and egg yolks—are known or suspected of being cancer-causing agents, under certain experimental conditions. The point is that nearly everything we use or create in our increasingly complex society has been suspected, by someone, of being a potential hazard. To forego the many benefits of these products of modern technology—simply on the basis of suspicions that are unsupported by medical evidence, would be a great and unwarranted disservice to the American public.[92]

If there was one image used repeatedly in industry advertisements it was that of the asbestos-clad firefighter emerging from the flames. The word 'Safety' constantly appeared in the industry's publications; so too

[89] Selikoff to John Higginson, IARC, 18 May 1973. Selikoff complained to Sheldon Samuels, AFL-CIO (letter, 29 May 1973), about 'the extraordinary selection of an industry representative to give the only address at the conference dinner (paid for by the asbestos industry!), in which he regretted the fact that some must die in the industrial utilisation of asbestos. But, since society needed this valuable material, we had to be brave and accept this unfortunate necessity.'

[90] 'Biological Effects of Asbestos: Report of the Advisory Committee on Asbestos Cancers to the Director of the International Agency for Research on Cancer, Lyon, 5–6 October 1972', *Annals of Occupational Hygiene* 16 (1973), pp. 9–17.

[91] Hill & Knowlton, 'The Situation Concerning the Asbestos Industry', 22 May 1979.

[92] Hill & Knowlton, 'White Paper', October 1966.

did 'Health'. This strategy of selling a hazardous product on its health and safety benefits was not unique—Rosner and Markowitz have highlighted a similar strategy in advertising lead paints and lead water pipes[93]—but no industry made more of this aspect than asbestos. Sadly, firemen would also eventually die of mesothelioma.

The industry line can be sampled in a series of major newspaper supplements on asbestos that appeared in the leading English newspapers in the 1960s and early 1970s. The articles were sponsored by the asbestos companies (which regarded them as no more than vehicles for advertising) and plugged asbestos use heavily. For example, *The Times* (28 November 1967) puffed up the 'unquenchable mineral', and emphasized the way in which it was able to offer fire protection on land and sea, and how transport depended on it. The man in the fireproof asbestos suit was ubiquitous in the photo-spreads, but there were no pictures of sick or dying workers. The authors in many of these supplements were identical and most of the articles were written by asbestos industrialists. In another asbestos supplement that ran in *The Financial Times* on 27 January 1972, T&N director Harry Hardie told everybody how asbestos was the 'fibre for safety and protection'.[94] A section on health was written by Walter Smither, who did not identify himself as Cape's physician, but as a 'medical consultant, Asbestosis Research Council'. According to Smither, 'it should be emphasised that there is no danger of environmental pollution from the use of asbestos products'. His article sat next to a boosterish feature on sprayed asbestos—a process already banned in America.

James Hardie's advertisements in 1960s and early 1970s stressed the virtues of water conservation (arguing that its pipes kept at bay the havoc wrought by drought) and also the safety aspects of its asbestos brake linings.[95] In France, the CPA, like the AIC, issued reassuring leaflets which emphasized the safety benefits of asbestos, its usefulness, and the need for its continued use. In 1976, the French asbestos industry placed full-page advertisements in *Le Monde* and other newspapers, stressing that asbestos was '*un material naturel*', no more dangerous than any other natural resource such as the salt in the sea: 'The problems posed by asbestos are nothing, compared to the immense services that it renders each day, without one even knowing it. Let us learn to live with asbestos.'[96]

[93] G. Markowitz and D. Rosner, *Deceit and Denial* (2002).
[94] Another major supplement was published in *The Financial Times*, 19 July 1974.
[95] B. Carroll, *'A Very Good Business'* (2001), pp. 122, 135.
[96] 'À Propos de L'Amiante', *Le Monde*, 17 November 1976, p. 8.

The need to continue to live with asbestos was, according to the industry, unavoidable. This was because asbestos was unique. Ignoring the fact that substitutes such as fibre glass, carbon, and cellulose had been around for decades and that governments were now recommending them, the industry argued that asbestos had no competitors and even if it had there would be problems. An industry insider commented: 'Football managers dream of having substitutes on the bench of the same calibre as their first-team stars. It seldom happens. So it is with asbestos alternatives. In addition to performance and cost criteria there is the need to consider health effects.'[97]

It was certainly true that scientists were concerned about the possibility of similar fibres to asbestos causing cancer—even Selikoff was initially wary of substitute materials. On the other hand, it was soon obvious that man-made minerals did not cause fibrosis. According to Wagner and Doll, there was also no evidence that they caused cancer. In 1980, Wagner argued that the use of glass fibre and mineral wools was 'in the interest of health', if the production of small-diameter fibres could be avoided. But this was where substitutes scored heavily: they could be engineered to have fibre dimensions large enough to avoid them lodging in the lungs.[98] In this way, the 'dividend' of researching the carcinogenic characteristics of asbestos could be 'spent' on man-made fibres. Some asbestos companies already made substitutes and a few industrialists, such as Stephan Schmidheiny, took the opportunity to go one step further and shed asbestos.[99] But many in the industry set their face against substitutes. Scientists, many of whom were funded by industry, showed little interest and remained obsessed with asbestos. Between the 1960s and early 2000s, one of the leading UK occupational health journals, *Annals of Occupational Hygiene* (mouthpiece of the BOHS), featured over 200 articles and contributions on asbestos. The corresponding number for articles on man-made fibres was only a quarter of that total.

The industry's defence has recently been resurrected and elaborated by Rachel Maines in her book *Asbestos and Fire*, which emphasizes the historic merits of asbestos in fire protection and the trade-offs between asbestos health hazards and fire risks.[100] Like the industry's propaganda in the 1970s, these safety aspects of asbestos are overdrawn. Asbestos

[97] W. Penney, AIA/NA, Annual Meeting and Industry-Government Conference, 18–19 September 1984, Twin Bridges Marriott Hotel, Arlington, VA.

[98] J.C. Wagner, G. Berry, and F.D. Pooley, 'Carcinogenesis and Mineral Fibres', *British Medical Bulletin* 36 (1980), pp. 53–6.

[99] S. Schmidheiny, 'My Path—My Perspective' (2006).

[100] R. Maines, *Asbestos and Fire* (2005).

had proved remarkably utilitarian and it had provided fire protection: but drawing a connection between asbestos use and the thousands of lives supposedly lost (or saved) from fires or other accidents is simplistic. Multiple factors are involved in fire mortality, as in other aspects of industrial hazard. These include building code changes, fire alarms, sprinklers, occupancy limits, and improved design of buildings for rapid exit. As regards fire protection in buildings, concrete encasement and man-made fibres have proved as reliable as asbestos in preventing conflagrations, which is why it has been relatively easy for modern skyscraper builders to dispense with the mineral. Historically, the majority of asbestos used has been in a/c building panels, roof sheets, and water pipes in which the fibre is used as a non-corrodible and tough reinforcer rather than as a fire-protector. To be sure, technical problems sometimes arose in the introduction of substitutes in some products (e.g. in brake linings), but these were not insurmountable. The main reason that the companies were reluctant to abandon asbestos was cost and profits: concrete encasement was calculated to be at least two-and-a-half times more expensive than asbestos, and ceramic fibres twenty times more expensive. No one factored in the associated health costs or the fact that asbestos had been often mined under appalling labour conditions.

The relatively rapid introduction of substitutes for asbestos offers the most pointed refutation of the argument that asbestos was unique and indispensable.[101] Neither US nor European cities have been consumed by fire; substitutes (which Maines almost completely ignores, apart from implying that they are as hazardous as asbestos) have proved eminently suitable; and there has been no epidemic of substitute-related disease.

Tragically, despite its resources and media skills, the industry did not publicize the hazards of asbestos. Documents produced in legal discovery show that the industry actively campaigned *not* to have asbestos marked with warning labels about its carcinogenic potential. In Britain, a label was eventually issued in the 1970s, which consisted of the letter 'a'—displayed prominently—and beneath it the words, 'Take Care'. The label warned that 'breathing asbestos dust can damage health. Observe the safety rules'. The wording lacked punch. But the AIA was not in favour of even these anodyne words, especially when T&N argued later that the word 'cancer' should be included to shield the group from legal liability. In 1980,

[101] For the industry line, see C. Moir, 'Asbestos: Why There is No Substitute', *Observer*, 29 August 1982.

Etienne van der Rest, who was chairman of the AIA and a member of the Belgian Eternit group, wrote anxiously to T&N to say 'how disappointed' he was to see the cancer warning: 'This could have as a consequence that the EEC would decide either to adopt the label as it is, either to encourage a skull and crossed bones label, or some other carcinogenic logo still to be created.'[102] As it transpired, in 1983 the EEC went no further than the bland statement 'asbestos dust is dangerous to health', which as Ruers and Schouten highlight was a clear sign of the influence of the Eternit group (the biggest producer of a/c in the EC) and also the AIA, which lobbied hard against a cancer warning.[103] The AIA lamented that it was no longer possible to prove that asbestos was not a carcinogen, but nevertheless believed it should be exempted from labelling because it was supposedly a 'special carcinogen'. This was on the bizarre basis that asbestos cancers had a long latency and there was a threshold level.[104]

Inevitably, the industry's opponents would begin to use the media themselves and journalists would become more critical. Swetonic had once comforted his colleagues by telling them the 'good news... that despite all the negative articles on asbestos-health... very few people have been paying attention'. He looked forward to the day 'when the press ceases to print anything about asbestos at all'.[105] This proved a forlorn hope and by the end of the 1970s the media tide was moving against the industry. The hollow reassurances of the lobby groups were questioned by journalists who were becoming familiar with the asbestos problem. Medical correspondents for *The Times,* Oliver Gillie and Peter Gillman, questioned the assertions of the AIC, pointing out that they were part of a massive publicity campaign and that the information was misleading.[106] In 1982, the UK industry was hit by a public relations disaster, when a one-and-half-hour television documentary 'Alice—A Fight for Life' ripped apart the myths surrounding asbestos and emphasized the dangers of environmental exposure and the toxicity of all types of asbestos. By focusing on a single mesothelioma sufferer, Alice Jefferson, and her slow death it brought home to ordinary people the cost of ARDs. In particular, 'Alice' exposed the compensation scandal that left many victims and their families not only bereft but in poverty. Its heaviest punches were reserved for the asbestos companies, especially the industry leader T&N (which

[102] E. Van der Rest to Harry Hardie, 5 March 1980.
[103] Ruers and Schouten, *Tragedy of Asbestos*, p. 27.
[104] AIA Minutes of Executive Meeting at RAF Club, Piccadilly, London, 8 November 1983.
[105] Swetonic, 'Why Asbestos?'
[106] 'Asbestos Safety Campaign Advert Claims Too Much', *The Times*, 4 July 1976.

had declined to take part in the programme). 'Alice' was shown in the US and in Europe, where the documentary won prizes. It was recognized as a path-breaking work that had put Britain (and other countries) on notice that asbestos was a major hazard. No one would ever look at asbestos in quite the same way again.

Some in the industry became dispirited. 'I have little doubt that we are losing the PR media battle with the anti lobby', wrote a T&N manager, 'especially following the Alice programme. Whatever efforts the AIC are making may be "silent but effective"—but where is the evidence of their impact? I can't see it'.[107] But the industry's public relations campaign had not been without effect: it bought the industry time and ensured that the decline in sales was not as precipitous as it might have been. During those crucial decades between the 1960s and 1980s, many individuals were needlessly exposed to dust. Moreover, the public relations campaign was combined with an equally effective strategy of exploiting and moulding the science of ARDs—a subject to which we will now turn.

[107] S. Marks, private and confidential memo to JBH, 19 October 1982.

5

The Chrysotile Defence

> There is no convincing evidence that chrysotile is more, or less, hazardous than any other asbestos fibres. There is ample evidence that all types of asbestos are associated with the risk of fatal lung scarring (asbestosis), lung cancer, pleural and peritoneal mesothelioma and gastro-intestinal cancer.
>
> Irving Selikoff to Paul C. Formby
> Vancouver, 18 December 1972.

> We've gone too far in eliminating asbestos—I mean the less dangerous white type, which carefully handled does more good than harm.
>
> Sir Richard Doll interviewed by Anna Wagstaff
> 'Richard Doll: Science Will Always Win in the End'
> *Cancer World* (December 2004), pp. 28–34, 32.

In reviewing the science it is important to remember that asbestos is such a toxic material that even relatively trivial exposure can result in serious or fatal injury. For that reason, one might have expected physicians and allied scientists to have led the campaigns against the mineral and against the companies that produced it. Yet, as we shall see, not only was the medical profession's reaction to the asbestos hazard often feeble, but scientists have also been among the industry's most strident defenders. There are two reasons why that was so: corporate suppression and intimidation meant that criticism of the industry came at a price. Another factor was the convergence of the economic, political, and social interests of the scientific establishment and commerce. Careers could be made from industry-sponsored research. No one commissioned research on behalf of asbestos workers.

The publication in 1960 of J.C. Wagner's paper on mesothelioma suggested there was no safe way of mining asbestos or manufacturing asbestos products. That discovery on South Africa's crocidolite fields threatened the survival of the entire industry. It was the third crisis the industry had faced during the twentieth century—the first was over asbestosis in 1930, the second, over asbestos-related lung cancer in 1940. However, the third was the most serious. In each instance, the industry leaders developed techniques which enabled them to survive. They ranged from intimidating medical researchers, suppressing evidence of risk, colluding with state and regulatory authorities, and generating favourable publicity. In South Africa and Canada, identifying what the industry knew about ARDs is made more difficult by the incomplete archival record.[1] As we have seen, it is also made difficult by the nature of the science. The medical literature was politicized from the early 1930s and for more than half a century the leading manufacturers influenced and even corrupted medical discovery. In defending the industry against mesothelioma, the industry's most potent weapon was the manipulation of science from within to promote doubt about the mineral's toxicity.

The Underbelly

Beneath the international conferences, the Royal Commissions, public inquires, and medical forums, lay a shadowy world where the rules of debate were very different. That world of intimidation and bribery was always hidden from public view and like so much of the asbestos story it has only come to light as the result of litigation. Nowhere were the shadows as dark as in the field of medical research.

Wagner's discovery in 1960 presented the South African industry with a dilemma. It was impossible to engineer dust out of the mills and still produce asbestos cheaply. Therefore the choice facing Cape Asbestos, Eternit, and Gefco was either to shut down the mines or suppress knowledge of disease. Having chosen the latter, those companies worked hard to keep information about mesothelioma secret. To retain a dwindling global market for crocidolite it was imperative that no further research into mesothelioma was conducted in South Africa.

Conditions in South Africa were unique. By 1960, mining had existed for over sixty years, and in the Northern Cape sizeable communities had experienced occupational and environmental exposure to a single type of fibre, crocidolite. The area was isolated and there were no other

[1] See Sources, pp. 276–83.

industries to compromise the results of an epidemiological survey. The Northern Cape was probably the best place in the world to study the effects of asbestos. Unfortunately because of apartheid, South Africa was also the country where industry had its greatest success in suppressing the knowledge of risk. The major companies were so successful in stifling debate that the mining of blue asbestos continued unabated between the 1960s and 1980s. Not only did it continue, but output actually rose! That history raises the question as to how medical knowledge could have become so unhinged from public policy.

Following the publication of Wagner's paper in April 1960, the Government Mining Engineer, Tony Gibbs, asked Ian Webster, the senior pathologist at the PRU to coordinate a survey of mesothelioma.[2] After some discussion it was agreed that a project in the Northern Cape would be funded by the asbestos industry, supported by a small grant from the South African Cancer Association.[3] The Survey was conducted between November 1960 and February 1962. By the following April, an interim report had found an alarmingly high incidence of mesothelioma and evidence that the disease was associated with asbestos.[4] Pollution in the mining towns was so extreme that residents were contracting asbestosis, an occupational disease, from environmental exposure.

Instead of closing the mines the industry shut down the survey and attacked the PRU scientists.[5] The director of the PRU, Dr I.G. Walters, later remarked that the project had been ended because of the industry's concern that it would be unable to recruit labour to its mines.[6] The Survey results were never published and Drs Jennifer Tallent and Bill Harrison, who fifteen years later conducted research at the PRU/National Centre for Occupational Health (NCOH) into mesothelioma among those same mining communities, were never told about that earlier work. It was as if it had never existed.[7] So complete was the secrecy surrounding mesothelioma that Guy Wilson and Jacoba Schnyders, who during the 1960s

[2] Jock McCulloch interview with Professor Ian Webster, Braamfontein, Johannesburg, 4 December 1997.

[3] 'Report on the Progress of Mesothelioma Survey as at 30 April 1962', PRU, South African Council for Scientific & Industrial Research (SACSIR), Johannesburg, unpublished, p. 2. NCOH Papers.

[4] 'Report on the Progress of Mesothelioma Survey', 1962, pp. 3–4.

[5] 'PRU Annual Report for Year Ending 31 March 1963', SACSIR.

[6] Covering letter from Dr I.G. Walters, Director of the PRU, 23 June 1964, in 'Field Survey in the North West Cape and at Penge in the Transvaal (Asbestosis and Mesothelioma)', PRU, Report No 1/64, SACSIR, 1964.

[7] Jock McCulloch interviews with Dr Bill Harrison, Killara, Sydney, 15 October 1999; and Dr Jennifer Tallent, Fish Hoek, South Africa, 16 December 1999.

and 1970s held middle management positions with Cape Asbestos—the world's major producer of crocidolite—were unaware of the existence of the disease until more than a decade later.[8] Wilson would often take his children on visits to the mines during summer holidays—something he told Jock McCulloch that he would never had done had he known about mesothelioma. That knowledge was even further removed from the men, women, and children who worked on the mines.

The Survey was known to the secretaries of the Departments of Mines, Treasury, and Trade, and the matter was discussed at Cabinet level in 1962.[9] In its internal correspondence, the Department of Mines recognized that any form of exposure to asbestos was 'extremely dangerous' for those who worked on the mines or who lived near mills. It acknowledged that in parts of the Northern Cape a large percentage of the population was suffering from ARDs.[10] The correspondence shows the Department's determination that the South African public should not learn about the dangers of asbestos. It feared that if trade unions in England and other countries became aware of mesothelioma it would destroy the market for South African fibre.

The response to medical discovery is often complex and it took time for Wagner's paper to change occupational health practices in Organization for Economic Cooperation & Development (OECD) states. But change did come, and within less than a decade new regulations in the UK brought an end to the use of crocidolite in British factories. In South Africa, the government threw its weight behind the industry and employers behaved as if ARDs did not exist. The same work and living conditions on the mines, described by Dr Kit Sleggs in 1948 and about which the Department of Mines was well aware, continued. What did change in South Africa was the science. After 1960, no creative research on asbestos disease came to public attention.

The lack of research from South Africa was noted by the scientific community. During the International Conference on Pneumoconiosis held at Johannesburg in 1969, Dr John Gilson, then director of the Pneumoconiosis Research Unit in the UK, commented of South African research: 'What surprises me a little is that here you have got three of

[8] Jock McCulloch interviews with Guy Wilson, Johannesburg, 5 July 1999; and interview by Engela Venter, with Willemina Jacoba Schnyders, Bloomfontein, 7 April 1999.

[9] Memo from Secretary for Mines, Pretoria, to Secretary for the Treasury, Pretoria, 4 July 1963; 'Research in Connection with Prevention of Asbestosis and Mesothelioma and the Prevention and Control of Dust in and by Asbestos Mines and Asbestos Mills'. TES Vol. 930B F5/389.

[10] Memo from Secretary for Mines.

the principal types of asbestos being mined. You have, it seems to me, a unique opportunity to look at the effects of the mining and milling. This seems an absolutely unique opportunity and yet we have not been given the results of such epidemiological studies.'[11] Over the next eight years, nothing changed and at the Asbestos Symposium held in Johannesburg in 1977 Professor J.C. McDonald from McGill University remarked: 'The highest priority for future research lay in South Africa, where case control studies were strongly indicated.'[12] Prior to 1990, the only notable work from the NCOH was a study by Drs Bill Harrison and J. Tallent of miners at Kuruman. That study was essentially a footnote to Wagner's 1960 paper.[13]

Professor Ian Webster was one of the most distinguished pathologists of his generation. In 1970 he became Director of the PRU (soon renamed the NCOH). He held that post until his retirement in 1983. Webster was one of the founders (alongside Selikoff) of the global occupational health organization, the Collegium Ramazzini, and he was the author of over 150 scientific papers. He was at the PRU when his brother-in-law Chris Wagner began research into ARDs. After participating in the autopsy of a miner named Bill Fello in 1956, Webster was convinced of the link between mesothelioma and asbestos, a conclusion from which he never wavered.[14] Webster helped set up the Survey of the Northern Cape in 1962, which confirmed his initial fears about crocidolite.

In 1962, Webster and the senior scientist with Cape Asbestos, Dr Richard Gaze, both read the Survey of the Northern Cape. Two years later they attended Selikoff's 1964 conference in New York City. During days of intense debate about mesothelioma, in which they participated, both men remained silent about those results. Their silence was all the more significant because the Survey had in the most dramatic fashion confirmed Wagner's work on mesothelioma and environmental exposure. An International Union against Cancer meeting held concurrently agreed to establish a number of national panels so that diagnostic criteria for mesothelioma could be standardized and results compared. On his return to Johannesburg Webster immediately ran into industry opposition. As a result, it took two years to set up a panel. By the end of the 1970s, the panel had collated a total of 2,334 cases. At that point, the

[11] H.A. Shapiro (ed.), *Pneumoconiosis Proceedings of the International Conference Johannesburg 1969* (1970), p. 262.

[12] H.W. Glen (ed.), *Asbestos Symposium Johannesburg, 1977* (1978), p. 86.

[13] J.M. Talent et al., 'A Survey of Black Mineworkers of the Cape Crocidolite Mines' in J.C. Wagner (ed.), *Biological Effects of Mineral Fibres* (1980), pp. 723–9.

[14] This and the following paragraphs draw on Jock McCulloch's interviews with Professor Ian Webster, Braamfontein, Johannesburg, 3, 4, 5 December 1997.

Medical Research Council (MRC) in South Africa decided it was not a 'research project' and, without consultation, closed it down. It was an important decision for without a register the incidence of mesothelioma could not be properly gauged. Today there is still no register in South Africa.

Following the suppression of the PRU Survey, all meaningful research in South Africa ceased. In its place the Department of Mines and the MRC established a web of committees and sub-committees to monitor the asbestos problem. They included the Executive Committee of the Asbestosis Research Project, the Sub-Committee on Cumulative Dust Exposure Records, the Asbestos Publicity Committee, the Editorial Committee of the National Research Institute for Occupational Diseases (within the MRC), the Sub-Committee on Asbestosis, and the Reconstituted Committee on Cumulative Dust Exposure Records. That administrative maze stifled innovative research.

Under the aegis of the MRC in July 1969, the Executive Committee of the Asbestosis Research Project was set up to review all asbestos research. The committee was chaired by the Government Mining Engineer, Tony Gibbs, and included Ian Webster and two industry representatives, Fritz Baunach and Justin MacKeurtan. The minutes show that industry fought any suggestion that asbestos was hazardous or that occupational disease was rife on the mines. It was even more aggressive in rejecting the conventions that asbestos causes cancer and mesothelioma. When Ian Webster became director of the NCOH in 1970 he was drawn into more intense conflict with the industry and in particular with the South African Asbestos Producers Advisory Committee (SAAPAC).

As international opposition to the use of South African fibre intensified, in May 1974 the South African Minister for Mines established the Asbestos Publicity Committee to counter what he termed 'wrong and prejudiced publicity about asbestos health hazards, especially against blue fibre'. The committee's brief was to monitor the extent of 'propaganda' against South African asbestos; refute and correct 'propaganda' through statements in the press or in scientific publications; keep a close check on local and overseas research; and make recommendations to the Minister.[15] The chair of the committee was Tony Gibbs and its members included Ian Webster, the Deputy Government Mining Engineer, Dr Isserow, and Fritz Baunach. Most of those on the

[15] Letter from Secretary for Mines to Professor I. Webster, Director National Research Institute for Occupational Diseases, Johannesburg, undated (May 1974).

committee had no research experience and the terms of reference and the surviving minutes reveal a crude mix of science and politics. In his personal notes, Ian Webster refers to it as the Asbestosis Propaganda Committee.

At one of its first meetings, the committee discussed the development of posters warning miners about the hazards of dust.[16] The industry did not want asbestos mentioned on the posters and the issue soon turned into a political struggle. Baunach warned that South African (black) workers were far less sophisticated than those in Europe or the US and that if black trade unions were legalized at some time in the future the posters would only cause trouble.[17] Webster fought hard to have effective posters displayed at mines but he had no allies. After some months of debate the Department of Mines decided that the posters would emphasize the dangers of smoking rather than asbestos. The phrase 'asbestos dust' was replaced by the broader term 'mineral dust'. Webster's role in the debate brought a rebuke from the Department which reminded him of the need to 'guard against alarmist attitudes'.[18]

The SAAPAC acting through Fritz Baunach had an open door to the MRC and the Deputy Secretary of Mines. It also had direct access to the Minister.[19] Baunach claimed that the mesothelioma problem was 'nothing but a political vendetta against South Africa', that it was really all about opposition to apartheid.[20] Fritz Baunach complained to the Department of Mines that Webster was wasting public money and he demanded the right to review personally all PRU publications on asbestos.[21] The MRC eventually agreed and during Ian Webster's period as director the industry exercised a veto over NCOH research. According to Webster: 'All the materials, and all the letters, reports and publications. And I can testify as all publications, all publications should be submitted to the asbestos industry. And that on the instructions of the president of the Medical Research Council.'[22] That system lingered after Webster retired as director in 1983. Ian Webster gave a paper at a Sydney conference on occupational disease in 1987, highlighting the low levels of exposure that could cause

[16] Minutes of Meeting of the Publicity Committee of the Department of Mines, 9 May 1974.

[17] Letter from Fritz Baunach, Asbesco Mine Services, to Mr T. Gibbs, GME, 3 October 1974.

[18] Letter from L. Jooste, Senior Manager, Base Minerals Division, Dept. of Mines, to Professor Ian Webster, Director National Research Institute for Occupational Diseases, 31 July 1974.

[19] Deposition by Dr Ian Webster, 2 November 1996, Sandton, Johannesburg, in Re: Asbestos Personal Injury, Civil Action Cases, Arrington Lead No. 93-9-114, 36, p. 43.

[20] Deposition by Dr Ian Webster, pp. 43–4.

[21] McCulloch interview with Webster, 4 December 1997.

[22] Deposition by Dr Ian Webster, p. 143.

injury. Once again he fell foul of the SAAPAC and his paper was never published.[23]

Harassment at the NCOH continued after 1983 when Professor J.C.A. (Tony) Davies replaced Webster as director. The industry was still fighting the same battle it had begun twenty-five years earlier and was just as vigilant in attacking its critics. In July 1986, Davies gave a paper on asbestos at a conference in Israel. In his presentation he made reference to a 20 per cent incidence of lung cancer among a cohort of deceased miners. It was hardly a surprising figure given the work conditions in South Africa. Baunach wrote in protest that Davies' paper was misleading and he rejected outright the claim that asbestos is a carcinogen.[24] Baunach then cautioned Davies that both the falling demand for South Africa asbestos and declining company profits were due to the grossly exaggerated reports of a health hazard. He went on: 'We (the industry) have never in any way attempted to interfere with the scientific programmes initiated by Prof. Webster and his colleagues.'[25] This time Baunach circulated a copy of his response to Dr Retief, the Director General of the Department of Health. The correspondence, which would have seemed bizarre to any medical researcher outside of South Africa, dragged on for months.

To prevent the NCOH from publishing original research the industry questioned every advance in knowledge, thereby creating doubt and with it a counter-position which could be used as a reference point. The SAAPAC was so successful that in the 1980s key research committees in Johannesburg were still debating whether or not asbestos is a carcinogen (something settled at least thirty years before elsewhere). That in turn rendered South African science irrelevant on the international stage. No lie was unthinkable and in addressing the public the industry linked criticism of asbestos to wider political agendas. In particular, the SAAPAC abetted by the Department of Mines spun the fantasy that anti-South African sentiment over apartheid lay behind the opposition to South Africa's crocidolite.

The industry fostered close relations with government and placed its representatives on key committees. From that position it could monitor publications and eventually established the right of veto. The industry also developed a method for dealing with individual critics. In writing to

[23] Deposition by Dr Ian Webster, pp. 129–30.

[24] Letter from F. Baunach (SAAPAC), Johannesburg, to J.C.A. Davies, Director NCOH, 13 October 1986.

[25] Letter from F. Baunach (SAAPAC), Johannesburg to J.C.A. Davies, Director NCOH, 29 April 1987.

opponents the SAAPAC, for example, would create an audience by circulating correspondence to adjacent authorities. It exhausted researchers by bogging them down in lengthy correspondence and intimidated those in bureaucratic positions by making accusations of unprofessional conduct to their superiors. That method was used against Ian Webster and his successor Tony Davies. It was also used against researchers at the University of Cape Town. The suppression created a gulf in the knowledge of risk between those in the workplace on the one side (who knew nothing of mesothelioma) and senior industry management and those in the Department of Mines on the other (who had access to the 1962 Survey). We do not know how many men, women, and children had their health ruined as a result.

Whiter than White

Until the 1960s, the industry had devoted little energy or interest to occupational health, but mesothelioma signalled a change. The industry suddenly acquired a research agenda and ARDs became not only a workplace problem, but also one that involved the laboratory. There the research took an unexpected turn. The centre-piece in the industry's strategy since 1960 has been the formulation of the fibre-specific theory—the so-called 'chrysotile defence' or 'amphibole hypothesis'. Simply put, these state that white asbestos is benign and that the amphiboles alone are responsible for ARDs, especially mesothelioma. Chrysotile, therefore, if it is handled in a 'controlled' fashion can still be used and the world can continue to enjoy the mineral's prodigious benefits—especially in the developing world, where any health risks are outweighed by the flow of clean water through a/c pipes. The Canadian industry, in particular, remains adamant to this day that chrysotile can be used safely and is a valuable material.

Obviously, chrysotile could only prevail if it was shown that it did not pose the same health hazards as crocidolite. As we have seen, during the 1960s and 1970s the industry began a massive research and public relations effort to save the industry. Funding was poured into front organizations, such as the AIA/NA. By 1971, the AIA/NA budget was almost $300,000, to be spent on monitoring medical conferences and papers, and then initiating 'lines of action'.[26] One tactic was to 'Start to

[26] Minutes of Meeting of Board of Directors, AIA/NA, 24 August 1971, p. 5.

tell the chrysotile story and discredit other fibers'.[27] American dollars were channelled towards Quebec Asbestos Mining Association (QAMA) and Institute of Occupational and Environmental Health (IOEH) at McGill University in Montreal. Eighty per cent of Canadian asbestos was mined in Quebec, with Thetford Mines at its centre. Canada had 40 per cent of the world chrysotile market, making it the world's largest producer of that mineral, with yearly shipments by the mid-1960s of 1.5 million tonnes. With this business under threat, there was no shortage of money available and by 1972, QAMA had expended over Can$2 million on research projects.[28]

The global industry was fortunate that blue asbestos, which was only mined in South Africa and at a single site in Australia, comprised a small fraction of the market. In the US, only trivial amounts of all varieties of amphiboles had been imported into the US before the 1940s; and even by the early 1970s annual imports of crocidolite stood at 18,000 tonnes compared with 840,000 tonnes of chrysotile, a mere 2 per cent.[29] On the other side of the Atlantic, during the twentieth century Britain imported 5 million tonnes of chrysotile, but only 150,000 tonnes of crocidolite (3 per cent). Chrysotile producers successfully protected these markets, while the demand for amphibole in the OECD states fell sharply. In 1969, the UK led the way in ostracizing crocidolite by persuading manufacturers to voluntarily cease importing it. In Pretoria, the Department of Mines believed, with some justification, that Canadian producers were publicizing the dangers of amphiboles in the hope of gaining a greater market share for their own asbestos.[30] Amphibole production in South Africa ended in 1996, leaving chrysotile as the only type of asbestos still used.

The survival of the chrysotile industry has been remarkable, since there was nothing in the pathbreaking research by Wagner (1960), Selikoff (1964), or Newhouse (1965) to suggest that chrysotile is benign. One of Wagner's original cases of mesothelioma was in a man with occupational exposure to chrysotile, while the cohorts used by Selikoff and Newhouse had exposure to a mixture of fibres including significant quantities of chrysotile.[31] The initial response by the scientific community to Wagner's paper was to note the association with crocidolite, but assume that all

[27] AIA/NA, Items for Discussion, 23 March 1973.

[28] 'IOEH: A Review of Background and Projects Sponsored', 1974.

[29] I.J. Selikoff and D.H.K. Lee, *Asbestos and Disease* (1978), p. 56.

[30] *Report of the Department of Mines for the Year Ending 31 December 1978* (Pretoria, 1978), p. 7.

[31] One of Wagner's original thirty-three cases was of a man who had worked in a brake repair shop, where he would only have come into contact with chrysotile.

types of asbestos were hazardous. In 1964, scientists at the New York conference rejected the Canadian claim that 'that only crocidolite is concerned with these tumours'.[32] In Britain, experts also agreed that 'it is highly improbable that only one type of fibre is always responsible' for mesothelioma.[33]

The asbestos industry had one story about chrysotile for private consumption and another for the public. While the public was told that white fibre was harmless and did not cause mesothelioma, legal discovery shows that this was something the industry itself did not believe. In August 1967, American physician Dr George Wright, who was an industry consultant and chair of the scientific committee of the IOEH, wrote a confidential review of its work. He noted the clear evidence linking lung cancer and asbestos—evidence that had led to the formation of the IOEH. Wright observed that there was also evidence that all types of asbestos cause mesothelioma.[34] He warned that the data were so strong that asbestos could be outlawed by governments or simply boycotted by consumers. In the following year, QAMA accepted that the 'three types of asbestos are in the same boat... [and] ... one cannot ignore that with proper circumstances the same fibrogenesis and malignancy apply to amosite and chrysotile'.[35] This conclusion was unavoidable, as cases of pleural mesothelioma and lung cancer had been identified among Canadian miners and other workers in the late 1940s and early 1950s. In 1949, for example, the 63-year-old treasurer of the Asbestos Corporation (a company member of QAMA) had died of 'extensive malignant mesothelioma'. His only exposure since 1920 had been as an office worker.[36] Not only did the industry know about such cases, but it suppressed publication of the evidence.[37]

The Quebec mines were the global centre of chrysotile production making disease rates among miners and their communities crucial to the industry's survival. Following Wagner's research, QAMA reinvoked its old defence that Canadian working conditions and products were safer than

[32] International Union Against Cancer, 'Report and Recommendations of the Working Group on Asbestos and Cancer', *ANYAS* 132 (1965), pp. 706–21.

[33] J.C. Gilson, 'Asbestos Cancer: Past and Future Hazards', *Proceedings Royal Society of Medicine* 66 (1973), pp. 395–403.

[34] Minutes of QAMA Special Summer Meeting, Murray Bay, Quebec, 8–11 August, 1967.

[35] Minutes of QAMA Special Meeting..., 28, 29 March 1968, Grand Bahama Hotel and Country Club, G.B.

[36] Saranac autopsy file sent by Paul Cartier, Case IM-573.

[37] W.E. Smith, 'Surveys of Some Current British and European Studies of Occupational Tumor Problems', *Archives of Industrial Hygiene and Occupational Medicine* 5 (1952), pp. 242–62; B.I. Castleman, *Asbestos: Medical and Legal Aspects* (2005), pp. 95–8.

elsewhere.[38] The lynchpin in that defence was Dr Corbett McDonald, a 46-year-old occupational health physician from England. McDonald was based at McGill University in Montreal, though initially he had little familiarity with the asbestos mining industry. The catalyst for his interest was his attendance at the 1964 New York conference. Immediately after that meeting—encouraged by Gilson, Wagner, and Knox—McDonald approached Johns-Manville and broached the idea of an epidemiological study of Canadian mining. According to McDonald, this was something which could 'only be done in collaboration with the industry'.[39] It was a well-targeted approach, since Johns-Manville was QAMA's main funder at a time when the latter was searching for 'authoritative publicity'.

In early 1965, the Canadian government made mesothelioma a registerable disease, thereby forcing QAMA to take some initiative. The federal government offered to finance a research programme, but QAMA was reluctant to accept as it would have lost control of the results.[40] Corbett McDonald offered an alternative. In July 1965, he submitted a research proposal to QAMA. He described the evidence for an 'association' between asbestos and cancer as 'not conclusive', and urged the need for urgent clarification of the issue because of the world's increasing dependence on 'this valuable mineral'. The result in 1966 was QAMA's funding of the IOEH, as described in Chapter 4.[41] It was the IOEH that was the main beneficiary of the $2 million expanded by QAMA.[42]

Those funds enabled McDonald to become the leading Canadian authority on ARDs. For QAMA he proved an ideal choice. McDonald, like Richard Doll, had the tact to work closely with industry, while never publicly criticizing it. He may have lacked Selikoff's charisma and his team was always too small to emulate the promethean output of Mount Sinai. But in the context of Canadian asbestos mining, McDonald's research was to prove most valuable, not least because McDonald was so sceptical of the idea that chrysotile could cause mesothelioma and persistently searched for other explanations for the incidence of that disease amongst miners and their families.

[38] M. Greenberg, 'Re. Call for an International Ban on Asbestos: Trust Me I'm a Doctor', *AJIM* 37 (2000), pp. 232–4; M. Greenberg, 'A Report on the Health of Asbestos, Quebec Miners', *AJIM* 48 (2005), pp. 230–7.

[39] McDonald to Kenneth W. Smith, J-M Corporation, 27 November 1964.

[40] Record of Discussions between John Waddell and Dr. John Beattie re. QAMA Research Institute of Occupational Health, Montreal, at Rochdale 14 December 1965 and with K.V. Lindell, New York, 15 December 1965.

[41] 'IOEH: A Review of Background and Projects Sponsored', 1974.

[42] 'IOEH: A Review of Background and Projects Sponsored', 1974.

With the help of the mining companies, McDonald was able to identify roughly 11,000 miners and millers who had worked in the industry. He then collected the available dust data. By 1971, the McGill team was ready to publish its results. McDonald's data suggested that levels of asbestosis and cancer were very low, so much so that it seemed that workers had a lower mortality than the corresponding population of Quebec. A few mesotheliomas had been found, but as McDonald and his team put it, Canadian miners and millers had 'nothing like the experience of the American insulation workers'.[43]

This finding was nothing if not well timed. Publication virtually coincided with the OSHA Labor Hearings in Washington in 1972, when the whole subject of an American safe threshold was being debated (see Chapter 4). In contrast to the British standard of 2 f/cc, McDonald argued that workers in chrysotile mines and mills could be safely exposed to 5–9 f/ml on average during a working life and an even higher threshold for a shorter employment period.[44] The hearings were heated, as the various experts could be questioned by their adversaries. Johns-Manville's Matt Swetonic watched enthralled at the protagonists 'ripping each other's guts out in a public forum'.[45] It was not a happy experience for McDonald, who sought to distance himself from his backers by arguing that the initiative for his survey came from the federal government and that it was McGill that subsidized his research. But his critics would have none of it. They flagged his industry connections and probed what they perceived as flaws in his scientific methods.[46] Dust counts before the 1960s were scanty and unreliable and McDonald's team made clear that they had conducted relatively few dust counts themselves. It was also questionable whether the Quebec study had allowed enough for under-reporting of ARDs and the effect of past suppression of data by QAMA. Critics of the study also highlighted that many of the workers examined had quite low dust exposures, and that many had worked in open-air pits (such as the giant Jeffrey Mine), while others had not worked or lived long enough to develop ARDs.

Inevitably, Selikoff was a sceptic. Jack Solon, vice-president of Johns-Manville, attended presentations by Selikoff in Toronto in 1973 and then

[43] J.C. McDonald et al., 'Mortality in the Chrysotile Asbestos Mines and Mills of Quebec', *Archives of Environmental Health* 22 (June 1971), pp. 677–86.

[44] Statement by J.C. McDonald, US Dept. of Labor Hearings on a Standard of Exposure to Asbestos Dust, Washington DC, 14–17 March, 1972, p. 13.

[45] M.M. Swetonic, 'Death of the Asbestos Industry' in J. Gottschalk (ed.), *Crisis Response* (1993), p. 305.

[46] P. Brodeur, *Expendable Americans* (1974), pp. 129–34.

told senior figures in the company: 'One fact of particular import for the industry—came through loud and clear. Dr Selikoff is going to fight hard on the chrysotile, crocidolite, amosite matter. He said quite directly that until there is much more positive evidence that chrysotile is less dangerous than crocidolite and amosite, he is not going to "buy" it.' He went on: 'This [refuting Selikoff] is one of our key projects now because it is important to us to refute him quickly. . . . The chrysotile matter is too important to us and we intend to hit hard with every available argument.'[47]

McDonald himself tackled some of the critics. He penned an angry letter to David Kotelchuck, an American environmental health expert, after the latter had argued that the asbestos problem was caused not by scientific ignorance, but by 'economic self-interest—that of the industry and the scientists who do its bidding'.[48] McDonald replied: 'We have gained nothing from [the project], except the chance to do important work', and told Kotelchuck:

> [L]ife for many honest medical scientists—of which our group at McGill is only one, has been made hell by people like you, Brodeur and the like. No doubt this is your intention; I suppose you believe that people like us should be driven out of their jobs or out of their minds. I would like you to know that you are being very successful. Competent and sensitive people cannot take this kind of abuse.[49]

However, more criticism of McDonald was to follow. At the end of 1974, a group of trade unionists, led by Paul Formby, invited a Mount Sinai team under Dr Bill Nicholson (1930–2001) to undertake an independent clinical study of Thetford miners.[50] Unlike the McGill study, it focused on asbestos mine and mill workers with at least twenty years of employment in chrysotile mining and milling. Nicholson found a much higher incidence of lung cancer and asbestosis than in the McGill cohort, though the risk of death from mesothelioma was evidently much less than in the factory workers and insulators that Selikoff had previously examined. They found only one Quebec mesothelioma.[51] The Selikoff study featured on a Canadian radio (CBC) programme in 1975 and immediately provoked a

[47] Memo: Report on the Recent Presentation in Toronto by Dr Selikoff. F.J. Solon, Denver, 18 May 1973.

[48] D. Kotelchuck, 'Asbestos Research', *Health Advisory Center*, No. 61 (Nov/Dec 1974), pp. 1–6, 20–7.

[49] McDonald to David Kotelchuck, 10 March 1976.

[50] L. Tataryn, *Dying for a Living: The Politics of Industrial Death* (1979), pp. 15–60.

[51] W.J. Nicholson et al., 'Long-Term Mortality in Experience of Chrysotile Miners and Millers in Thetford Mines, Quebec', *ANYAS* 330 (1979), pp. 11–21.

strike. McDonald was inevitably drawn into the dispute, because his own research had shown relatively little risk for miners. When interviewed on CBC, he claimed that he was invited in 1965 by the federal government and the International Union against Cancer to study asbestos miners. He denied that his research was an industry initiative, but he was unsure how much money the industry had contributed to his work. He did not think it would be 'proper to sound an alarm, because the alarming situation is largely due to past circumstances'; and believed that the Mount Sinai study was 'quite unnecessary'.[52] However, Selikoff was not finished. Militancy next erupted at the Baie Verte chrysotile mines in Newfoundland, where another Selikoff study in 1976 had highlighted hazardous working conditions and significant ARDs.

Selikoff's activities in Canada ended abruptly when the Quebec industry was nationalized, but before then trade union agitation at Thetford had triggered a federal inquiry. Published in 1976, it found that dust control was very poor and the lead investigator, Judge René Beaudry, found it 'shocking' that in some plants workers were still handling asbestos fibre with their bare hands.[53] He went on to comment on employers: 'They have kept available information about the dangerous effects of asbestos dust away from the workers and the unions.'[54] A similar picture was drawn by an Asbestosis Working Group in Ottawa, which suggested that McDonald's statistics on mesothelioma were underestimated.[55] These reports concluded that the industry was more concerned with medicalizing the asbestos problem through the compensation system and epidemiology than by spending money on protecting workers from the dust. Revelations were also emerging of poor working conditions at the Canadian plants of Johns-Manville in Scarborough, T&N in Montreal, and Bendix Automotive in Windsor.[56]

However, the McGill team, with industry resources to encourage them, were nothing if not dogged. Critics at the time and for some years afterwards never quite dislodged their basic finding that miners and millers had a lower risk of ARDs, which in turn suggested that chrysotile was a relatively safe fibre. The asbestos industry had found its counterpoint to

[52] Interview with McDonald, Midday Magazine, CBC Radio, 7 March 1975.

[53] R. Beaudry, G. Lagace, and L. Jukau, *Rapport Final: Comité d'Étude sur La Salubrité dans L'Industrie de L'Amiante* (1976).

[54] Beaudry, *Rapport*, p. 381.

[55] Subcommittee on Environmental Health (Department of National Health and Welfare), *Report of the Asbestosis Working Group* (Ottawa, 15 February 1976).

[56] R. Storey and W. Lewchuck, 'From Dust to DUST to Dust: Asbestos and the Struggle for Worker Health and Safety at Bendix Automotive', *Labour/Le Travail, Journal of Canadian Labour Studies*], 45 (2000), pp. 103–40.

Selikoff. It had once argued that Selikoff was unscrupulous in extrapolating from the insulators' experience, but QAMA had no scruples about extrapolating the Canadian experience for its own ends. QAMA was able to attack the validity of the media reports by arguing that leading scientists had shown that its miners were supposedly healthier than the rest of the Quebec population.

McDonald's research was important in other ways. It was cited in government inquiries and commissions where it gave weight to employers' claims about the safety of asbestos. It shed the industry's past (by apparently confirming all those traditions that Canadian fibre was 'safe'), deflected attention from the present, and heaped the blame on amphiboles. At the UK government inquiry in 1976 (the Simpson Committee), the industry drew on McDonald's work to claim that within certain thresholds white asbestos was harmless. The government accepted that chrysotile 'rarely caused mesothelioma' and that this 'favourable' point should be used to frame policy.[57] In Canada in 1984, the tensions between trade unions, a concerned public, and the industry were resolved when a Royal Commission in Ontario recommended a ban on crocidolite and amosite, but endorsed the use of chrysotile if there was adequate dust control. The Commissioners relied heavily upon McDonald's work and in their three-volume report his voice is prominent.[58] The industry was well pleased with the Commission's findings, which exonerated chrysotile and laid the blame for disease upon the distant past when exposure levels were so high. The Commission also blamed amphiboles imported into Ontario from South Africa for mesothelioma among Canadian workers. In effect, the Commission truncated the domestic use of asbestos, while giving a green light for Canada to continue to mine and export chrysotile to the developing world.

By the early 1980s, the chrysotile defence was in robust health, helped by scientific work on asbestos and cancer. Scientists had begun with renewed vigour to examine the dust content of lungs and the effects of different fibres on laboratory animals. It was not easy. Counting millions of fibres in a few grams of lung is as difficult as it sounds; so, too, is administering fibres to rats and hamsters. Yet many thousands of laboratory hours were spent in doing exactly that during the 1970s and 1980s, especially by the industry's research bodies such as the ARC. As the results

[57] Health & Safety Commission, *Asbestos. Vol. 1: Final Report of the Advisory Committee* (1979), p. 62.

[58] *Report of the Royal Commission on Matters of Health and Safety Arising from the Use of Asbestos in Ontario* (1984); cf. 'Asbestos Mining', vol. 1, pp. 173–8.

became available, scientific opinion began splitting and fragmenting like a piece of chrysotile. One finding from the animal bio-essays was that man-made fibres (which some of the firms already produced) were also potentially carcinogenic. Overall, however, strong support for the chrysotile defence emerged. Mearl Stanton, an American pathologist, conducted experiments on rats which suggested that the carcinogenicity of asbestos (and other) fibres depended on their dimensions and durability (biopersistence) rather than physicochemical properties. Crocidolite seemed to fit the picture on the 'Wanted List': the fibres are relatively long (at over five microns), thin, and highly durable.[59] Amphiboles are also more likely than chrysotile to be found in the lungs of dead asbestos workers (mainly because chrysotile fibres are more easily dissolved in the lungs and more inclined to fragment, making them difficult to see even with an electron microscope). These findings sustained the chrysotile defence.

The defence was used in industry advertisements, where chrysotile was pictured as a 'curly, soft' fibre that was easily cleared from the lungs (in contrast to the deadly looking needle-like fibres of crocidolite). Inevitably, as the number of personal-injury actions gathered pace in the 1980s, the chrysotile defence was used increasingly in courts. Arcane arguments developed at inquests and in courtrooms about what type of fibre—and how many fibres per gram of lung tissue—had been retrieved from workers' lungs. It was often a gruesome charade, that took no account of the distress suffered by relatives, but the strategy was pursued relentlessly by insurers and defence lawyers. One victims' group in the UK found it worthwhile to invest in an electron microscope to defend such cases.

The chrysotile defence, though, had to contend with some awkward facts. McDonald's research suggested that lung diseases in miners and millers were not common: however, his team (and another group in British Columbia) *had* found several mesotheliomas in Quebec workers. In the atmosphere of the 1970s, even one was too many. Such cases appeared to shoot down the chrysotile defence, thereby setting the chrysotile industry a major problem.[60] However, the results of the IOEH research continued to be reassuring. McDonald and his McGill colleagues initially suggested that 'contaminants' in the Canadian environment, rather than

[59] M.F. Stanton et al., 'Relation of Particle Dimension to Carcinogenicity in Amphibole Asbestos and Other Fibrous Materials', *Journal of the National Cancer Institute* 67 (1981), pp. 965–75.

[60] A. Churg et al., 'Lung Asbestos Content in Chrysotile Workers with Mesothelioma', *American Review of Respiratory Disease* 130 (1984), pp. 1042–5, found six deaths from mesothelioma in ninety consecutive autopsies of Québec miners and millers.

chrysotile, were to blame for tumours. Organic and synthetic oil fell first under suspicion, followed by polyethylene in asbestos storage bags. McDonald then alleged that the cause of many Quebec mesotheliomas was imported Australian crocidolite that had been used during the Second World War for the manufacture of gas masks.[61]

Perhaps the most intriguing suggestion for the carcinogenicity of chrysotile was that many of the observed mesotheliomas in chrysotile workers (especially miners and millers in Canada) were due to amphiboles. Since the 1960s, it had been known that Canadian ore contained blue fibrous riebeckite (crocidolite), yet researchers and asbestos companies initially ignored this. By the late 1980s and 1990s, however, it was suggested by McDonald and others that tremolite (which typically constituted about 1 per cent in commercial-grade chrysotile) was the cause of mesothelioma in Canadian workers.[62] Further, that by identifying tremolite-free ore, then chrysotile mining could continue. It seemed a persuasive theory, especially since tremolite was often found in the lungs at *post mortems* on mesothelioma victims.

The apotheosis of the chrysotile defence was reached in 1997. An issue of the *Annals of Occupational Hygiene* (the journal of the British Occupational Hygiene Society that had orchestrated the two-fibre threshold) showcased the completion of McDonald's Quebec cohort study. It was a memorable occasion in other ways. In an unorthodox gesture, the journal invited one of the McGill scientists, Doug Liddell, to write a guest editorial. Since Liddell was one of McDonald's co-authors, he was in effect asked to editorialize on his own work. The opportunity proved too much for Liddell, who from a safe distance launched a vicious attack on Selikoff (now conveniently dead) and on the Mount Sinai physicians. Many years of simmering rage and frustration boiled to the surface as Liddell dubbed the Americans as 'The Lobby' and argued that they were motivated by 'menace' and 'intense *malice*'. The article contained an even more bizarre assertion that chrysotile is 'essentially innocuous, except possibly in textile manufacture', and that this was the view of most scientists in the field.[63] The McGill articles showed an excess of lung cancer and thirty-eight mesotheliomas, but this was attributed characteristically

[61] A.D. McDonald and J.C. McDonald, 'Mesothelioma after Crocidolite Exposure during Gas Mask Manufacture', *Environmental Research* 17 (1978), pp. 340–6.
[62] A. Churg, 'Chrysotile, Tremolite, and Malignant Mesothelioma in Man', *Chest* 93 (1988), pp. 621–8.
[63] F.D.K. Liddell, 'Magic, Menace, Myth and Malice', *Annals of Occupational Hygiene* 41 (1997), pp. 1–12.

to tremolite not chrysotile.[64] Another reassurance was that only certain 'central' mines in Thetford had a tremolite problem; 'peripheral' mines had a much lower tremolite component, which could be removed in processing to provide an amphibole-free product. By the end of the twentieth century, then, the McGill scientists had presented the world with a product that was whiter than white.

Chrysophiles and the Doctrine of Controlled Use

The fibre debate has been characterized by McDonald as one between 'chrysophiles' and 'chrysophobes'.[65] The chrysophiles grew in influence during the 1980s. When the Environmental Protection Agency (EPA), moved to phase out and ban virtually all asbestos products in the US, the Canadians joined with American industry interests in opposing it. In 1991, US courts overturned the ban, ensuring that as this book goes to press asbestos is still not banned in the US.[66]

In the early 1990s, chrysophiles were in the ascendancy at some conferences, where complaints were heard about the 'uniting of several quite different minerals under the term *asbestos*'.[67] Between 1986 and 1993, the influence of the Canadians has also been detected in meetings of the International Program on Chemical Safety (IPCS), where products such as asbestos-cement (a/c) were cleared for continued use; and in meetings and reports under WHO auspices.[68] At such meetings, the chrysotile defence joined hands with the concept of 'controlled use', which meant that by good health and safety practices the risks of using chrysotile would be negligible. This meant that society could enjoy the 'dividend' of using asbestos fire-retardants and friction products. Besides McDonald and his

[64] J.C. McDonald and A.D. McDonald, 'Chrysotile, Tremolite and Carcinogenicity', *Annals of Occupational Hygiene* 41(1997), pp. 699–705; A.D. McDonald et al., 'Mesothelioma in Quebec Chrysotile Miners and Millers: Epidemiology and Aetiology', *Annals of Occupational Hygiene* 41(1997), pp. 707–19; F.D.K. Liddell, A.D. McDonald, and J.C. McDonald, 'Dust Exposure and Lung Cancer in Quebec Chrysotile Miners and Millers', *Annals of Occupational Hygiene* 42 (1998), pp. 7–20.

[65] J.C. McDonald, 'Unfinished Business: The Asbestos Textiles Mystery', *Annals of Occupational Hygiene* 42 (1998), pp. 3–5.

[66] Environmental Working Group. *Asbestos: Think Again* (2004), pp. 31–2.

[67] G.W. Gibbs, F. Valic, and K. Browne (eds.), 'Health Risks Associated with Chrysotile Asbestos', Report of Workshop in Jersey, Channel Islands, 14–17 November 1993, *Annals of Occupational Hygiene* 38 (1994), pp. 399–646.

[68] B.I. Castleman, 'The Manipulation of "Scientific Organisations": Controversies at International Organisations Over Asbestos Industry Influence', *Annals of the Global Asbestos Congress—Past, Present and Future* (CD-ROM, 17–20 September 2000, Osasco, Brazil).

colleagues, proponents of this argument have included at various times: Sir Richard Doll, Dr Peter Elmes (former director of the Pneumoconiosis Research Unit in the UK), Dr Robert Murray (once medical adviser to the Trades Union Congress in the UK), Dr Kevin Browne (a former physician of Cape Asbestos),[69] Professor Jean Bignon (a leading French lung specialist), American physician Hans Weill, and Dr Christopher Wagner.

Ironically, most of these individuals were once seen as threats to the industry and had sometimes encountered considerable opposition, including intimidation. Ultimately, however, their reaction to the asbestos problem was conservative. For example, Doll developed a working relationship with T&N, even occasionally acting as litigation consultant. In 1970s, his belief that the risk of lung cancer in the Rochdale factory had been eliminated was found to be far too optimistic. Nevertheless, he was unshaken in his view that workers could continue to be exposed to chrysotile within the safe thresholds that he and his fellow scientists had devised. By the 1980s, he was complaining about the 'unjustified threats of prosecution to which industrial companies are increasingly subject'.[70] His transparency, however, was to be questioned at the end of his career, when it was revealed that his Oxford college had received a substantial donation from T&N and that he had accepted a long-standing and secret retainer from the chemical giant Monsanto.[71] Meanwhile, Doll's protégé Julian Peto, who has perhaps done as much as any other epidemiologist to alert the world to the future mesothelioma epidemic, in a controversial speech in 2005, attacked the anti-asbestos lobby as 'hysterical' and 'completely unscientific'.[72] Peto argued that substantial exposure to chrysotile meant 'really little risk' and that the resolution of asbestos problems in the developing world was no concern of the West.[73] His chrysophile views came as a shock to some in the ban-asbestos movement, who had counted him as a supporter.

[69] K. Browne, 'Chrysotile: Thresholds of Risk', Paper presented at an International Seminar on Safety in the Use of Chrysotile Asbestos: Basis for Scientifically-Based Regulatory Action, Havana, Cuba, 12–13 September 2000.

[70] R. Doll, 'Occupational Cancer: A Hazard for Epidemiologists', *International Journal of Epidemiology* 14 (March 1985), pp. 22–31.

[71] G. Tweedale, 'Hero or Villain? Sir Richard Doll and Occupational Cancer', *IJOEH* 13 (2007), pp. 233–5.

[72] J. Peto, 'Asbestos Cancer Deaths in the UK: The Past, Present, and Future', Lane Lecture, Manchester University, 23 November 2005.

[73] G. Tweedale, 'What You See Depends on Where You Sit: The Rochdale Asbestos Cancer Studies and the Politics of Epidemiology', *IJOEH* 13 (January/March 2007), pp. 70–9.

Dr Peter Elmes (1921–2003) became a T&N consultant and joined the industry-controlled ARC. Some of his reports on factory and mining conditions were trenchant, but he did not publicize them and the dusty conditions that he saw did not influence his chrysophile views. He defended the 'safe-use' of asbestos and thought that switching to substitutes was 'risky and expensive'. He worried more about the closure of factories than ARDs. He agreed that asbestosis was a disease that could mean early retirement at sixty, but added comfortingly that 'life may not be appreciably shortened and death from unrelated disease is common'.[74] Elmes was to prove useful to the industry in US property damage litigation (see Chapter 7), where in the mid-1980s he gave evidence at the request of defendant asbestos firms, such as US Gypsum. His testimony supported every industry position—and beyond. He described asbestosis as a non-life-threatening condition ('it very, very seldom even shortens life appreciably'), argued that 'the two-fiber standard is probably unnecessarily stringent', and even equivocated over whether heavy chrysotile exposure could cause asbestosis.[75]

Some idea of the allegiances of Robert Murray (1916–1998) can be divined from the fact that when he was at the TUC he passed copies of Selikoff's letters to his medical and company friends at T&N. He successfully sued one of his foremost critics, Alan Dalton, for libel. Dalton's offence was to argue that Murray had pro-industry views that were contrary to his position in the trades' union movement.[76] After the trial, Murray was recruited by the AIA as a consultant. Like Elmes, he testified at the request of defendant asbestos companies, where he argued that they had been 'unfairly picked upon'.[77] Always a *bête noirie* to the industry's critics for the Dalton episode, Murray did not endear himself to them for his unwavering belief that asbestos had benefits for the developing world in terms of cheap asbestos building materials and water pipes.

In France, Professor Jean Bignon had initially supported activists in their struggle against asbestos. In 1977, Bignon wrote a protest letter to the French government about the industry's *White Book: Asbestos, the Truth*. Bignon rebutted its assertions that asbestos was a safe product by

[74] P.C. Elmes, 'Current Information on the Health Risk of Asbestos', *Royal Society of Health Journal* 96 (1976), pp. 248–52.

[75] Elmes' testimony, *County of Anderson, Tennessee (Head of Education and Superintendent of Schools) v. US Gypsum and National Gypsum*, US District Court for the Eastern District of Tennessee Northern Division, 4 March 1985.

[76] A. Dalton, *Asbestos Killer Dust* (1979).

[77] Murray testimony, *Milton Jenkins v. Fibreboard* Corp., Superior Court of the State of California for the County of Los Angeles, 12 July 1993, p. 102.

emphasizing that cancer was a serious problem, that all asbestos types caused it, that substitutes were safer, and that France had no dust regulations. Bignon warned of the serious consequences over the next thirty years, but the letter was not made public at the time.[78] However, in 1982, when the industry organized its lobby group the CPA, Bignon joined it. He shed his opposition to asbestos and lent his support to the CPA's doctrine of 'usage controlé'. Bignon was joined by his protégé Professor Patrick Brochard (another lung specialist) and Etienne Fournier, a toxicologist and a member of the Academie de Medicin. They appeared to legitimize what was essentially a public relations campaign. EU measures to end asbestos use were blocked in France and by 1990s the country was the biggest user of asbestos in Western Europe.

Bignon continued to support the CPA into the 1990s.[79] In June 1995, Bignon and Brochard praised the CPA as having done 'good work during the last thirteen years'.[80] When the media exposed the CPA, *Le Monde* published an editorial, 'CPA: Le *Mea Culpa* of Jean Bignon', in which Bignon lamented, 'we should have collectively opposed the creation of the CPA'.[81] Bignon said he knew nothing of the CPA's funding and blamed the inertia of the authorities for not regulating the problem properly. But Bignon's critics were unsympathetic and journalist François Malye attacked him for remaining silent so long. He quoted Bignon as stating that the French ban was political and not scientifically based, partly due to 'agitateurs'.[82] When a French Senate report reviewed the whole episode, it described the role of the medical community and the government as 'ambiguous'.[83]

Perhaps no one epitomized this scientific ambiguity more than Christopher Wagner. At first glance, no one should have been more antagonistic towards the magic mineral. Wagner had paid a high price for his research. Following the publication of his article on mesothelioma, a rumour circulated in Johannesburg that the mining companies were threatening to have him shot.[84] Wagner was under such pressure that

[78] Bignon's letter is reproduced in R. Lenglet, *L'Affaire de L'Amiante* (1996), pp. 231–5; and F. Malye, *Amiante: Le Dossier de L'Air Contaminé* (1996), pp. 273–80.

[79] Malye, *Dossier*, p. 118.

[80] M.G. Deriot and M. J.-P. Godefroy, *Rapport [No 37]...sur...Le Bilan et Les Consequences de La Contamination par L'Amiante* (Senat: Session Ordinaire de, 2005–6), p. 83.

[81] Quoted in B. Hopquin, 'Amiante: Vingt-cinq Ans D'Intox', *Le Monde*, 22 April 2005.

[82] Malye, *100,000 Morts*, p. 132.

[83] Deriot and Godefroy, *Rapport*.

[84] Deposition of Dr Ian Webster, 12 November 1996, in Re: Asbestos Personal Injury Cases, Arrington Lead No 93-9-114 In the Circuit Court of Jones County, State of Mississippi, Second Judicial District, p. 135.

two years later he accepted a position at Llandough Hospital in South Wales in the UK, where the government's PRU was based. Wagner came at the invitation of PRU head Dr John Gilson and he worked there until his retirement in 1988. When he returned briefly to Johannesburg in 1966, he was confronted by senior figures from 'the South African medical establishment' who accused him of damaging the local economy.[85] Wagner enjoyed a distinguished career at Llandough and in 1985 he received the Charles S. Mott Prize for 'the most outstanding recent contribution related to the causes and ultimate prevention of cancer'. When he died in June 2000, the *Guardian* rightfully described him as an outstanding international authority on asbestos-related cancer.[86] It is ironic that the obituary also noted that Wagner had left the world a safer place.

At Llandough, Wagner continued to work on ARDs and for more than twenty years his views were orthodox. In November 1967, he presented a paper at the QAMA-sponsored Second Antigua Conference on the Biological Effects of Asbestos. Wagner spoke about animal experiments that he had conducted with blue, brown, and white fibre. Wagner was surprised that chrysotile from Thetford produced the highest rates of mesotheliomas. He then repeated the experiments and got the same results. He even found tumours with low doses of chrysotile. Wagner noted that scientists in Cincinnati had found the same. He told an audience that included Drs Gilson, Selikoff, and Vigliani: 'So it looks as though this particular type of chrysotile [Thetford] is a nasty thing to have in one's pleural cavities.'[87] In a follow-up study, Wagner and his colleague Geoffrey Berry found more mesotheliomas in rats with intra-pleural injections of white than with blue or brown fibres.[88] Wagner produced similar results in a subsequent study.[89] A large-scale inhalation study by Wagner in 1974 found that chrysotile was the most carcinogenic of the fibres.[90]

With such a weight of experimental evidence, Wagner rightfully concluded that all types of asbestos cause all three ARDs. In 1979, as a

[85] Jock McCulloch interview with Dr Chris Wagner, Weymouth, Dorset, 22 March 1998.

[86] K.M. Connochie, 'Chris Wagner' , *Guardian*, 1 July 2000. See also *Independent*, 4 July 2000, and *Daily Telegraph*, 11 July 2000.

[87] Second Antigua Conference on the Biological Effects of Asbestos 20–23 November 1967. Quebec Asbestos Mining Association, Vol. 3, p. 92. Transcript Marked: 'Restricted Copy, No. 27'.

[88] J.C. Wagner and G. Berry, 'Mesothelioma in Rats Following Inoculation with Asbestos', *British Journal of Cancer* 23 (1969), pp. 578–81.

[89] J.C. Wagner, G. Berry, and V. Timbrell, 'Mesotheliomas in Rats Following the Intrapleural Inoculation' with Asbestos', H.A. Shapiro (ed.), *Pneumoconiosis: Proceedings of an International Conference, Johannesburg, 1969* (1970), pp. 216–19.

[90] J.C. Wagner et al., 'The Effects of the Inhalation of Asbestos in Rats', *British Journal of Cancer* 29 (1974), pp. 252–69.

panel member of the International Agency for Research on Cancer (IARC), he endorsed its finding that chrysotile causes mesothelioma.[91] Wagner's position was in line with that of the IARC, the EPA, the US Department of Health & Human Services, and NIOSH. Less than twelve months later, Johns-Manville's senior health and safety adviser Paul Kotin wrote to Wagner asking if he had any evidence of the hazards of low exposure. Wagner's reply can hardly have pleased Kotin: 'We did produce 2 mesotheliomas in rats after a single day's exposure', Wagner wrote, 'and we did get a few mesotheliomas with the Canadian chrysotile'.[92] He went on to comment that he was unsure how regulatory authorities in the US would interpret his results. He underlined those views in a medical journal in 1980, when he stated that his experiments had 'cast some doubt on the epidemiological evidence that crocidolite is much more hazardous than chrysotile as far as mesotheliomas are concerned'.[93]

Paradoxically, as evidence linking chrysotile to mesothelioma continued to accumulate, Wagner changed his mind. That *volte-face*, which put Wagner at odds with the leading researchers in the field, occurred in 1986 when Wagner published a remarkable paper that proclaimed the 'innocence of chrysotile to humans'.[94] He was aware that his own animal experiments showed the opposite, but argued that this was due to contaminated fibres. He went on to say that commercial treatments could now straighten chrysotile fibres to render them safe, while a search was being made for economic methods of removing 'contaminants'. No such treatments have ever been found, but henceforth Wagner regarded chrysotile (in McDonald's words) as 'close to harmless' and his views on amosite also mellowed.[95] This proved congenial to the big British and American companies and their lawyers, with whom Wagner reached a *modus vivendi*. At their request, he began testifying in court.

In May 1990, Wagner gave evidence in a case involving the US conglomerate Raymark (formerly Raybestos-Manhattan).[96] Under oath, he

[91] IARC Monographs on the *Evaluation of the Carcinogenic Risk of Chemicals to Humans* (September 1979), p. 23. See also J.C. Wagner et al., 'The Comparative Effect of Three Chrysotiles by Injection and Inhalation in Rats' (1980), pp. 363–73.

[92] Wagner to Paul Kotin, Johns-Manville, 17 April 1980.

[93] J.C. Wagner and G. Berry, 'Carcinogenesis and Mineral Fibres', *British Medical Bulletin* 36 (1980), pp. 53–6.

[94] J. C. Wagner, 'Mesothelioma and Mineral Fibers' [Charles Mott Prize paper], *Cancer* 57 (15 May 1986), 1905–11.

[95] *Independent*, 4 July 2000.

[96] J.C. Wagner deposition, *Claude Cimino v. Raymark Industries*. US District Court for the Eastern District of Texas, Beaumont Division Civil action No. B-86-0456-CA, 30 May 1990, pp. 1–245.

endorsed the three pillars of the industry position on mesothelioma: the disease is always dose-related, even heavy exposure to chrysotile does not cause mesothelioma, and 20 per cent of mesotheliomas are not caused by asbestos. He also disputed the toxicity of amosite.[97] Under cross-examination, Wagner admitted to providing monthly reviews of the current literature to a lawyer named Bruce Shaw, but he could not recall how much he was being paid for his services. He admitted to having sent Shaw and his colleagues drafts of work he was doing with another scientist, E.B. Ilgrin, thereby suggesting the research was commissioned.[98] Wagner denied that in the previous year he had worked for any asbestos company or that he had ever worked for any US asbestos manufacturer.[99]

In 1991, Wagner served as a member of the American Health Effects Institute (HEI), which was running an Asbestos Literature Review Panel on asbestos and cancer.[100] The panel's eighteen members included several experts noted for their conservatism (the HEI was a government-industry-sponsored affair). Wagner was the only one to dissent from the panel's finding that all types of asbestos, including chrysotile, cause mesothelioma. In a statement appended to the final report, Wagner rejected that orthodoxy.[101] He also argued that there was a background rate of mesotheliomas unrelated to asbestos exposure, and emphasized instead, the role played by passive smoking and radon.[102] Wagner's dissent in such an important forum helped perpetuate debate about the toxicity of chrysotile.

Wagner's precise relationship with the asbestos companies might have remained a mystery, but for an unexpected legal battle between two asbestos giants. It involved Owens-Illinois (O-I), which had used large quantities of amosite and chrysotile, and T&N. The case, which involved a $1.6 billion damages claim, began in a Texas district court in June 2000. One of the issues at stake was O-I's level of knowledge about the dangers of asbestos, and in particular the company's attempts to improperly

[97] Wagner deposition, *Cimino v. Raymark*, in which he stated (pp. 79–80) that 'chrysotile does not cause mesothelioma', but that amosite can, 'probably with very heavy dosage'.

[98] Wagner deposition, *Cimino v. Raymark*, pp. 114–19. In that article with Ilgren, Wagner argued, in effect, that chrysotile does not cause mesothelioma. See E.B. Ilgren and J.C. Wagner, 'Background Incidence of Mesothelioma: Animal and Human Evidence', *Regulatory Toxicology and Pharmacology* 13 (1991), pp. 133–49.

[99] Wagner deposition, *Cimino v. Raymark*, p. 123.

[100] The Health Effects Institute, *Asbestos in Commercial Buildings: A Literature Review and Synthesis of Current Knowledge* (1991).

[101] 'Statement by Dr. J. Christopher Wagner concerning some of the conclusions of the Literature Review Panel, August 27th 1991' (1991), pp. S2–1 to S2–3.

[102] For an account of Wagner's dissent, see R. Stone, 'No Meeting of Minds on Asbestos', *Science* 254 (16 November 1991), pp. 928–31.

influence the medical literature. An affidavit presented in defence of T&N by Paul Hanly, a New York attorney, involved Wagner. According to Hanly, in December 1987 Hanly met attorney Bruce Shaw, who at that time was representing both T&N and O-I. Hanly recalled that Shaw had informed him that 'O-I had been paying Dr Wagner $6,000 per month for some period of time irrespective of whether Dr Wagner did any work for O-I'.[103]

Hanly agreed to attend a meeting in the UK with Shaw and Wagner to discuss mesothelioma. According to Hanly: 'Mr. Shaw privately confirmed to me after the meeting that he was trying to persuade Dr Wagner to say or write publicly that only crocidolite asbestos was an undeniable cause of mesothelioma, and that the role of amosite asbestos was non-existent or at most minimal. At the time my lay view was that this position was scientifically unsupportable.' For that reason Hanly thought Shaw would be unable to persuade Wagner: 'I was therefore quite surprised when, in a paper written two years later, Dr Wagner wrote in a published paper that the evidence was overwhelming that the main cause of mesothelioma was crocidolite asbestos—and that amosite asbestos was implicated in just a few cases. The paper fails to mention any financial support from O-I.'[104] However, documents tendered during the Texas case reveal that from 1986, O-I made regular payments to Wagner through its legal firm Nelson, Mullins, Riley & Scarborough. The documents also reveal that the arrangement continued for more than fifteen years.[105] His employment was never acknowledged by O-I and certainly not by Wagner, either in his publications or at the numerous conferences Wagner attended during the period of his association with O-I. It was an association he even denied under oath.

It is equally significant that Wagner's stance shifted at a time when the evidence linking all types of asbestos to mesothelioma had become overwhelming. Wagner was in a position to influence the reception of knowledge and it appears that his brief with O-I was to keep alive doubt regarding the link between chrysotile (and amosite) and mesothelioma. If so, it was a role he played with some success.

By the end of this life, Wagner had completed the journey from pariah to a man with whom the industry could do business. The last

[103] Affidavit of Paul J. Hanly, Jr, In ... *Owens-Illinois, Inc., v. T&N, Ltd.*, in the US District Court for the Eastern District of Texas, Marshall Division, CA No.2-99CV01117-DF, 24 January 2000, p. 10. See also Amended Affidavit of R. Bruce Shaw in Response to Affidavit of Paul J. Hanly Jr, in *Owens-Illinois v. T&N*, 16 February 2000.

[104] Hanly Affidavit, pp. 10–11.

[105] Documents tendered before *Owens-Illinois, Inc. v. T&N*, Ltd in the US District Court for the eastern District of Texas, Marshall Division CA No.2-99CV01117-DF.

conference Wagner attended was in Montreal in 1997, when he joined fellow chrysophiles in extolling the benefits of white asbestos. The Canadian government was among the sponsors. The published volume included a dedication by Sir Richard Doll to Robert Murray, who had died soon after attending the conference. One of Wagner's contributions was a paper which cast doubt on environmental hazards and even attempted to undermine the classic 1965 work of Molly Newhouse.[106] Wagner's co-author was none other than Cape Asbestos's old physician Kevin Browne—an ironic bedfellow in view of Wagner's previous experience with that company. Wagner opined: 'Where we all went wrong was in always referring to asbestos in a generic term. If we stuck to chrysotile as asbestos, and talked about the amphiboles, we would not be in the ghastly mess we are in at the moment.'[107]

The Chrysophobes

The idea that chrysotile was innocuous was a difficult concept to sell to the growing number of victims, who did not care about the scientific niceties of which type of asbestos and what fibre dimensions had caused their cancers. Asbestos activist groups did not like the concept either. Many scientists also had their doubts about the supposed harmlessness of chrysotile.

In practice, the distinction between chrysotile and crocidolite was misleading. In the same way that the terms 'white', 'blue', and 'brown' are not an accurate description of fibre colour (chrysotile looks sea-green rather than white), the geological distinction between them is often blurred. The mine at Penge in South Africa's Limpopo Province was the world's only source of amosite. Yet the seams of amosite were invariably mixed with crocidolite and both fibre types were milled together. As a result anyone using amosite downstream was inadvertently exposed to crocidolite.[108] As we have seen, Canadian chrysotile deposits contain small amounts of amphiboles, such as tremolite. A geological survey conducted in the late 1950s, documented the presence of crocidolite in Quebec's chrysotile

[106] K. Browne and J.C. Wagner, 'Environmental Exposure to Amphibole-Asbestos and Mesothelioma' in R.P. Nolan et al. (eds.), *The Health Effect of Chrysotile Asbestos* (2001), pp. 21–8. On Newhouse, see Chapter 7.

[107] Wagner comments in Nolan et al., *Health Effects of Chrysotile*, p. 67.

[108] Jock McCulloch was given this information by former miners during a visit to Penge in November 2002. See also A.L. Hall, *Asbestos in the Union of South Africa* (1930).

mines.[109] In addition to the variable seams, the distinction between fibres was blurred during manufacture. Pressure pipes and long-span sheeting were made from white asbestos, but such products would warp if taken from the moulds 'green', or not fully set.[110] To overcome that problem, manufacturers added small amounts of crocidolite to a/c mixtures thereby making it possible to work 'green-strength' materials. Soviet industry also enhanced the quality of inferior domestic chrysotile by using crocidolite as a kind of 'broad-spectrum antibiotic'.

Despite the certainties of the McDonald school, establishing the toxicity of asbestos by fibre types is difficult. Exposure in the workplace was usually to a mixture of fibres and when asbestosis first appeared in the 1920s it was found in factories that processed all types of asbestos. Asbestos-related lung cancer began occurring with increasing frequency between the 1930s and 1950s in factories that processed blue, brown, and white fibre. Research on mesothelioma by fibre types was further hampered by the extended latency period, the lack of reliable data on usage in the past, and the relative rarity of mesothelioma amongst the general population.

The McGill research showed the problems of a fibre-specific theory. McDonald's results, which were at odds with the work of Pedley and Stevenson (described in Chapter 3), were taken as an exoneration of chrysotile because the incidence of ARDs was supposedly low. The Mount Sinai physicians had also found the incidence of mesothelioma in Quebec quite low, but they had another explanation (apart from the innocence of chrysotile) for this anomaly. Bill Nicholson believed that it was due to the atypical exposures of Quebec workers, when compared with asbestos textile and insulation workers. This was primarily because asbestos fibre bundles were not extensively broken down in the mining and milling stages (compared with, say, asbestos textile production, where the fibres were extensively refined and manipulated). This meant that Quebec workers escaped exposure to the thinnest and most carcinogenic fibres.[111] This hypothesis received strong support, when studies appeared of an asbestos textile spinning plant in Charleston, South Carolina. Although only chrysotile had been used, a marked excess of lung cancers (including

[109] A. De, 'Petrology of Dikes Emplaced in the Ultramafic Rocks of South Eastern Quebec' (Princeton University PhD, 1961).

[110] Jock McCulloch interview with Pat Hart CEO, Griqualand Exploration & Finance Company, Braamfontein, Johannesburg, 6, 7 July 2001.

[111] W.J. Nicholson, 'Comparative Dose-Response Relationships of Asbestos-Fiber Types: Magnitudes and Uncertainties', *ANYAS* 643 (1991), pp. 74–84.

some mesotheliomas) was apparent.[112] McDonald could offer no reason for this and regarded the textile findings as an 'unexplained mystery'.

Critics were not impressed, either, with the view that any problems with chrysotile must be due to contaminants. The idea that oils and plastics bags played a role in asbestos-related cancers was quickly discounted by experimental studies. Nor was there evidence that crocidolite was imported into Canada for gas-mask production during the Second World War and it was most certainly never imported from Australia, as McDonald supposed.[113] A more obvious explanation for the crocidolite fibres found in the lungs of deceased Quebec miners lies in the presence of crocidolite in the Quebec ore.

Many scientists, too, regarded the 'tremolite hypothesis' as deeply unconvincing and noted that the mining companies had never before made public the presence of such contaminants (which would inevitably have compromised the claim that Canadian asbestos was harmless). To most, it merely underlined the truism that asbestos was never a single entity and that to single out one variety as safe was reckless. As one critic remarked: '[A]ttributing mesothelioma production to this amphibole contaminant... is clearly suspect. If tremolite cannot be removed from chrysotile via industrial processing, the whole issue of tremolite contamination appears academic at best.'[114] Obstinately, McGill scientists argued that by careful screening and selection of mines, the production of chrysotile with 'minimal' contamination was feasible. But the Canadian industry has never been able to guarantee such a 'clean' product; and nor have Canadian epidemiologists.

Certainly, McDonald was correct in asserting that chrysotile is much more biologically and chemically reactive than amphiboles, so that chrysotile fibres tend to dissolve and divide into a multitude of small fibres that are removed from the lung more easily than amphiboles. However, the science was never as clear cut as the industry and its researchers suggested, mainly because the mechanisms by which any type of asbestos induce cancer are still not fully understood. American asbestos legislation has never differentiated between fibre types. In 1986, the OSHA in America reviewed the published scientific evidence and concluded that all

[112] See, for example, J.M. Dement et al., 'Exposures and Mortality Among Chrysotile Asbestos Workers. Part II. Mortality', *AJIM* 4 (1983), pp. 421–33.

[113] The only possible source of Australian crocidolite was the Wittenoom mine, which did not export fibre during the period in question. See J. McCulloch, *Asbestos: Its Human Cost* (1986), pp. 70–100.

[114] M. Huncharek, 'Asbestos and Cancer: Epidemiological and Public Health Controversies', *Cancer Investigation* 12 (1994), pp. 214–22.

fibre types, alone or in combination, had been observed in studies to cause lung cancer, mesothelioma, and asbestosis. It appeared that long fibres did more damage than short ones; on the other hand, scientists were far from confident that short fibres (under five microns) were not carcinogenic.[115]

In addition, a steady stream of epidemiological studies through the 1980s and 1990s from as far afield as Australia, Germany, and the US suggested that chrysotile did cause mesothelioma (aside from other ARDs).[116] Independent research at Zimbabwe's chrysotile mines found cases of asbestosis, lung cancer, and mesothelioma, though lack of data made it impossible to quantify the rates. In the developed countries, mesothelioma was found in individuals exposed only to white asbestos: namely, Italian mine and mill workers (where the mineral was uncontaminated with tremolite), brake mechanics and makers of friction products, and railroad workers. Evidence was also found that white asbestos caused mesothelioma in wine filter workers, metal workers, and even after environmental exposure. In 1987, an IARC working group underlined that chrysotile induced lung cancer and pulmonary mesothelioma and that there was no safe level.[117] Besides the IARC, American organizations such as the EPA and NIOSH regarded chrysotile as a causative agent for mesothelioma. In 1996, for example, NIOSH concluded that chrysotile should be treated with the same concern as amphiboles.

At the end of the twentieth century, McDonald still had his supporters. The BOHS had lamented how, after he had brought home the glad tidings about chrysotile, he and his colleagues 'had to endure a campaign of vilification which was motivated by those whose motives were often not scientific'.[118] But others had less sympathy for the sanguine Canadian views. In the 1990s, as Barry Castleman has shown, industry participation in bodies such as the IPCS and the WHO met with increasing resistance from scientists who demanded more objectivity in assessing health risks.[119] They were joined by asbestos action groups, which were forming in countries such as France, India, and Brazil on the wave of anger

[115] Selikoff and Lee, *Asbestos*, pp. 427–8.

[116] For an introduction to the large literature on which this statement is based, see R.A. Lemen, 'Chrysotile Asbestos as a Cause of Mesothelioma: Application of the Hill Causation Model', *IJOEH* 10 (2004), pp. 233–9; and G. Tweedale and J. McCulloch, 'Chrysophiles versus Chrysophobes: The White Asbestos Controversy, 1950s–2004', *Isis* 95 (2004), pp. 239–59.

[117] WHO/IARC, *IARC Monographs on the Evaluation of Carcinogenic Risks to Humans. Overall Evaluations of Carcinogenicity: An Updating of IARC Monographs vols 1–42, Supplement 7* (1987).

[118] 'The Quebec Asbestos Cohort', *Annals of Occupational Hygiene* 41 (1997), p. 1.

[119] Castleman, 'Manipulation'.

at the steadily rising mortality from ARDs. In 1999, the International Ban Asbestos Secretariat (IBAS) demanded a worldwide ban on all types of asbestos—a call echoed by the Collegium Ramazzini.[120]

The views of lay groups and scientists were beginning to converge. Pathologists began questioning the idea that amphiboles were uniquely dangerous and that white asbestos was harmless because it was removed more readily from the lungs (after all individuals had still been exposed to a carcinogen, even if the culprit had disappeared). Questions were also raised about the emphasis placed on burdens of fibre in the lung itself: after all, mesothelioma occurred in the pleura and it had been known since the 1980s that chrysotile had a predilection for that area—exactly what one would expect if it caused mesothelioma.[121] When pathologists examined mesothelioma fibre burdens under the microscope, they certainly found amphiboles—but also chrysotile! They were therefore unable to exclude a role for it in causing the malignancy.[122] Subsequent studies of mesothelial tissues by analytical electron microscope showed that in many tumours the major fibre type identified was chrysotile; moreover, that many of the fibres were 'short' (under five microns).[123] This suggested clearly that no precise fibre length could be identified that is without biological activity (especially since short fibres inevitably outnumber the long).[124] Selikoff was unmoved by this debate, because, as he pointed out: 'most individuals, who are exposed to fibres of one length, are simultaneously exposed to others of different lengths, and this is certainly true if we consider repetitive, accumulated exposures'.[125]

By the mid-1990s, therefore, a marked reaction had set in against the chrysophiles, with some even suggesting that chrysotile was mainly

[120] This group of occupational health physicians was founded by Dr Irving Selikoff at Mount Sinai School of Medicine in 1982. See P.J. Landrigan, 'Asbestos—Still a Carcinogen', *NEJM* 338 (28 May 1998), pp. 1618–19; and articles re. 'Call for an International Ban on Asbestos', *AJIM* 37 (2000).

[121] Y. Suzuki and N. Kohyama, 'Translocation of Inhaled Asbestos Fibres from the Lung to Other Tissue', *AJIM* 19 (1991), pp. 701–4. See also Y. Suzuki and S.R. Yuen, 'Asbestos Fibers Contributing to the Induction of Human Malignant Mesothelioma', *ANYAS* 982 (2002), pp. 160–76.

[122] V.L. Roggli, P.C. Pratt, and A.R. Brody, 'Asbestos Fiber Type in Malignant Mesothelioma: An Analytical Scanning Electron Microscope Study of 94 Cases', *AJIM* 23 (1993), pp. 605–14.

[123] Y. Suzuki and S.R. Yuen, 'Asbestos Tissue Burden Study on Human Malignant Mesothelioma', *Industrial Health* 39 (2001), pp. 150–60. See also Y. Suzuki and S.R. Yuen, 'Asbestos Fibers Contributing to the Induction of Human Malignant Mesothelioma', *ANYAS* 982 (2002), pp. 160–76.

[124] R.F. Dodson, M.A.L. Atkinson, and J.L. Levin, 'Asbestos Fiber Length as Related to Potential Pathogenicity: A Critical Review', *AJIM* 44 (2003), pp. 291–7.

[125] Selikoff to W. C. Bocim (IBM), 4 April 1990.

responsible for mesothelioma.[126] Even those who did not go so far nevertheless refused to give chrysotile a clean bill of health. The emerging consensus that chrysotile causes mesothelioma was highlighted by several publications in the late 1990s.[127] In 1998, the IPCS, under the joint sponsorship of the UN Environment Program, the ILO, and the WHO, published a monograph devoted entirely to chrysotile. It concluded: 'Exposure to chrysotile asbestos poses increased risks for asbestosis, lung cancer and mesothelioma in a dose-dependent manner. No threshold has been identified for carcinogenic risks. Where safer substitute materials for chrysotile are available, they should be considered.'[128]

These developments coincided with a Canadian legal challenge, conducted through the World Trade Organization (WTO), to the French decision to ban chrysotile (see Chapter 8 for a fuller discussion). The Canadian industry was now a shadow of its former self. In 1999, Canada produced 345,000 tonnes of asbestos (over 18 per cent of world output), making it the second largest producer after Russia, with most of the fibre destined for the developing world. The industry employed about 1,500 in various mining jobs (compared with 6,000 in 1964). But the Canadian industry still thought it worthwhile to try and get the French ban on asbestos overturned as a way of advertising chrysotile's beneficial properties to its diminishing customer base. This presented another opportunity for chrysotile defenders and their opponents to lock horns: one charging unfair trade practices, the other scientific corruption and misinformation.[129]

Corbett McDonald emerged from his retirement to support personally the Canadian government's position in the French WTO case. At the end

[126] A.H. Smith and C.C. Wright, 'Chrysotile Asbestos is the Main Cause of Pleural Mesothelioma', *AJIM* 30 (1996), pp. 252–66; L.T Stayner, D. Dankovic, and R.A. Lemen, 'Occupational Exposure to Chrysotile Asbestos and Cancer Risk: A Review of the Amphibole Hypothesis', *American Journal of Public Health* 86 (1996), pp. 179–86.

[127] A. Tossavainen, 'Asbestos, Asbestosis and Cancer: The Helsinki Criteria for Diagnosis and Attribution', *Scandinavian Journal of Work and Environmental Health* 23 (1997), pp. 311–16; MRC Institute for Environment and Health, *Chrysotile and its Substitutes: A Critical Evaluation* (2000), p. 4. See also MRC Institute for Environment and Health, *Fibrous Materials in the Environment: A Review of Asbestos and Man-Made Mineral Fibres* (1997); P.T.C. Harrison et al., 'Comparative Hazards of Chrysotile Asbestos and Its Substitutes: A European Perspective', *Environmental Health Perspectives* 107 (1999), pp. 607–11.

[128] IPCS, *Environmental Criteria 203: Chrysotile Asbestos* (Geneva, 1998), p. 94. Posted at: www.inchem.org/documents/ehc/ehc/ehc203.htm

[129] B.I. Castleman and R.A. Lemen, 'The Manipulation of International Scientific Organizations', *IJOEH* 4 (1998), pp. 53–5; B.I. Castleman, 'WTO Confidential: The Case of Asbestos. World Trade Organization', *International Journal of Health Services 32* (2002), pp. 489–501. See also R.A. Lemen, 'Challenge for the 21st Century—A Global Ban on Asbestos' (CD-ROM, 17–20 September 2000, Osasco, Brazil).

of his career, he stood by the results of his research, which he contended showed that chrysotile could be used safely. But the dispute was resolved in favor of the French in 2000, when the WTO confirmed the carcinogenic risk of chrysotile fibres, thereby cutting the ground from under the Canadian industry and its supporters.[130] The WTO judged that there was no safe threshold for chrysotile and that the concept of controlled use could not be realized. Chrysotile was not whiter than white, after all.

Conclusion

The WHO and the WTO did not end the scientific disputes. They were too entrenched and continued unabated into the new millennium. However, the arguments—against the backdrop of a steady drip of government bans against *all* forms of asbestos—were looking somewhat tired. Selikoff and Nicholson had died; McDonald had retired; and the arguments had been heard before. But reputations were at stake; so too were the remnants of Canada's chrysotile trade with the developing world. So in 2001 the arguments over the banning of asbestos were aired yet again in the *Canadian Medical Association Journal* amongst the familiar opponents. The editors of the *CMAJ* limply suggested that what was needed was another panel of experts to debate the problem.[131] In 2003, Dr David Egilman, an activist physician in America and a long-time student of the chrysotile defence, struck a heavy blow against the whole edifice of the McGill University studies. He published an article that ridiculed the studies as the 'Anything But Chrysotile' argument.[132] To Egilman, it was nothing more than a strategy that was used by the industry to expand market share and avoid liability.

A historical perspective on the debate is useful. There has always been a perceived spectrum of risk with asbestos. Amphiboles have traditionally been regarded by asbestos workers as the most dangerous fibres, a conclusion borne out by experience. The industry's greatest health disasters at Hebden Bridge and Armley in the UK, Wittenoom in Australia, and Kuruman and Prieska in South Africa have involved amphiboles.

[130] WTO, 'European Community—Measures Affecting Asbestos and Asbestos-Containing Products'. Report of the Appellate Body, WT/DS135/AB/R, 12 March 2001; Report of the Panel, WT/DS135/R, 18 September 2000.

[131] Editorial, 'A Ban on Asbestos: Is Now the Time?', *CMAJ* 164 (20 February 2001), p. 453.

[132] D. Egilman, C. Fehnel, and S.R. Bohme, 'Exposing the "Myth" of ABC, "Anything But Chrysotile": A Critique of the Canadian Asbestos Mining Industry and McGill University Chrysotile Studies', *AJIM* 44 (2003), pp. 540–57.

Subsequent research confirmed what workers suspected. But chrysotile has never been regarded as risk-free either by labour or independent scientists. Abundant evidence is available to show that chrysotile can cause mesothelioma: and, of course, it causes lung cancer and asbestosis, a fact which has often been lost in the chrysotile debate. The present scientific consensus is clear and has been for some time: amphiboles are more potent as a cause of ARDs than chrysotile, but chrysotile is an undoubted carcinogen and certainly not 'innocuous'. As if to underline that point, the UK banned chrysotile in 1999 (with some temporary exemptions), which meant the disappearance of asbestos cement. A year earlier the Council of Europe recommended that all forms of asbestos (i.e. white) should be banned in all Member States—a recommendation that became law in 2005. By then bans had been instituted in South America (Chile and Brazil) and Australia.

Ironically, Canada's occupational health record now looks no better than other industrialized countries. It has failed abysmally to provide safe working conditions for its workers—the consequences of which in terms of ARDs are still being felt. The National Institute for Public Health in Quebec produced a report in 2004 on the *Epidemiology of Asbestos-Related Diseases in Quebec.*[133] Between 1982 and 1996, it concluded that 832 Quebec residents, 655 men and 177 women, were diagnosed with mesothelioma. The rate for men was calculated to be 9.5 times greater than for the rest of the country and the rate for women was two times greater. Not surprisingly, opponents of asbestos use asked: if a socially and industrially advanced country like Canada cannot provide safe working conditions for workers in the asbestos industry, what about those developing countries to which Canadian fibre was exported?

The controversy about fibre type looks increasingly to have been largely a product of industry manipulation. Robert Proctor has coined the term 'the social construction of ignorance' to describe the strategy used by tobacco and asbestos companies in maintaining doubt as to the dangers of their products.[134] In 1987, during a heated correspondence initiated by the SAAPAC, Tony Davies reflected on these parallels between asbestos and tobacco. He concluded: 'Big industry is powerful and has no hesitation in enlisting big guns to slam people like me into the ground.'[135]

[133] Institut National de Santé Publique du Quebec, *Report of the Epidemiology of Asbestos-Related Diseases in Quebec* (Quebec: July 2004). Posted at: http://www.inspq.qc.ca/pdf/publications/293-EpidemiologyAsbestos.pdf

[134] R.N. Proctor, *Cancer Wars* (New York, 1995), pp. 8–9.

[135] Letter from Davies, Director NCOH, Johannesburg, to Director-General, Department of National Health, 29 May 1987.

It was a comment that could have been made by any director of the PRU/NCOH over the previous twenty-five years. In addition to suppressing medical discovery and intimidating researchers, the industry had also patronized science to get the result it wanted. So long as there was a counter-discourse about chrysotile, the industry could argue that the banning of asbestos was not justified. Having arguably influenced the research process, it then ruthlessly used those results against plaintiffs in court. As David Egilman has asserted, besides 'promoting the marketing and sale of asbestos, these [Quebec] studies have had a tremendous effect on litigation from the 1960s onwards'.[136] The effect was apparent in other directions. Canada maintained its chrysotile output above 1 million tonnes until 1981 (with the mid-1970s being peak years at over 1.5 million tonnes a year). Thereafter there was a steady decline, but even so in the twentieth century two-thirds of Canadian asbestos (41 million tonnes from a total of 61 million tonnes) was mined *after* 1960. Over forty-five years after Wagner's famous paper, the Quebec chrysotile industry still survives.

The co-opting of many members of the medical community played its part in that. The latter had shown that it sat more comfortably with industry and government than with critics of the industry or victims' groups.[137] Scientists like Doll praised the benefits of industry, but they rarely looked closely at the casualties. No one, apart from Selikoff, tried to launch programmes designed to look for better medical treatments for cancer. As far as we can see, the industry's hiring of consultants aroused little critical comment amongst epidemiologists and physicians. Again, no one apart from Selikoff seems to have been perturbed by corporate influence or by the sight of some academics parading themselves, as Selikoff put it, 'with silken banners waving in the air streams of corporate jets, pound and dollar insignia interwoven'.[138]

In contrary ways, the careers of Professor Ian Webster, Dr Chris Wagner, Sir Richard Doll, and Professor Jean Bignon illustrate the vulnerability of their profession. These were successful scientists, who were damaged to a greater or lesser extent by an industry determined to survive. Doll's asbestos research was not the ornament it should have been to a promethean career. Bignon found himself uncomfortably positioned between industry and victims. Murray was increasingly reviled. Webster

[136] Egilman, 'ABC', 541.

[137] J. Huff, 'Industry Influence on Occupational and Evironmental Public Health', *IJOEH* 13 (2007), pp. 107–17.

[138] Selikoff to M. Greenberg, 17 June 1983.

spent his career fighting a losing battle in Johannesburg for the right to conduct research. Wagner was forced to leave South Africa in 1962, because of the mining industry's opposition to his work and then spent the last two decades of his career in the secret employ of an asbestos manufacturer. When one of the authors interviewed Wagner at his home in Weymouth, England, shortly before his death, he complained that from the mid-1950s the industry set out to frustrate scientific discovery and gave only minimal cooperation to researchers like him.[139] He added that some time in the 1970s the whole scientific endeavour was hijacked by lawyers and the press, and he expressed regret that he had ever worked on asbestos disease. It was not such an odd lament and others may have felt the same.

[139] J. McCulloch interview with Wagner, 22 March 1998.

6

Hiding the Elephant of Compensation

Once you know that you've got the bug [asbestosis], you're in trouble. You know that you're ill and that sooner or later you just won't be able to earn money any more. You'd think that you wouldn't have any more problems thrust on you then. You'd be wrong there. Your troubles have only just started.

Arthur Rhodie, a Glasgow lagger,
Interviewed in Laurie Flynn, *Asbestos: The Dust that Kills in the Name of Profit*
(London, 1974), pp. 3–4.

I have used the word *VICTIMS* deliberately. I know some people do not like it, but we believe that is exactly what we are. We are not dying of natural causes. We are dying because of a product that made large profits for big companies... but we are not passive, helpless victims. We have a good, strong and healthy anger inside us.

Ella Sweeney
'Asbestos Diseases Foundation of Australia, Inc.'
Paper presented at
Global Asbestos Congress, Osasco, Brazil
17–20 September 2000

In 1918, at the New York Hippodrome, the great magician and escapologist Harry Houdini staged an audacious trick; he led onto the stage a large elephant called Jennie. Having introduced her to his audience, Houdini enticed Jennie into a specially constructed cabinet and then closed the doors. The orchestra worked up a suitable finale, whereupon the maestro pulled open the doors to reveal that Jennie had vanished. The elephant had not really disappeared, of course: she was simply hidden behind a mirrored screen.

Ever since asbestos-related diseases (ARDs) were diagnosed and accepted by governments as industrial illnesses, the asbestos industry has been attempting to hide its own elephant—mounting compensation claims. That analogy is not entirely inappropriate. At the end of the 1990s, one American judge famously referred to the 'elephantine mass' of asbestos claims that needed resolving.[1] These claims were in the hundreds of thousands and were registering on a national scale. Soon after his re-election, President George W. Bush began implementing his own plans to hide the asbestos elephant. In early 2005, he descended into the heart of the 'judicial hellhole' of Madison County, Illinois (so named by its detractors because of the size of asbestos awards there) to lament the scores of bankrupted asbestos companies, and 'frivolous' lawsuits launched by greedy lawyers. The Republican solution was a national superfund of $140 billion, bankrolled by manufacturers and insurers that would replace litigation by no-fault compensation, thus banishing all those distressing headlines of astronomic damages.

We are all familiar with those headlines. They tell of notable victories (or sometimes defeats) in the courts, where victims are either denied damages or awarded millions of dollars of compensation. In 2003, Roby Whittington, an Indiana steel worker with mesothelioma, won a Madison County verdict of $250 million against US Steel. No matter that such awards are invariably greatly reduced in private settlements after the trial.[2] Such events catch the public eye, loaded as they are with emotional issues and able, it seems, to reveal all the key actors: workers, industrialists, lawyers, judges, regulators, physicians, and insurers. Despite the media attention, however, the issue of compensation is still poorly understood. Few people are aware of the legal mechanics of compensation for industrial disease; even fewer will be familiar with the physical and emotional suffering caused by ARDs. As one American attorney put it: 'The ravages of asbestos disease, especially mesothelioma, are almost unimaginable to the uninitiated.'[3]

The word 'compensation' itself is filled with ambiguities, for how can one compensate someone for illness and death? As one miner in Libby, Montana, complained, shortly before he died: '[T]hey [W.R. Grace] will never have to pay like we did, because it won't cost them their lives.'[4] Yet

[1] Justice David Souter, *Ortiz v. Fibreboard*, 1999.

[2] The Whittington award of $50 million compensation/$200 million punitive damages—a record in American asbestos litigation—was later settled for under $50 million.

[3] Shephard Hoffman to G. Tweedale, 2 July 2000.

[4] Les Skramstad (who died of asbestosis in 2007) quoted, *Great Falls Tribune*, 10 February 2005.

compensation is central to the asbestos story. First, monetary compensation remains an individual's *only* form of restitution. As one mesothelioma widow remarked after the death of her 33-year-old husband: '[T]aking some of [Johns-Manville's] profit away from them is the retaliation available to me.'[5] Second, compensation for many people is an economic necessity, especially since ARDs can strike individuals in their prime. The fact that America has no medical insurance scheme should be noted. Third, compensation can act as a financial penalty for an industry, thereby improving safety standards and leading perhaps to the substitution of hazardous products. Fourth, compensation generates statistics: it is often how we 'see' occupational disease and so it fundamentally shapes our perceptions. Finally, there is the moral dimension, for as one veteran asbestos campaigner remarked: '[J]ust compensation is one of the hallmarks of a civilised society'.[6] Unfortunately, companies have little interest in paying out for most of these reasons. As this chapter will show, asbestos companies have made strenuous efforts to avoid paying *any* compensation.

Compensation Before 1964

Workers were dying from asbestosis before 1914. However, the first known compensation case occurred in 1924, after the death of a T&N worker in Britain. Her name was Nellie Kershaw, a Rochdale asbestos spinner, who died at the age of 33. The case was a harbinger. It involved a young woman dying in poverty, a bereft husband with a young son, and a grisly troop of lawyers, coroners, doctors, and hard-nosed company men rejecting begging letters.[7] At that time, asbestosis was not officially compensated. Even by the Second World War, only a handful of countries offered any compensation for the disease.

Britain was the first to introduce a national compensation scheme specifically for asbestosis in 1931. In the US, there was no federal scheme, and although many states in America had workmen's compensation laws, most provided little coverage for occupational diseases. By 1939 only eleven of forty-eight American states specifically listed asbestosis.

[5] N. Rossi, *From This Day Forward* (1983), p. 293.

[6] A. Dalton, 'Adding Insult to Injury for the Asbestos Victims' (letter), *The Guardian*, 23 January 2001.

[7] I.J. Selikoff and M. Greenberg, 'A Landmark Case in Asbestosis', *JAMA* 265 (20 February 1991), pp. 898–901.

In 1936, Germany introduced a compensation scheme that involved government accident insurance/prevention institutions known as Berufsgenossenschafts [BG]. Membership was compulsory for asbestos companies, but the BG protected the industry from claims and provided occupational health guidelines.[8] Until recently, very little was known about how these schemes operated and how many claimants were involved, but recent research has uncovered at least part of the picture for Britain and America.

Britain's pioneering Asbestosis Scheme provided payments for sufferers, combining it with medical monitoring and dust regulations. But benefits—set by the government, but paid initially by the companies—were paltry and strictly circumscribed. For example, the scheme only applied to a few 'scheduled' workers in the industry (mainly in the asbestos textile factories) and many limitations were built into the scheme. For example, workers who had left the industry were excluded. Not surprisingly, the industry's compensation costs were miniscule.[9]

In America, asbestosis compensation cases began in 1927 when a Massachusetts asbestos textile worker filed a claim. A group of Johns-Manville workers launched claims in 1929 and by the mid-1930s at least sixty-nine more state compensation suits were brought against the company. During the 1930s, claims were filed against other leading manufacturers, such as Raybestos-Manhattan. The latter compensated about twenty cases in the 1930s, with average payments between $2,000 and $3,000, sometimes less. But it was no more than a trickle of cases. The low numbers reflected the lack of state coverage and the fact that the American system was adversarial, involving the court system. Insurers rarely accepted liability without a legal fight. It also reflected the stringency of compensation legislation. Here, too, many claims were 'statute-barred', so that workers could not claim if they had left the industry more than a year or had worked in more than one state. This was an especially cruel limitation for insulators with their itinerant work practices.

The fact that many cases were dismissed or quietly settled also hid the extent of ARDs. In 1933, Johns-Manville settled eleven New Jersey factory workers' claims for asbestosis, on condition that the lawyer involved filed no more claims. Corporate pressure meant that New Jersey did not introduce compensation for asbestosis until 1951. Companies were not passive

[8] X. Baur and A.B. Czuppon, 'Regulation and Compensation of Asbestos Diseases in Germany', G.A. Peters and B.J. Peters (eds.), *Sourcebook on Asbestos Diseases* (1997), pp. 405–19.

[9] G. Tweedale, *Magic Mineral to Killer Dust* (2001), pp. 69–119.

victims of the compensation system; they were involved through their political and economic influence in the formulation and implementation of regulations. So, too, were the insurance companies. Company pronouncements at that time give a flavour of the attitude towards compensation. Vandiver Brown, Johns-Manville's corporate attorney, attended one of the formative meetings of the Industrial Hygiene Foundation in Pittsburgh in 1935, and heard speakers state that the 'strongest bulwark against future disaster for industry is the enactment of properly drawn occupational disease legislation'.[10] As Brown summarized it, this would 'eliminate the jury and . . . the shyster lawyer and the quack doctor', and also provide a forum for evidence that might cast doubt on disability. According to Brown, the 'worst Workmen's Compensation Commission is preferable to the best jury'.[11]

In Canada, state compensation was available in Ontario and Quebec after 1943, but there was no state medical surveillance of workers and compensation was sought only as a last resort by the most disabled, because there was no extra payment if a worker lost his job.[12] This ethos was compounded by the companies, which had to pay compensation. At Johns-Manville in Asbestos, Quebec, the medical director was Dr Kenneth W. Smith. In 1948, Smith conducted a survey of 708 mill workers and found lung disorders rife, with at least fifty-nine cases of asbestosis. To deflect compensation claims, Smith suggested moving the men to less dusty areas and did not inform workers of his diagnosis, 'for it is felt that as long as the man feels well, is happy at home and work, and his physical condition remains good, nothing should be said . . . [and] . . . the company can benefit by his many years of experience'.[13]

Smith was involved with the same 'hush-hush policy' at compensation meetings at Johns-Manville's plant in New Jersey in the late 1950s. Notes of some of these meetings have survived. In the case of John Hudak, a 58-year-old worker with asbestosis, Smith feared an impending claim: 'I see no reason to bring a man in like this [for health counselling], it is dangerous.' A colleague agreed: 'Now take that Blanik woman, she's very nervous. If she is called in, she will get hysterical and I'm sure you will have a claim on your hands.'[14] Smith suggested to one colleague that they should 'purchase a small and inexpensive shredding machine' to

[10] Brown memo to M.F. Judd, Raybestos-Manhattan, 22 January 1935.

[11] Brown comments, 6th Saranac Symposium (1950), pp. 567–8.

[12] Information courtesy of Professor Katherine Lippel.

[13] K.W. Smith, 'Industrial Hygiene—Survey of Men in Dusty Areas', 1949. Typescript.

[14] Notes of Johns-Manville compensation meetings.

destroy medical records that might be used against Johns-Manville in compensation cases.[15]

During and after the Second World War, more countries introduced benefits for asbestosis: Italy in 1943, France in 1945, Czechoslovakia in 1946, Japan in 1947, the Netherlands in 1949, Belgium in 1953, Switzerland in 1953, Austria in 1955, and Norway in 1956. Not only was asbestos compensation slow to be introduced worldwide and full of limitations,[16] but the legislation proved inflexible in another crucial respect. Compensation covered asbestosis. However, from the 1930s it was increasingly apparent that lung cancer was a risk for asbestos workers. Yet the only countries that introduced compensation for asbestos-related lung cancer were Germany in 1943 and Czechoslovakia in 1948.[17] Even as late as 1964, as Selikoff's conference got underway, these were still the only countries in the world to compensate asbestos-related lung cancers.[18] The failure of America and Britain to extend compensation is striking, especially since these countries had recognized the lung cancer link.

In terms of financial restitution, compensation for ARDs before 1964 hardly scratched the surface, aside from the moral issue. Some historians have suggested that compensation for industrial injury and disease raised costs, forcing companies to reform.[19] This never happened in the asbestos industry, especially in America. Even in successful claims, the top benefit would usually be well under half a worker's average weekly income, since most workers were classed as low-income earners. Often benefits were so paltry that they were hardly worth pursuing. Industrialists certainly feared workmen's compensation, especially in America during the silicosis scare in the 1930s (see Chapter 3). But by 1964, it had long been shown that the fear of exorbitant costs was without foundation, partly because industry was so successful in shaping compensation statutes to its own ends. For thousands of workers disabled by silicosis and asbestosis, the workmen's compensation statutes enacted in America in the 1930s and 1940s amounted to 'a swindle, in which men and women, who had

[15] Dr Kenneth W. Smith to C.H. Grote, Manville plant, 26 February 1963.

[16] One important limitation was that in the US and UK, military personnel (such as in the Navy yards) were barred by the government from taking action for personal injuries.

[17] R. Proctor, *The Nazi War on Cancer* (1999), pp. 107–13.

[18] The abject failure of workmen's compensation in cancer cases is highlighted by B.I. Castleman, *Asbestos: Medical and Legal Aspects* (2005), p. 195. During the 1960s and 1970s, New York State compensated only about five cases a year of occupational cancer.

[19] M. Aldrich, *Safety First: Technology, Labor, and Business in the Building of American Work Safety 1970–1939* (1997), p. 97; C.C. Sellers, *From Industrial Disease to Environmental Health Science* (1997).

been unsuspectingly robbed of their health, were cheated out of just recompense for their loss of earning power, their suffering and, in many cases, their early death'.[20]

Junkyard Dogs: Compensation American-Style

One of the most striking features of the situation before the 1970s was the virtual immunity of manufacturers from litigation. In North America and Europe, it was rare for victims to sue asbestos companies under the common law. This was because compensation was specifically designed to block legal action, by making it impossible for plaintiffs for follow both routes. In America, the Supreme Court in 1917 removed workers right to sue their employers in return (supposedly) for a fair compensation system in which they would not have to prove liability and would receive adequate benefits.[21] Even when it became easier to sue, legal limitations meant that claims had to be launched within only a few years of exposure.[22] Given the latency of ARDs, this effectively meant that asbestos claims were statute-barred.

The biggest problem in launching a common law action was that sick workers needed to prove company negligence. Before the 1960s, this was very difficult. Individual workers and trade unions lacked detailed knowledge of ARDs and access to company documentation. Even if they had, the legal bar was often set impossibly high. In France, workers had to show what was known as 'inexcusable fault', which amounted to a deliberate and reckless act by employers that was virtually impossible to prove in court.[23] In Belgium, too, a civil claim had to show the employer's intention to cause injury and disease. In whatever country, companies could either simply plead ignorance or counterclaim that the worker had been negligent. Workers lacked the resources to fight powerful industrial interests and many were too sick to fight for long. ARDs (especially mesothelioma) imposed their own cruel time limitations and offered plenty of opportunity for delaying tactics by defendant companies and insurers.

[20] P. Brodeur, *Outrageous Misconduct* (1985), p. 23.

[21] As Selikoff later remarked on the failure of this idea: 'the quid has disappeared and the quo has remained'. Selikoff to Hon. Jack B. Weinstein, Chief Judge, United States District Court, 21 April 1986.

[22] N.J. Wikeley, *Compensation for Industrial Disease* (1993).

[23] A. Thébaud-Mony, 'Justice for Asbestos Victims and the Politics of Compensation: The French Experience', *IJOEH* 9 (July/September 2003), pp. 280–6.

The New York conference in 1964 was vital, because it gave plaintiffs a weapon they lacked—information. Moreover, Selikoff's data were derived from the unions, which meant that his studies could not be controlled by the industry. This development was made more profound by its context. In 1965, the American legal system finally recognized that workers' compensation was useless for those injured by occupational diseases with long latency and where exposure had often been by multiple employers. A special rule was introduced to widen the liability net. This was the law of strict liability, which made sellers of products liable for a 'failure to warn' ordinary consumers and users of an 'unreasonably dangerous product'. No longer would plaintiffs need to prove cases against their employers under workmen's compensation law or attempt to sue companies for negligence or breach of warranty. They could now launch legal actions directly against the numerous makers and sellers of asbestos-containing products, who fell outside the workmen's compensation regime.

Borel v. Fibreboard et al. (1973), which involved an insulator Clarence Borel who died of mesothelioma, was the first test case for the applicability of the new product liability law. It set a legal landmark. 'Disaster has struck!', commented the insurers for the companies involved, correctly divining its import.[24] *Borel* began the longest running mass tort litigation in US history, though the slide into litigation began slowly. In 1974, a Dallas-based lawyer Fred Baron launched an important class action suit against Pittsburgh Plate Glass (PPG) Industries, which operated the Tyler plant in Texas—a case that was later settled for $20 million in 1977, thereby making millionaires of the lawyers involved. In 1978, there were enough asbestos lawyers around (about 150) to form the Asbestos Litigation Group, which was designed to pool knowledge and resources. Initially, however, the industry was able to construct a state-of-the-art defence (see below), which when implemented by skilled defendant attorneys such as Lively Wilson, drew first blood. In the late 1970s, Ron Motley, a brash Southern lawyer, tried his first cases against Wilson and lost.

Before 1980, only about 950 asbestos cases were filed in the US federal courts (no corresponding figures are available for state courts), but in the early 1980s the potential of product liability law was more fully revealed. By 1982, 21,000 claims had been filed and lawsuits were appearing at the rate of 6,000 a year. The cost of settling them was also rising, so

[24] Alexander & Alexander, Borel claims report, 22 October 1973.

that the average cost for Johns-Manville for each case was approaching $50,000. Insurers began projecting ultimate costs up to $40 billion. By 1988, 100,000 claims had been filed. American asbestos litigation, which created a whole branch of the legal industry, was gathering steam.

It was based on product liability law, which proved to have a tremendous reach. Even if firms became bankrupt, usually other sellers or distributors of asbestos could be targeted along the supply chain. Steve Kazan, who launched some of the key claims against Johns-Manville, explained the process:

> In the early years you went after obvious defendants who were most involved. As the manufacturers of insulation disappeared, you turned to higher hanging fruit. Then the distributors—not [culpable] in a religious or moral sense, but legally [so]—disappear. So you go after the installers and applicators. Then come the retail hardware stores.[25]

By naming as many asbestos entities as possible in a claim, a plaintiff would increase the chances of success. In the 1980s, a claim would usually list about twenty defendants; in the next decade, sixty to seventy defendants might be listed. By 2002, over 8,000 defendants had been named in litigation. Companies only distantly related to the asbestos industry were pulled into court. For example, the company that manufactured the filters for Kent cigarettes during the 1950s was hit by several claims after workers and even their relatives developed asbestosis and mesothelioma. The cigarette manufacturer itself, Lorillard, also settled claims after smokers alleged that the crocidolite filters had caused their mesothelioma. Bigger companies were caught in the product-liability net, too. Halliburton—the oil-services company formerly run by US Vice-President Dick Cheney—was hit by 300,000 asbestos claims due to the liabilities of Dresser Industries, a maker of pipeline couplings that was acquired by Cheney in 1998. Hardly a sector of American industry has been untouched. Ford, Dow Chemicals, Union Carbide, General Electric, Shell, Texaco, MetLife, and leading railway companies have become embroiled in asbestos litigation.

The ability of the American legal system to produce litigation seemed endless. Producers who were sued issued cross-complaints against other producers to 'lay-off' their own liability. At the end of the 1980s, Owens-Corning prepared a three-volume book of asbestos product labels, so that

[25] Steven Kazan, quoted in *Houston Chronicle*, 3 October 2004.

plaintiffs could identify other possible causes of their asbestos exposure. The strategy reduced Owens's liability and added dozens more defendant companies to the litigation.

The other key feature of American litigation was the use of a jury, whose working-class members were likely to sympathize with the sufferings of a sick worker on the witness stand, especially in certain state jurisdictions. Plaintiffs' lawyers could indulge in what was known as 'forum shopping' by searching for the most favourable jurisdiction to try cases. New York, Ohio, and Texas became well-established asbestos trial states, with Mississippi particularly favoured. By 2002, a fifth of all pending asbestos cases were in Mississippi. Jurors in such states were characterized by their eagerness to award, not only compensation, but also punitive damages to register their moral outrage at reckless or 'outrageous misconduct'. Most cases were settled before trial, but that in itself was partly a reflection of the corporate fear of juries.

Litigation was facilitated by legal discovery, which obliges adversaries in litigation to exchange relevant documentation. Many of the early legal actions against the industry had foundered on the fact that the extent of company knowledge was a closed book. In 1978, however, Raybestos-Manhattan produced the Sumner Simpson papers (see Chapter 3), which enabled Motley to win an order for a retrial of an asbestos case he had lost, after a judge agreed that the correspondence revealed a pattern of denial and suppression. The case was soon settled. Lawyers like Motley also recruited researchers to plough through old state records. They discovered that the leading companies had been settling compensation cases for insulators for decades. Executives and company physicians were also deposed to help build up the picture of conspiracy.

Individual workers could now theoretically have their day in court. This development also transformed the status of asbestos attorneys. A book is yet to be written about these individuals, but when it is, the roll call might include the following: Karl Asch, Robert Sweeney, Scott Baldwin, Peter Angelos, Gene Locks, Paul Gillenwater, Larry Madeksho, Conard Metcalf, Joe Rice, Walter Umphrey, Richard Glasser, Mike Moore, Dick Scruggs, George Kilbourne, Randy Bono, Shepard Hoffman, Mark Lanier, Roger Worthington, and Jim Walker. The net earnings of these men would run into billions of dollars. For example, since 2000, Randy Bono, the 'king of asbestos litigation' in Madison County, was reported to have settled some 600 mesothelioma cases at up to $3 million each, netting over $1 billion in settlements (defendant lawyers say more). Since lawyers usually take 30–40 per cent of awards, Bono's share was obviously substantial.[26]

Many regard such individuals as modern versions of the dreaded 'ambulance chaser' (a phrase first coined in New York City in the 1890s), only now they have metamorphosed into something worse—the junkyard dogs, which clamp onto the flesh of hapless companies. It is a caricature that trial lawyers have not entirely discouraged. Some have become politicians and the American Trial Lawyers' Association (ATLA) has become a staunch contributor to the Democrats. John Kerry's running mate in the 2004 presidential election was multimillionaire personal-injury lawyer John Edwards, whose campaign finance chairman was Fred Baron, one of the richest and most successful of the asbestos attorneys. The ATLA, which was formed in 1946 as no more than a handful of lowly workmen's compensation attorneys, is now the group American conservatives love to hate.

Perhaps the most publicized asbestos attorney is Ron Motley, who was born in 1944 in Charleston, South Carolina, the son of a gas station owner. Flamboyant, quick-tongued, and immensely energetic, Motley rode the wave of asbestos litigation in the 1980s when he built up a national network of local lawyers. Motley's speciality was consolidated cases, where local lawyers would do the groundwork and Motley would arrive just before the trial. His *modus operandi* has been likened to that of a brain surgeon: 'First other doctors tested the patient and stabilised his vital signs. Then one morning, Motley walked in, cut him open, fixed his brain, and went home.'[27] Motley's career and lifestyle is legend: his enormous wealth has garnered island mansions off the South Carolina coast, yachts, jets, and dogs named after different types of asbestos. His practice, Motley Rice, was eventually powerful enough alongside fellow asbestos attorneys Mike Moore and Dick Scruggs to take on the tobacco industry. Motley was sympathetically portrayed in a film *The Insider* (1999) about the tobacco whistleblower Jeffrey Wigand. The huge American tobacco settlement ($246 billion) in 1998 is reputed to have netted Motley Rice fees of $2–3 billion. Little wonder that the asbestos lawsuit industry has been characterized by its critics as Trial Lawyers Inc.—a vast money-machine creating revenues twice those of Coca Cola.[28]

[26] P. Hampel, 'Madison County: Where Asbestos Rules', *St Louis Dispatch*, 18 September 2004.

[27] D. Zegart, *Civil Warriors: The Legal Siege on the Tobacco Industry* (2000), p. 31.

[28] Manhattan Institute, *Trial Lawyers Inc: A Report on the Lawsuit Industry in America* (2003).

Developments Outside America

The *Financial Times* in the UK felt that American litigation was pure theatre, with swaggering trial lawyers, sheepish corporations that paid up without complaint, and Robin Hood juries dishing out cash as a form of redistributive justice. Still, it had to admit that 'nothing can thrill like asbestos'.[29] The thrills were more limited elsewhere, with asbestos litigation a pale reflection of its American counterpart.

Coincidentally, the first significant court action against a UK asbestos company was won in the early 1970s at about the same time as Borel. Thereafter the number of civil actions against the leading UK companies increased. Claims had to be directed against the employer—there was no product liability in the UK—and not until 2000 could solicitors adopt American-style conditional fee arrangements (in other words, no win no fee). Cases were settled by judges, with no juries and no punitive damages. Legal aid from the government was sometimes available, but in the UK the loser of a lawsuit paid the winner's costs and legal discovery was a tepid affair compared with the US, with court documents not made available to the public. The high awards typical in American cases, where the average mesothelioma payment was approaching $2 million by 2001, were not seen in Britain. The £1million paid to the widow of plastic surgeon James Emerson and the £4 million to the widow of company director Anthony Farmer were well publicized, but only because they were so rare. The Farmer payment was a British record for mesothelioma, but both amounts were only so large because both men were in their prime and high earners. Damages in Britain are based on loss of income and earning power, so that typical awards for, say, mesothelioma ranged between £50,000 and £100,000 during the 1990s and early 2000s. Before the 1990s, payouts were much less and were trivial compared with most American settlements.

By the 1980s—nearly a century after the first adverse health effects of asbestos had been noted—the only other countries, apart from America and the UK, where litigation registered was in Australia and Japan. In Australia the first cases were launched against the Colonial Sugar Refining Company (CSR), the owners of the Wittenoom mine. In 1977, Cornelius Mass became the first worker to sue CSR (renamed Midalco), but he died of mesothelioma aged 41 before the case could be tried. CSR and the state insurers immediately stopped compensation payments to his widow

[29] *Financial Times*, 14 May 2002.

and three children. In 1979, Jane Joosten, the 51-year-old ex-company secretary at Wittenoom, also sued CSR for mesothelioma—a case that was soon dismissed by the judge as simply 'sad misadventure'. The case was appealed, but tragically Joosten, having battled mesothelioma for three years, died only two hours before the court session started. No compensation was paid either to her or her husband, who later developed pleural plaques. Finally, in 1985 the first mesothelioma legal action was won by Harold Pilmer, who sued Melbourne hardware company McPherson's and won A$225,500. In 1988, a Wittenoom case at last succeeded against CSR. With the relaxation of the statutes of limitations and the impact of entrepreneurial legal firms such as Slater & Gordon, the pace of litigation began to quicken. Hundreds of writs were issued, including 250 against CSR alone.[30]

In Japan, civil actions for damages were rare: only a handful of cases were launched after the late 1980s, with no environmental actions and none for product liability. Most cases emanated from the shipyards, particularly from Yokosuka City near Tokyo, which was well known for its shipbuilding industry and American naval base. In 1988, eight former workers in Yokosuka sued Sumitomo Heavy Industries Ltd, thus beginning a long legal action that was only settled in the workers' favour in 1997. Since then, other cases have been brought by workers and their families against the Japanese and American governments, with varying success. The Japanese government has attempted to get cases thrown out on technicalities or has denied negligence. A complicating factor is liability: Japan claims that the US is responsible for reimbursement, while America argues the opposite.

By the 1990s, litigation was appearing in the Netherlands and Belgium. In the former, about a thousand claims were submitted by two specialist lawyers. Most were settled out of court after long battles through the usual thicket of statutory limitations.[31] In Belgium, the first legal action against employers was brought in 1996 by Luc Vandenbroucke, a maintenance worker with mesothelioma. In Belgium, the legal obstacles are severe: an employer is not liable, even for gross negligence, and victims cannot claim if twenty years have passed since exposure. Vandenbroucke lost and subsequently only a handful of personal-injury cases have been

[30] B. Hills, *Blue Murder* (1989), pp. 76–87. See also M. Peacock, *Asbestos: Work as a Health Hazard* (1978).

[31] P. Swuste, A. Burdoff, and B. Ruers, 'Asbestos, Asbestos-Related Diseases, and Compensation Claims in the Netherlands', *IJOEH* 10 (April/June 2004), pp. 159–65.

brought.[32] Similarly, in Switzerland class-actions are not recognized in the courts and, in any case, workers have only ten years in which to file a claim against their last employers. In Italy few ARD cases have reached the courts, largely because they hinge on criminal charges of manslaughter and unpremeditated injury—charges that are difficult to prove and usually end in jail sentences not monetary awards.[33] One such case was launched in 1987 against Eternit in Casale Monferrato by the unions, the local council, and victims' groups. It was not concluded until 1993, when the Supreme Court convicted the management for manslaughter and unpremeditated injuries and awarded L7 million to about 1,700 people. Victims in France were also belatedly successful: in 2002, the Supreme Court in France changed legal precedent by upholding thirty judgements of 'inexcusable fault' against employers: for the first time the latter were found to have an absolute obligation to ensure workers' safety and so were liable for ARDs. French civil courts also found in favour of a group of shipyard workers and in 2004 thirteen former miners at Eternit's Corsica asbestos mine were awarded €1.5 million ($1.92 million) after an action by the French national health services and the victims.[34] But no charge against the employers has ever been laid in a French criminal court.

It should be emphasized that most countries still relied heavily on some form of state social security. Even in America, workmen's compensation was not quite moribund, despite the ferocity of the junkyard dogs. In Canada, workers' compensation for ARDs remains the norm and it is still almost impossible to sue employers for damages relating to workplace asbestos exposure. In Britain, those with ARDs could claim Industrial Injuries benefit or claim under the Pneumoconiosis Act, which provided modest lump sum damages for those who were unable to sue their employer because the firm was no longer in business. In the leading European countries, various funds and state insurance schemes offered payments for ARDs. The German BG has already been mentioned. Other schemes include: the Swiss Accident Insurance Organization (SUVA); the Belgian Occupational Diseases Fund (FMP) and its adjunct the Asbestos Fund (AFA); the Dutch Institute for Asbestos Victims (IAV); the Italian National Institute of Insurance for Professional Illness & Injury (INAIL); and the Fond D'Indemnisation des Victimes de L'Amiante (FIVA)

[32] S.Y. Nay, 'Asbestos in Belgium: Use and Abuse', *IJOEH* 9 (July/September 2003), pp. 287–93.

[33] P. Comba, E. Merler, and R. Pasetto, 'Asbestos-Related Diseases in Italy: Epidemiological Evidences and Public Health Issues', *IJOEH* 11 (2005), pp. 36–44.

[34] G. Meria, *L'Aventure Industrielle De L'Amiante en Corse* (2003).

in France.[35] In Australia, compensation was handled by various state schemes, such as the Dust Diseases Board in New South Wales. In New Zealand, a no-fault compensation scheme was handled by a government Accident Compensation Corporation (ACC).[36] By 2006, a special law in Japan tasked an Environmental Restoration & Conservation Agency (ERCA) with processing asbestos compensation claims.

These schemes, which usually prevented employers being sued, were designed to reduce the so-called 'second agony' of having to fight for compensation (the first agony being ARD itself).[37] Workmen's compensation certainly has benefits, not least for governments, insurers, and industry, because it avoids the 'transaction' or administrative costs of litigation (which in America are said to swallow half the bill for litigation, because of lawyers' fees). In Europe sick workers usually also have access to national health systems, which are free on demand, unlike the laissez-faire and costly medical systems in the US. However, state compensation is riddled with problems and inequalities.

The EU is founded on the concept of harmonization, but this does not extend to industrial-disease compensation: no universal compensation scheme exists or EU-wide registry for ARDs.[38] Usually monetary awards are far less than can be achieved by litigation. Even in Germany, payments only amount to two-thirds of salary and there are no awards for pain and suffering. One BG chairman said in 2003, without a trace of irony: 'it's safe to say that German industry is glad that the BGs provide a fitting and, above all, socially acceptable solution'.[39] In Switzerland, asbestos victims have been marginalized by SUVA, because most remain ineligible for compensation or are given very small awards. In Italy, the state insurer has a reputation for actively discouraging claims. Under INAIL asbestos claimants need to prove that they have worked at certain dust exposures for at least ten years, which means that thousands of claims have been rejected.[40] In France, FIVA's compensation payments did not start until

[35] See generally R. Best, 'Liability for Asbestos-Related Disease in England and Germany', *German Law Journal* 4 (1 July 2003), pp. 661–83; Y.R.K. Waterman and M.G.P. Peters, 'The Dutch Institute for Asbestos Victims', *IJOEH* 10 (April/June 2004), pp. 166–76; C. Manaouil, M. Graser and O. Jardie, 'Compensation of Asbestos Victims in France', *Medicine and Law* 25 (September 2006), pp. 435–43.

[36] T.E. Kjellstrom, 'The Epidemic of Asbestos-Related Diseases in New Zealand', *IJOEH* 10 (April/June 2004), pp. 212–19.

[37] Waterman and Peters, 'Dutch Institute'.

[38] European Forum of the Insurance Against Accidents at Work & Occupational Diseases, *Asbestos-Related Occupational Diseases in Europe: Recognition, Figures, Specific Systems* (2006).

[39] Klaus Hinne, Opening Address at European Asbestos Conference, Dresden, 2003.

[40] Comba, 'Asbestos-related Diseases', p. 41. See also E. Merler et al., 'Occupational Cancer in Italy', *Environmental Health Perspectives Supplements* (S2) 107 (May 1999), pp. 259–71.

after 2003. The Fund, to which employers contribute, has been criticized for protecting negligent employers from punishment. In the Netherlands, the IAV is only a mediator, without binding legal power and only deals with mesothelioma victims (the government does not pay for either asbestosis or asbestos-related lung cancer). In Belgium, which has one of the most reactionary systems in the developed world, the criteria for ARDs have always been impossibly strict. Self-employed Belgian victims are excluded from compensation and so too are civil servants and members of the armed forces.[41] The new AFA was supposed to ameliorate the harshness of this regime: however, one of its provisos was that beneficiaries of the Fund would give up any legal rights to sue their employers. Individuals with asbestos-related lung cancer and pleural plaques were excluded by the Fund. Likewise, until 2007 the self-employed and non-occupationally exposed were also excluded from ARD state benefits in the UK.

In Australia, some state schemes provide reasonable support—even in New South Wales funding medical research—but in others (notably Victoria) compensation is much less favourable and most workers who develop asbestos-related lung cancer never make a claim. In Tasmania, limitations mean that claims for pain and suffering end with the victim's death. In New Zealand, the government's accident scheme is not framed for occupational diseases, with ARD claimants caught between legislation that does not allow civil actions against employers if exposure occurred before 1974, yet provides instead a system riddled with delays, arguments, and cost-cutting. The government has traditionally paid a 'pension' of only A$68 a week. The ACC became embroiled in controversy over a revision to the law in 2002, which according to the ACC only allowed lump sum payments if exposure to dust had occurred after that date. The ACC argued that 'historic' exposure did not warrant compensation, then backed down, but then tried to claw the money back from families through the courts. One mesothelioma sufferer described the process as 'farcical, callous [and] cavalier'.[42] In 2006, the Court of Appeal decided that asbestos victims *were* entitled to lump sum payments, though not before several had died knowing that the issue was unresolved.

A common problem is that those afflicted with an ARD need to immerse themselves in complex administrative mechanisms of which they are either unaware or do not understand. In Japan, the low number of

[41] S.Y. Nay, 'Asbestos in Belgium: Use and Abuse', *IJOEH* 9 (July/September 2003), pp. 287–93.

[42] M. Johnson, 'Dying Man Awaits Cash Verdict', *New Zealand Herald*, 26 May 2005.

compensated cases has been linked with the fact that many victims are unorganized workers who are not informed about the compensation scheme and its application procedures—even though employers have a legal obligation to assist workers in initiating procedures for compensation.[43]

An overarching problem is that state compensation is dependent on government recognition of specific ARDs. This recognition can be extremely slow. Take a 'signature' disease such as mesothelioma. This was compensated in the UK after 1966 and in the Netherlands after 1968. But France, Germany, Spain, and Japan did not make it compensable until the late 1970s. In Canada, Belgium, Poland, Switzerland, and in some states in Australia, it was not compensable until the 1980s. In Italy, mesothelioma (without asbestosis) was not listed for compensation until 1994. In some Central and Eastern European countries, such as Czechoslovakia and Hungary, mesothelioma had not been officially listed as a specific occupational disease by the mid-1990s.[44] In Greece, pleural mesothelioma is not recognized officially as a compensable disease, even though claims can be made for peritoneal mesothelioma.[45] In some countries in the developing world, such as India, mesothelioma is still not recognized as an occupational disease.

Asbestos-related lung cancer has also only slowly been accepted for compensation. Many countries did not make this compensable until the 1980s—at least thirty years after its medical recognition. In Belgium, asbestos-related lung cancer was listed for compensation only in 1999. Often it was a requirement that asbestosis needed to be present alongside the lung cancer as evidence of asbestos exposure, despite medical evidence that asbestos could trigger lung cancer without asbestosis.[46]

Insulating Asbestos Companies from Compensation

For ARD sufferers, the callousness and inadequacy of compensation systems translated into poverty and social exclusion that has been well

[43] S. Furuya, Y. Natori, and R. Ikeda, 'Asbestos in Japan', *IJOEH* 9 (July/September 2003), pp. 260–5.

[44] P. Brhel, 'Occupational Respiratory Diseases in the Czech Republic', *Industrial Health* 41 (2003), pp. 121–3; E. Fabianova et al., 'Occupational Cancer in Central European Countries', *Environmental Health Perspectives* 107 (S2) (May 1999), pp. 279–82.

[45] Panagiotis Behrakis presentation, European Asbestos Conference: Policy, Health and Human Rights, Brussels, 22–23 September 2005.

[46] Not until 2005 did the UK government compensate lung cancer where no asbestosis was present, but where there was evidence of substantial asbestos exposure.

Table 6.1. Asbestos Victims' Action Groups

Date	Name	Country
1978	Society for Prevention of Asbestosis & Industrial Diseases (SPAID) [Occupational & Environmental Diseases Association after 1996]	UK
1979	White Lung Association	US
1979	Asbestos Diseases Society of Australia	Australia
1983	Hull Asbestos Action Group	UK
1986	Clydeside Action on Asbestos	UK
1987	Ban Asbestos Network Japan (BANJAN)	Japan
1989	Associazione Eposti Amianto (AEA)	Italy
1990	Asbestos Diseases Foundation of Australia	Australia
1990	Asbestos Diseases Association	New Zealand
1991	Ban Asbestos Network (BAN)	International
1992	Clydebank Asbestos Group	UK
1993	Merseyside Asbestos Victims Support Group	UK
1993	Gippsland Asbestos-Related Diseases Support (GARDS)	Australia
1994	Greater Manchester Asbestos Victims Support Group	UK
1994	Association Nationale de Défence des Victimes de L'Amiante (ANDEVA)	France
1995	Dutch Committee of Asbestos Victims	Netherlands
1995	Brazilian Association of Workers Exposed to Asbestos (ABREA)	Brazil
1996	Concerned People Against Asbestos (CPA)	South Africa
1998	Asosiacon de Extrabajadores de La Nicalit (AEXNIC)	Nicaragua
1999	International Ban Asbestos Secretariat (IBAS)	International
2000	Belgian Asbestos Victims Group (ABEVA)	Belgium
2001	Association of Asbestos Victims (ACHVA)	Chile
2002	Ban Asbestos Network India (BANI)	India
2002	Association Against Asbestos—Program for Study of the Occupational Risks of Asbestos (AFA-PEART)	Peru
2002	Justice for Asbestos Victims	UK
2002	Comité d'Aide et d'Orientation des Victimes de l'Amiante (CAOVA)	Switzerland
2003	Ban Asbestos Canada	Canada

described in Johnston and McIvor's study of the asbestos tragedy in Scotland.[47] It was a scenario that was to be endlessly repeated in cities and towns around the world that had processed asbestos. Ultimately, the 'healthy anger' generated by the failures of the compensation system resulted in the appearance of the first victims' action groups and eventually an anti-asbestos lobby.

The first groups appeared in Britain and the US—the largest end users of asbestos—in the 1970s (see Table 6.1). Similar groups then appeared in developing countries during the 1980s and by the year 2000 others had materialized as far afield as South America and South Africa. In 1991, an international federation known as the Ban Asbestos Network

[47] R. Johnston and A. McIvor, *Lethal Work* (2000). See also J. Leneghan, *Victims Twice Over* (1994).

(BAN) was formed. During the 1990s, as the movement gathered force, a global peregrination of meetings took in cities as far afield as Milan and Sao Paulo. In 1999, the IBAS was established to distribute and monitor information. The key individual in its operation is Laurie Kazan-Allen (sister of the American lawyer Steve Kazan), who since 1990 has published the *British Asbestos Newsletter*.

The industry claimed that anti-asbestos groups were filled with left-wing subversives. But in almost every instance, the driving force behind the foundation of these groups has not been a revolutionary agenda—unless one calls fairness revolutionary—but the physical and emotional damage caused by asbestos. In trying to heal that damage, such groups networked with lawyers and trade unions, lobbied governments, sought the help of the media, tried to raise awareness of asbestos hazards, and, above, all tried to comfort victims and help them gain financial recompense.[48] For the first time, scientists, government officials, and company executives faced placards, media campaigns, and groups of angry dockers and laggers. A string of publications by activists such as Nancy Tait, Alan Dalton, Henri Pézerat, and George Wragg offered an alternative discourse on asbestos.[49]

In taking on the asbestos industry, however, the victims' groups faced a difficult task. Paul Brodeur set the asbestos tragedy within the context of the Cold War and what he termed the medical–industrial complex.[50] If one accepts the reality of that complex, then obviously the efforts of victims' action groups would be strongly resisted. Corporations could defend themselves and had formidable resources. They could erect legal defences in court to defeat claims in the early stages; or they could pool their resources and outspend the opposition. They could try to recoup their costs from insurers, which in turn might also erect defences to defeat claims. They could sell off businesses or attempt to shift their assets; or they could take shelter from litigation by seeking bankruptcy. If all this failed, then an appeal to the government could be made to bail them out. Whichever defence was tried, the clock was always ticking on the sick and dying, ensuring that delay itself became a strategy.

[48] J. Flanagan and T. Whitson, 'Asbestos Victims Support Groups in England', *IJOEH* 10 (April/June 2004), pp. 177–9.

[49] N. Tait, *Asbestos Kills* (1976); A. Dalton, *Asbestos Killer Dust: A Worker/Community Guide* (1979); Collectif Intersyndical Securite des Universites-Jussieu CFDT, CGT, FEN, *Danger! Amiante* (1977); G. Wragg, *The Asbestos Time Bomb* (1995).

[50] P. Brodeur, *Secrets: A Writer in the Cold War* (1997).

In the 1970s, the asbestos companies used, with some temporary success, the so-called state-of-the-art defence, originally concocted by Johns-Manville and the insurance industry. The basic tenet was that the companies and their insurers knew little about the dangers of asbestos until 1964 and so were innocent.[51] In 1978, Johns-Manville told its shareholders, 'there simply is no fault on the part of J-M'.[52] The industry also argued that some injuries should be non-compensable; that exposure to some products (such as asbestos tiles and a/c panels) was not harmful enough to warrant compensation; and that the type of asbestos that was usually encountered (chrysotile) was harmless. Given that smoking had such a close interaction with lung cancer, the argument that most asbestos cancers were caused by smoking could also be thrown into the mix. However, by the end of the 1970s legal discovery had blown apart this defence. Even had it remained viable, so many claims were made that it was impossible to defend and adequately examine each one. In addition, most companies had insurance cover, so there were compelling reasons to settle. Although claims were rising, the number of defendants was also rapidly increasing. Thus claims lodged against at least a dozen or more manufacturers offered additional protection.

However, at the beginning of the 1980s it was apparent that the liabilities of the asbestos companies, the construction products industry, and the insurance industry were going to be enormous. In 1982, the Union Asbestos Rubber Company (UNARCO) filed for bankruptcy. At that time, this was not considered a particularly significant event, as its assets and insurance were limited. But in the same year, the mighty Johns-Manville sought relief in bankruptcy—an event that sent shockwaves through the American economy (and is considered below). As litigation progressed through the 1980s and 1990s, other trends appeared. Defendants began settling mass claims in their thousands to cut their total liability and save money on transaction costs. Leading defendant companies teamed up, alongside their insurers, to form claims facilities (such as the Wellington Group in 1985) to pool resources and settle claims together. Meanwhile, in 1991 the judiciary attempted to concentrate all cases in a single federal court. When this proved unworkable, in the following year the leading asbestos companies tried to cut a novel $1.3 billion deal with a group of plaintiffs' attorneys, led by Ron Motley and Gene Locks, to settle thousands of present and *future* claims. The scheme, known as *Georgine*,

[51] This argument is the centrepiece in P. Bartrip, *Beyond the Factory Gates: Asbestos and Health in Twentieth-Century America* (2005).

[52] J-M Corporation, *Annual Report* (1978), p. 27.

became highly controversial and was eventually rejected by the Supreme Court.[53]

A key player in the medical–industrial complex was the insurance industry. It was important for two reasons. First, the insurers conducted their own risk assessments of asbestos, making them an indirect party to dangerous working conditions. As one would expect, this knowledge was often highly sophisticated. It is often mentioned that an official of Prudential Life in New Jersey observed that many carriers declined to insure asbestos workers for life insurance even as early as 1918. Second, insurers were in line to pay out heavily when claims against the companies began to escalate after the 1970s. Although companies are named on writs, usually these are simply passed to insurers, who become the real line of defence against payment. For those claiming compensation, the insurers (presuming they could be found) presented yet another, if often hidden, obstacle. Experience would show that they could fight as tenaciously as the asbestos companies when large claims were at stake.[54]

A major problem for the insurance companies was that, although they had known for decades that asbestos workers were working in dusty conditions and dying prematurely, they had largely kept silent while their underwriters happily wrote policies for workmen's compensation and comprehensive general liability. Many of these policies would prove flawed and loosely worded, but nobody complained because the workmen's compensation system was so one-sided that few claims were ever accepted. After the 1970s, however, the situation became serious for the insurance industry. Internal insurance industry documents in that decade show the leading insurers formulating the 'state of the art' defence and unanimously agreeing to deny liability for ARDs.[55] At first, the insurers and the asbestos companies worked together; but later a bitter struggle broke out between them as to who should pay. Complicated arguments developed as to when insurance coverage should run: from when the claimant was exposed to asbestos or from when ARDs were diagnosed or discovered? The ensuing arguments about 'triggers' and 'manifestation' seem arcane to outsiders, but on such questions have hinged billions of dollars in compensation. When insurers were hit by the costs, they

[53] S.P. Koniak, 'Feasting While the Widows Weep: Georgine v. Anchem Products Inc.', *Cornell Law Review* 80 (May 1995), pp. 1045–158; Castleman, *Asbestos*, pp. 748–9.

[54] On insurers, see Brodeur, *Outrageous Misconduct*, pp. 183–211.

[55] American Insurance Association Meeting (Enterprise Liability Study Group), New York, 20 May 1977; Travelers Insurance Company, Memo of Meeting of Discussion Group: Asbestosis. New York, 21 April 1977.

claimed that the asbestos industry had not alerted them to the risk. It was a bizarre argument for an industry that was supposed to analyse risk.

The huge growth in the number of defendants meant that hundreds of millions of dollars worth of untapped insurance coverage became exposed. As these defendants were often very peripheral, no policy inclusions had generally been written into their coverage. Like a fire storm consuming oxygen, US litigation sucked in Lloyd's of London, which had been the insurer of last resort for many of the biggest American companies, including Johns-Manville. In 1973, one executive apparently remarked: 'asbestosis is going to change the wealth of nations. It will bankrupt Lloyd's and there is nothing we can do to stop it'.[56] By the end of the 1980s, Lloyd's indeed teetered on the brink of collapse. It had operated as a kind of private club, which had accepted members ('Names') on condition that their personal liability was unlimited. Unlimited asbestos claims combined with unlimited liability meant bankruptcy for many Names. Some reacted by suing the company's internal syndicates for allegedly hiding the problems of asbestos while signing them up to expand Lloyd's membership. The case, *Lloyd's v. Jaffray*, was dismissed in 2000, but claims that Lloyd's tried to hide its asbestos liabilities—even from its chief executives—have persisted. Meanwhile, a reinsurance vehicle, Equitas, was set up to absorb Lloyd's huge pre-1993 exposures. Between 2001 and 2005, it paid out about $3 billion in 35 major asbestos settlements, before Equitas was taken over in 2006 by Berkshire Hathaway—the US investment company owned by billionaire Warren Buffet. The latter already owned the rump of Johns-Manville, besides an investment in another bankrupt asbestos company, USG. The $7 billion (£3.8 billion) deal will only be profitable if Berkshire can limit its payouts to claimants.

In Britain, the insurance industry and defendant firms stated in 2004 that they faced up to £20 billion in compensation costs over the next thirty years. Ironically, both insurers and victims faced the same problem—overexposure to asbestos. The difference was that insurers were able to launch a powerful and orchestrated campaign to limit their liabilities.

In 2001, in the so-called Fairchild court case, the insurers first challenged the entitlement of most mesothelioma victims to compensation by arguing that if an individual had been exposed to dust by multiple

[56] Ralph Rokeby-Johnson, quoted in D. McLintock, 'The Decline and Fall of Lloyd's of London', *Time*, 21 February 2000.

defendants, then no one should pay because no one could prove which dust, or more particularly which single fibre, had caused the cancer. It was unlikely that any jury would have accepted this absurdity, but in Britain such cases are settled by a judge and he ruled in favour of the defendants. The ruling, which meant that an employer could admit to exposing a worker to dust and not pay any compensation, caused outrage. It was overturned after an appeal to the House of Lords, but not before it had caused delays and great distress to victims and relatives. The ABI tried to pre-empt the appeal by settling the cases at the last moment, so that the original judgement would stand. One of the plaintiffs, Doreen Fox, whose husband had died of mesothelioma, bravely turned down £115,000 so that the appeal could go ahead. But about 2,000 mesothelioma sufferers had died during the judicial process, with no hope of compensation for either themselves or their dependants.

After this rebuff, the insurers simply regrouped and in 2005 contested compensation for pleural plaques, by arguing that calcified patches on the pleura of the lungs caused by asbestos were not a compensable injury and that the modest damages payable under English law (£2,500–20,000) should not be allowed.[57] The courts first reduced the amount payable in future awards to a maximum of £7,000 and then in 2006 abolished the payments entirely (even though plaques had been compensated since the early 1980s). It was calculated that this would save the insurers over £1 billion. Meanwhile, the insurers won a notable court decision involving 'bystander' exposure. In 2004, a judge awarded damages of £82,000 against shipbuilders Harland & Wolff, after Teresa Maguire, the wife of one of its workers, developed mesothelioma from dust brought home on his clothes. In the following year, the Appeal Court overturned the judgement on the basis that the firm could not have foreseen in the 1960s that Mrs Maguire would suffer personal injury from asbestos dust. Another victory followed in 2006, when in *Barker v. St Gobain/Corus* the insurers did a U-turn and now agreed that multiple defendants were liable. But they succeeded in persuading the courts that if there was more than one employer, compensation should be spread between defendant companies (previously full damages could be claimed against any one company). Since in practice many companies would be defunct, uninsured, or impossible to trace, this would have meant many families missing out on part or all of their compensation.

[57] Pleural plaques, which are a marker for asbestos exposure and for possible future diseases (such as mesothelioma), are not compensated under UK government benefit schemes. Legal redress is the only route.

The *Barker* ruling was so unfair that the government immediately reversed it, though this interrupted rather than halted the attempt by insurers to cut their costs. At the end of 2006, as if to demonstrate that no strategy was too unthinkable for the insurance industry, a number of companies began contesting mesothelioma payments using a 'trigger' theory. The argument used by the insurers was that liability should depend not on the date when the victim was exposed, but when the mesothelioma began to grow. Given the latency of mesothelioma, exposure usually occurs decades ago, while it can be argued medically that the tumour itself only begins proliferating a few years before the first symptoms. Happily for the insurers, in that time lag employers have often ceased trading and no insurance exists.

What could not be achieved in the courts, the insurers achieved by spinning off liabilities into shell companies that then declared insolvency.[58] Such strategies were part of an overall trend that had been apparent since 2001. In that period, while asbestos deaths escalated, UK insurers, with the help of pro-defendant court verdicts, had reduced significantly their bills and shrunk the parameters of compensation.

Bankruptcy Strategies and the Corporate Veil

Bankruptcy has been the favoured escape route from liability in America. In the US, companies do not even need to be insolvent to file for bankruptcy and while restructuring takes place they can enjoy protection from creditors, such as asbestos claimants. The escape route is an administrative procedure known as Chapter 11, which hit the front pages in 1982 when Johns-Manville used it to shelter from thousands of lawsuits. Because Johns-Manville was such a prominent American company (it had assets of $2 billion and was on the *Fortune* 500 list), its bankruptcy generated a heated controversy. As the *Harvard Law Review* accurately predicted: 'If successful, Manville's strategy will have a profound effect on all asbestos-related tort litigation.'[59] Since then, asbestos companies have been among the most consistent and controversial filers for Chapter 11. Seventeen more companies filed for bankruptcy in the 1980s, due to their

[58] The most notorious example was the insolvency of Chester Street Insurance in 2001. See 'Asbestos Compensation Endangered by Insurer Collapse', *British Asbestos Newsletter*, Issue 41 (Winter 2000–2001).

[59] Anon, 'The Manville Bankruptcy: Treating Mass Tort Claims in Chapter 11 Proceedings', *Harvard Law Review* 96 (March 1983), pp. 1121–42.

asbestos liabilities, followed by another twenty-four in the 1990s. This had a knock-on effect, as each bankruptcy increased pressure on other firms. Then in 1994 Congress enacted the Manville Amendments, which greatly increased the attraction of Chapter 11 for asbestos companies. The Amendments allowed companies to seek bankruptcy protection from *future* liability, if they can show that this liability exceeds the assets of the company. Bankruptcies then escalated, especially between 2000 and 2005, when forty companies used Chapter 11. This brought the grand total to almost eighty companies since 1982.[60]

For victims it was a disturbing trend. In Chapter 11, claims are immediately stayed, while lawyers and administrators haggle over the debts of a company that continues in business. Reorganization plans usually involve a trust for victims paid for by future profits and the insurers, who contribute in return for discharge from claims. But at least six years and sometimes more than ten years can elapse before companies emerge from bankruptcy and begin payments.[61] Johns-Manville took six years to create two independent trusts—a Personal Injury Trust (PIT) and a Property Damage Settlement Trust—as a source of payments for asbestos claimants. The Trusts have had a chequered history, with the PIT (the larger of the two) accused of poor administration. Castleman notes that the PIT paid $1.1 billion to settle claims between 1988 and 1994, while running up $179 million in expenses.[62] At first, many existing claimants received sizeable payments, but the PIT soon ran out of money and it is now so drained by the number of claims that it can only pay five cents in the dollar of each settlement. For mesothelioma, the PIT now pays successful claimants a paltry $10,000.

The rump of the original Manville company, disconnected from its previous history and asbestos claims, has prospered under the ownership of Warren Buffet and his Berkshire Hathaway group. Manville has current sales of about $2.5 billion. Not surprisingly, the Johns-Manville bankruptcy has been attacked as one of the 'greatest miscarriages of justice against workers in [American] history'.[63] Other Chapter 11 asbestos companies can also usually be found on their websites celebrating 'business as usual'.[64] Of course, it was imperative that companies did do well, since one of the ironies of Chapter 11 is that plaintiffs become stakeholders in

[60] G. Tweedale and R. Warren, 'Chapter 11 and Asbestos: Encouraging Private Enterprise or Conspiring to Avoid Liability?', *Journal of Business Ethics* 55 (November 2004), pp. 31–42.

[61] S.J. Carroll et al., *Asbestos Litigation* (2005), pp. 118–19.

[62] Castleman, *Asbestos*, p. 803.

[63] Brodeur, *Secrets*, p. 187.

[64] Environmental Working Group, *Asbestos Think Again* (2004).

the company. On the other hand, Chapter 11 was usually accompanied with dire warnings about claims forcing firms out of business with attendant job losses.

Chapter 11 can have international repercussions. In 1997, American car parts maker Federal Mogul (FM) took over T&N along with its asbestos liabilities. Four years later, FM filed for Chapter 11 bankruptcy, after being hit by the resurgence in American asbestos claims (though in addition the company was awash with debt from an overly ambitious acquisitions strategy). A familiar picture unfolded for those seeking compensation: although FM/T&N had substantial revenues of $6 billion a year, all litigation was frozen, with the company claiming that the estimated £9 billion of liabilities would soon swallow its $10 billion assets. However, the bankruptcy had some unusual features. The American claims (well over 100,000) were typically product-liability suits from non-T&N workers in the earliest stages of ARD, while in the UK the 400 claims involved former T&N workers with serious ARDs. Yet when FM/T&N's 130 UK companies were forced into administration in 2001, all UK asbestos claims also ground to a halt. Through no fault of the UK claimants, resolution of these claims had become inextricably ensnared with the horse-trading of attorneys and creditors several thousand miles distant. The withdrawal of compensation produced a particularly tragic situation for some. Former T&N worker Ken Hall was awarded a settlement of £70,000 before his death from mesothelioma in 2001, but the offer was then withdrawn.

FM/T&N emerged from Chapter 11 at the start of 2008, though in the UK a fund had already been established to pay UK victims a mere 20 pence in the pound. Yet the bankruptcy lawyers involved in Britain had racked up costs of £70 million. This was £1 million more than the cash set aside for compensation, with the hourly rate for the leading bankruptcy administrators, Kroll, hitting £425. Whatever the payout, scores of mesothelioma victims will have died without compensation from the company. In effect, T&N's asbestos liabilities had been taken 'offshore', which raises questions about the long-term rationale behind FM's bizarre decision to buy a British company loaded with asbestos claims. Concurrently, Cape plc succeeded in pushing through the English courts a £40 million compensation fund for *future* claims, even though Cape is still solvent. No one expects the money to cover all future liabilities, yet Cape is not legally obliged to top up the fund and admits future claims may not be paid. Not surprisingly, one critic of these exit strategies wrote: 'Viewed as pieces of an evolving jigsaw puzzle, [these] developments... produce a picture of a society where corporate survival takes precedence over life and death issues, common law principles and human rights.'[65]

Chapter 11 is not available for companies outside America, but there are other options. Companies can shield assets by exploiting limited liability law, which regards a company as having a fictional 'personality' that is separate from its members. This is the so-called 'corporate veil', which provides an incentive for risks to be externalized by the relatively simple mechanism of making one company a 'parent' and another a 'subsidiary'. Because a parent company is simply a shareholder, it is legally no more responsible for the unlawful behaviour of a subsidiary than would be a member of the public for the liabilities of a large public company in which he/she owns a single share. Not surprisingly, this veil allows modern corporations the opportunity to use complex and confusing corporate structures to distance themselves from liability.

Cape Asbestos, itself once part of the great South African conglomerates of De Beers and Anglo-American Corporation, provides a classic example of these tactics. In 1982, after Cape had been ordered in one American court verdict to pay $55 million in damages for exporting a lethal product without warnings, the company simply refused to pay and closed its US operations. Reluctant to give up the American market, however, Cape established a secret shell company in Liechtenstein through which Cape continued to ship blue fibre to America. In 1988, when American plaintiffs sued Cape in London (*Adams v. Cape*), the judge refused to pierce the corporate veil and hold Cape accountable, even though he accepted that Cape operated worldwide as a single economic entity and that some of the participants had given false testimony.[66]

In the 1980s, such strategies were utilized elsewhere. Leveraged buyout subterfuges were used to distance the Jim Walker Corporation (another *Fortune 500* company) from its wholly owned asbestos subsidiaries, Celotex and Carey Canada.[67] In America, the reaction of W.R. Grace & Company to its spiralling asbestos liabilities after the mid-1980s was to divest the company of over $4 billion in assets behind an elaborate series of restructurings and technical bankruptcy. The result was a near empty shell for asbestos-claimants and a tax-free exit with the assets for the

[65] L. Kazan-Allen, 'Tipping the Balance: Exit Strategies for UK Asbestos Defendants', *British Asbestos Newsletter* (Autumn 2006), p. 1.

[66] G. Tweedale and L. Flynn, 'Piercing the Corporate Veil: Cape Industries and Multinational Corporate Liability for a Toxic Hazard, 1950–2004', *Enterprise and Society* 8 (June 2007), pp. 268–96.

[67] Castleman, *Asbestos*, p. 747.

stockholders (though the bankruptcy court was later able to claw back about $1billion).[68]

In the late 1990s, James Hardie in Australia also restructured its business to deal with its growing asbestos liabilities. In 2001, it established a grossly inadequate A$293 million compensation fund, while simultaneously shifting its assets and corporate headquarters not to the US, where Hardie did most of its business, but to the Netherlands. This was done ostensibly for tax reasons. When the fund began running dry within only three years and the Dutch parent was revealed to be fireproof to claims (because Australia does not have an enforcement treaty with the Netherlands), a public outcry resulted in a New South Wales government commission. The report on the scandal in 2004 revealed a masterly use of the corporate veil to distance a company from liability.[69] Faced with a union boycott and a barrage of bad publicity, Hardie was forced in early 2007 to renegotiate the fund to cover liabilities set at A$4.5 billion over the next forty years. The company, nevertheless, managed to wring A$1.5 billion in tax concessions from the government, because it was an offshore company.

Another layer of complexity that works in the favour of multinational companies is caused by the fact that ARDs themselves can be transnational. In Italy, several mesothelioma cases have occurred in Italian migrants returning from work in Australia at the Wittenoom mine. The question arises: Who compensates them? ARDs have also been suffered by Italian workers who had worked for Eternit at Niederurnen in Switzerland, and then returned home. In 2001, the public prosecutor of Turin launched a criminal investigation into Eternit and approached the Swiss government. SUVA, the national accident insurer, initially refused to release its files and, though this was overruled in 2005 by the Swiss courts, the documents have yet to find their way to Italy. The Turin prosecutor is also investigating 2,000 further claims, and by 2006 the pressure on Eternit was beginning to tell. In October 2006, one week after a Swiss TV programme highlighted the disastrous record of another

[68] K. Rivlin and J.D. Potts, 'Not So Fast: The Sealed Air Asbestos Settlement and Methods of Risk Management in the Acquisition of Companies with Asbestos Liabilities', *New York University Environmental Law Journal* 11 (2003), pp. 626–61; J. Heenan, 'Graceful Maneuvering: Corporate Avoidance of Liability through Bankruptcy and Corporate Law', *Vermont Journal of Environmental Law* (2003).

[69] NSW Government: Cabinet Office, *Report of the Special Commission of Inquiry into the Medical Research and Compensation Foundation* (September 2004); B. Hills, 'The James Hardie Story: Asbestos Victims' Claims Evaded by Manufacturer', *IJOEH* 11 (2005): 212–14; G. Haigh, *Asbestos House: The Secret History of James Hardie Industries* (2006).

factory in Payerne,[70] Eternit established a SwFr1.25million ($1million) asbestos fund for former employees in Switzerland. It was evident that such a sum would never compensate most asbestos victims or dissuade Swiss victims from pursuing Eternit in the Glarus cantonal courts.

A Simple Truth

By the end of the twentieth century, asbestos claims in America were reaching a crescendo. By 1994, a quarter of a million claims had been filed within the preceding decade; by 2005, the total was about three-quarters of a million. Some plaintiffs' counsel increasingly took on large numbers of cases, including many clients who were not necessarily sick from asbestos, but who had medical evidence of pulmonary injury. These so-called 'unimpaired claims' (by the 'worried well', as some have called them) greatly boosted the numbers seeking compensation. As its final defence, some asbestos defendants resurrected an idea that had been popular at the end of the 1970s: a Congressional bail-out scheme that would end the litigation. Public Citizen, the consumer rights group, estimated that between 2003 and 2004 a group of leading companies spent $144 million in lobbying for such a bill—a campaign which would, if it succeeded, see their liabilities greatly reduced.[71] In 2005, their wish was granted when a Senate bill was drawn up that would establish a $140 billion asbestos compensation fund that would operate for thirty litigation-free years.

Coincidentally, a sudden welter of studies appeared suggesting that it was not ARDs that posed a problem, but litigation.[72] Grotesquely, the metaphors and adjectives that were usually applied to ARDs—sick, crippling, injurious, and victims—were now applied to companies and the economy. Reports commissioned by insurers suggested that it was the economy that was suffering, with projected asbestos claims of $275 billion and 55,000 lost jobs.[73] The *Wall Street Journal* depicted a 'Job-Eating Asbestos Blob',[74] while *Fortune* headlined asbestos as a '$200 billion

[70] TSR, 'Mourir Amiante, en Silence', 28 September 2006.

[71] Public Citizen, *Federal Asbestos Legislation: The Winners Are ...* (May 2005).

[72] M.J. White, 'Asbestos and the Future of Mass Torts', *Journal of Economic Perspectives* 18 (Spring 2004), pp. 183–204.

[73] J. Stiglitz, J.M. Orszag, and P.R. Orszag, *The Impact of Asbestos Liabilities on Workers in Bankrupt Firms* (2002). See also J.L. Biggs, 'Proposed Resolution Regarding the Need for Effective Asbestos Reform', American Academy of Actuaries Statement (10 July 2003).

[74] *Wall Street Journal*, 23 January 2002.

miscarriage of justice'.[75] One might regard these sums as a bargain, if the magic mineral was as indispensable to civilization as the industry and its supporters have always claimed. Amongst those supporters are historians Peter Bartrip and Rachel Maines, who have deplored the American wave of asbestos litigation. Maines believes that the explanation for the avalanche of litigation is the opportunism of American lawyers and argues that victims should be far more accepting. According to Maines: 'We have all received the societal benefits of asbestos as a fire-resistive material, and we are all responsible [so] ... it would be simpler and more humane to ensure that all Americans receive the health care they need without requiring them to call a lawyer who can find somebody to blame for their condition.'[76]

Such facile interpretations ignore why the large number of claims has arisen; nor do they discuss the complexities of asbestos compensation. This chapter has shown that two systems have basically developed to solve this question: first, workmen's compensation; second, litigation. The two systems are not always exclusive, though it can be seen that the former is more closely associated with European countries and the latter with the US. Interestingly, America moved from favouring a system of workmen's compensation to one which allowed litigation freer rein. In the early 1980s, it appeared that tort law would provide a solution as '*the* uniquely effective and indispensable means of exposing and defeating the asbestos conspiracy, providing compensation to victims, and deterring future malfeasance'.[77] Brodeur believed that the history of asbestos litigation demonstrated the triumph of common law against vested interests. It did triumph to some extent; indeed, one can argue that litigation was more effective than government regulation in ending asbestos use. However, the question of how to compensate remaining claimants is difficult. Litigation is now less popular with the US government. It ticks many of the boxes mentioned at the start of this chapter—especially regarding punishment for corporate wrongdoing—but clearly not enough money can be liberated from the system if everyone is compensated at the highest court rates, given (and this is an important qualification) that the same system allows corporations the escape routes of bankruptcy and the corporate veil.

[75] R. Parloff, 'The $200 Billion Miscarriage of Justice', *Fortune* 145 (4 March 2002), pp. 154–8, 162, 164.

[76] R. Maines, *Asbestos and Fire* (2005), p. 170.

[77] D. Rosenberg, 'The Dusting of America: A Story of Asbestos—Carnage, Cover-Up, and Litigation', *Harvard Law Review* 99 (1986), pp. 1693–706.

On the other hand, European-style state welfare is far from perfect. Europeans regard the American court system with either contempt or amusement, but victims of asbestos are as disadvantaged in European countries as they are in the US—sometimes more so.[78] Moreover, the lack of litigation ensured that levels of asbestos use continued to be higher in Europe than the US during the last decades of the twentieth century.

Whatever the country, much of the current noise in the compensation debate is being generated by insurers, conservative think tanks, and the financial press. Undoubtedly, compensation costs are high in some respects. Asbestos costs in the leading European countries are now absorbing an average of a quarter of the total cost of all occupational disease. In Germany, for example, in 2003 over €314 million was spent on asbestos compensation by means of insurance. In France, total costs between 2000 and 2020 are estimated at between €27 billion and €37 billion. But these figures are only so high because of the numbers negligently exposed, not the largesse of individual payments. Hardly anyone has ever bothered to set the debate in any historical context, by highlighting either the enormous wealth generated by the industry or the way in which an unjust compensation system helped bury the problem of ARDs. What is especially striking is that, in contrast to the thousands of scientific papers on the epidemiology of ARDs and the ample newsprint devoted to the travails of bankrupt companies, there is an almost complete absence of studies which explore objectively the financial impact of ARDs on national health systems or, particularly, on victims. One of the few that has (published in the 1980s), showed the drastic impoverishment that occurs when workers and their families are devastated by ARDs. The study concluded that

> those who benefit from the production and use of commodities, such as the producers and users of asbestos, have not paid the full cost of the commodities. They have, in effect, been subsidised by the workers who have died and by their survivors. Whether one wishes to measure social goals in terms of economic efficiency, morality, or some common-sense definition of what is just, compensation that fails to pay even the net loss to the survivors of dead workers is grossly inadequate.[79]

[78] S. Jasanoff and D. Perese, 'Welfare State or Welfare Court: Asbestos Litigation in Comparative Perspective', *Journal of Law and Policy* 12 (2004), pp. 619–39.

[79] W.G. Johnson and E. Heler, 'Compensation for Death from Asbestos', *Industrial and Labor Relations Review* 37 (July 1984), pp. 529–40. On the impact on health systems, see A. Watterson et al., 'The Economic Costs of Health Service Treatments for Asbestos-Related Mesothelioma Deaths', *Annals New York Academy of Sciences* 1076 (September 2006), pp. 871–88.

A more recent study has confirmed the failure of American workers' compensation to adequately compensate victims at the end of the twentieth century. The study identified substantial cost shifting from workers' compensation systems to individual workers, their families, private medical insurance, and taxpayers.[80]

These findings would come as no surprise to anyone who has read the thousands of asbestos claims' files generated by the asbestos tragedy. Of course, few have read them, since the published discourse on asbestos has been so dominated by established science, industry, and the government. We have been privileged to read many of these files. It is difficult to pick a representative letter—there are so many from which to choose—but the following unsigned letter amongst Selikoff's papers gives an insight into the problems of ARDs:

28 January 1986

Dear Dr Selikoff,

The year the doctor found asbestos cancer, my husband lived one year and three months. His medical bills came to $77,000. His insurance stopped when he died. The doctors did not want to testify that his illness was related to his work. I could not draw workmen's compensation. I worked and used up my sick leave days while my husband was ill. I fell a year after my husband's death and crushed a disc in my back. The doctor operated and wanted me to stay off from work, but I had to work to pay my doctor's bills. I had to go back to work before I was able. I am too young for Medicare and Social Security. My medicine runs to $100 a month. I now do have a lawsuit, but so far nothing has happened. I sure appreciate you for doing so much for the working man. Do you think anything will come from Washington to help the family left holding the bag?

The idea that litigation costs are ruining the American or any other economy is a myth and a misunderstanding of the scale and scope of compensation for industrial disease.[81] The simple truth is that there is not *enough* compensation. It is rare for any victim in Europe to get more than €100,000 and without a civil action, which is impossible in many countries; most payments do not even reach €20,000, even for mesothelioma. Actually, most victims in whatever country get *nothing*—a fact that is easily demonstrated.

In the UK, during the 1990s, well under a half of those with mesothelioma were not compensated by the government (the self-employed and

[80] J.P. Leigh and J.A. Robbins, 'Occupational Disease and Workers' Compensation: Coverage, Costs and Consequences', *The Milbank Quarterly* 82 (2004), pp. 689–721.

[81] Public Citizen, *Asbestos Cases in the Courts: No Logjam* (2006).

bystander cases being unable to claim benefits at all). For asbestos-related lung cancers the figures are even starker: currently these number between about 1,800 and 3,800 cases each year, but in 2004 a mere seventy-five were awarded compensation by the state.[82] In France, by the year 2000 annual mesothelioma deaths had surpassed a thousand a year, yet not many more than a quarter were compensated and there were wide regional differences.[83] In Germany, despite its vaunted social insurance scheme, since the 1980s there has been significant under-claiming of government compensation by both men and women with mesothelioma. For example, in 2002 there were more than 250 female mesothelioma deaths and only seventy-five claims (possibly due to the failure of doctors and coroners to ask about the exposure history of women with mesothelioma).[84] In Denmark, one study found that between 1994 and 2002, of the 695 registered cases of mesothelioma nearly half had not been reported to the government compensation board, despite a legal requirement to do so.[85] Italy's restrictive laws meant that only about 15 per cent of mesotheliomas were compensated before 1994—a figure that only improved slightly when the law was relaxed.[86] In Central and Eastern European countries, only trivial numbers of asbestos cancer cases have been compensated since the mid-1990s.

In Canada, 832 mesotheliomas were diagnosed in Quebec between 1982 and 1996, but the Workers' Compensation Board registered only 261 between 1967 and 1997.[87] In Japan, compensated asbestos-related lung cancers and mesotheliomas had reached 127 in 2004, but deaths due

[82] L. Kazan-Allen, 'A Comparative Review of European Asbestos Compensation' (2005). Unpublished typescript available on ANDEVA website.

[83] M. Goldberg, S. Goldberg, and D. Luce, 'Regional Differences in the Compensation of Pleural Mesothelioma as Occupational Disease in France, 1986–1993', *Revue Épidémiologie Santé Publique* 47 (1999), pp. 421–31.

[84] O. Hagemeyer, H. Otten, and T. Kraus, 'Asbestos Consumption, Asbestos Exposure and Asbestos-Related Occupational Diseases in Germany', *International Archives of Occupational and Environmental Health* 79 (September 2006), pp. 613–20. See also Hans Joachim-Woitowitz, 'Asbestos-Related Occupational Diseases—The Current Situation', European Asbestos Conference, Dresden, 2003.

[85] J. Hansen et al., 'Registration of Selected Cases of Occupational Cancer (1994–2002) with the Danish National Board of Industrial Injuries', *Ugeskr Laeger* 169 (30 April 2007), pp. 1674–8.

[86] Merler, 'Occupational Cancer'; F. Montanaro et al., 'Occupational Exposure to Asbestos and Recognition of Pleural Mesothelioma as Occupational Disease in the Province of Genoa', *Epidemiologia e Prevenzione* 25 (March/April 2001), pp. 71–6.

[87] L. Kazan-Allen, 'Canadian Asbestos: A Global Concern', *IJOEH* 10 (April/June 2004), pp. 121–43. See also B. Waller and L. Marrett (eds.), *The Occupational Cancer Research and Surveillance Project* (2006), p. 23; J. Brophy, 'The Public Health Disaster Canada Chose to Ignore', in *Chrysotile Asbestos: Hazardous to Humans, Deadly to the Rotterdam Convention* (2006), pp. 17–20.

to mesothelioma alone had hit 953 in the same year.[88] In 2006, after the 'Kubota shock' (when cases of bystander mesothelioma were revealed close to asbestos factories in Osaka), a law was introduced to provide aid to victims of ARDs not covered by workers' accident compensation: but in the first year of the scheme only 143 of the 1,653 applicants had received compensation, amidst widespread complaints at the slowness of the scheme and the fact that 20 per cent of applicants died of cancer before their claim was processed.[89] It should also be noted that the bystander and environmental cases described in Chapter 7 have rarely been compensated. In the UK and in Australia, not until the 1990s did victims have any success in taking companies to court. As the Maguire case demonstrates, it is a difficult and uncertain task.[90] Only in 2007 was the first environmental case in the Netherlands won against Eternit (for providing asbestos waste to a resident of Goor for paving a farmhouse yard). Even in America, it is not easy to win such cases.

This raises fundamental questions. If the compensation system is so inefficient at present in the developed world, what will happen to the large number of future victims as asbestos deaths and the prevalence of ARDs continue to rise? What happens also to compensation in the developing world?

[88] *Japan Times*, 9 October 2005.

[89] Mamoru Kishibe and Yoshiaki Takeuchi, 'Asbestos Relief Measures Lacking', *Yomiur Shimbuni*, 28 March 2007.

[90] In 2007, the British government belatedly allowed environmentally exposed individuals and the self-employed to claim modest state benefits for mesothelioma.

7

Don't Disturb the Dog: Asbestos in the Environment

> The floating fibres [of asbestos] do not respect job classifications. Thus, for example, insulation workers undoubtedly share their exposure with their workmates in other trades: intimate contact with asbestos is possible for electricians, plumbers, sheet-metal workers, steamfitters, laborers, carpenters, boiler makers, and foremen; perhaps even the supervising architect should be included.
>
> I.J. Selikoff
> 'Asbestos and Neoplasia'
> *JAMA* 188 (6 April 1964), pp. 22–6

> Paul Rudolph, an architect whose career epitomized the turbulence that engulfed American modernism in the 1960s, died yesterday at New York Hospital. He was 78 and lived in Manhattan. The cause was mesothelioma.
>
> *New York Times*
> 9 August 1997

In 1980, Dr Irving Selikoff gave a talk in California. He told his audience that ten years earlier a thirty-year-old New Yorker, Simon Beers, had consulted him at Mount Sinai Hospital because of a pain in his chest. Selikoff discovered that Beers had been born on Ross Street in Brooklyn, moved to Wilson Street, and after marrying at 16 had started work at Macy's. A biopsy showed that Beers was suffering from pleural mesothelioma. Selikoff recalled: 'The diagnosis was clear, because on the [disease] site was the kind of asbestos that tended to be used in shipyards. We made the diagnosis with a 15¢ map of New York. Here was Ross Street, here was

Wilson Street, across the way was the Brooklyn Navy Yard. He had gotten his asbestos in his mother's arms.'[1]

Selikoff related this tragedy to demonstrate his anxiety about the dangers of environmental exposure to asbestos. Such anxieties would have seemed fastidious, even faintly ridiculous, during the Industrial Revolution when many workers endured choking clouds of dust on a daily basis, often without complaint. The fear that one could become sick from contamination in the open air or inside ordinary buildings would have seemed bizarre. This is not to say, however, that environmental concerns about industrial pollutants were non-existent. Campaigns against smoke pollution began in the late nineteenth century in some American and European cities. Moreover, some industrial raw materials and products were recognized surprisingly early as dangers to the wider community. As Gerald Markowitz and David Rosner have shown in America, leaded petrol and lead paints had been recognized by the inter-war period at the latest as causing potential adverse health effects *outside* the factory setting. They believe that lead, 'the mother of all industrial poisons', should be regarded as the 'paradigmatic toxin that linked industrial and environmental disease in the first two-thirds of the twentieth century'.[2] If that is so, then we would argue that asbestos must be a prime candidate as the paradigmatic environmental polluter of the final third of that century.

Outside the Factory Walls

If there was a defining moment when asbestos in the environment became a public issue, it was during the New York conference on the Biological Effects of Asbestos in 1964. However, even before then there were disturbing signs that asbestos outside the factory gates might be an issue. Even in the era when asbestosis was recognized as the chief problem—say, between the 1920s and 1950s—it was recognized that ARDs were not simply confined to certain factory occupations. They occurred in ancillary occupations, such as insulation (lagging), where workers sawed, hammered, drilled, mixed, and sprayed asbestos—usually in confined and airless places in ships and buildings. These men were not, as the industry repeatedly told everyone, asbestos workers in the ordinary sense, but that did not prevent them from developing asbestosis and cancer. The first

[1] Selikoff Address at University of Southern California Convention on the Medical and Legal Aspects of ARDs, 31 January 1980.

[2] G. Markowitz and D. Rosner, *Deceit and Denial* (2002), p. 41.

asbestosis cases among insulators were documented in the early 1930s; so too were the first compensation claims, with Johns-Manville settling its first cases in 1935. Thereafter, the number of sick insulation workers began to increase steadily, as did the references in the medical literature. When Barry Castleman began searching this literature for his landmark book on asbestos, he had little trouble in filling several pages with documented cases of ARDs worldwide in insulation workers, dating from the 1930s.[3] He also found references to ARDs amongst product users, such as welders, boiler riveters, and plumbers. These cases, too, began to appear in the 1930s. By the Second World War, the industry had no reason to doubt that asbestosis was a problem *outside* its mines and factories.

Asbestos was affecting the environment in other ways. The mining areas in Canada, South Africa, Europe, Russia, and Australia were particularly dusty, but they received far less attention in the medical press before the 1950s because of their isolation. In the aptly named Asbestos township in Quebec, the open mines and dumps had become so extensive that asbestos became the mineral that ate the town. Internal Johns-Manville documents show that as early as the 1940s, trade unions and company physicians in Quebec were becoming concerned about the 'downwind risk' from environmental pollution. Asbestos was blowing around the mining townships like tumbleweed.[4] At Wittenoom in Western Australia, although the mine was 5 miles from the town, miners' families were heavily exposed to asbestos.[5] Tailings dressed the race course (where crocidolite pooled in horses' hoof-marks), the car parks, and the golf course and they were used around most domestic dwellings to suppress the red dust. According to the headmaster of Wittenoom Primary School, before 1966 asbestos tailings were used to surface the playground—in places up to 3 inches deep—and a small mound of tailings was piled within yards of the main school building.[6]

But it was in South Africa where major environmental problems occurred. Once asbestos was disturbed by mining, large areas of the Northern Cape and what is now the Limpopo Provinces became badly contaminated. One of the discoverers of mesothelioma, Paul Marchand,

[3] B.I. Castleman, *Asbestos: Medical and Legal Aspects* (2005), pp. 391–400.

[4] Elliott DeForest, Secretary Northwest Magnesia Association, 14 June 1943. Letter in J-M files, cited in '*Johns-Manville v. Home Insurance*, et al.: Judicial Council Co-ordination Proceeding No. 1072. Memo, 22 May 1986, pp. 46, 115.

[5] J. McCulloch, 'The Mine at Wittenoom: Blue Asbestos, Labour and Occupational Disease', *Labor History* 47 (February 2006), pp. 1–19.

[6] Letter from G. Duncan, Chief Administrative Officer, Education Department, Perth, to Commissioner of Public Health, Perth, 5 December 1966.

visited these areas in 1958 with Kit Sleggs and found an 'environmental mess'. He recalled that 'the land was blue for miles around the asbestos settlements. The mills indiscriminately spewed blue dust clouds over the countryside, and whenever the wind arose, a blue haze covered the dumps'.[7] He wondered how anyone could avoid the 'pestilence'. In 1962, officials at the Department of Mines in Johannesburg found that contamination of residential areas around one of the region's asbestos mills in Penge could be as high as 22 f/cc, while clouds of road dust churned up by passing lorries reached 25 f/cc (aside from clouds of additional dusty particles).[8]

This was the setting for the discovery of the link between asbestos and mesothelioma, which was to be publicized at Selikoff's New York conference. One of Wagner's key findings was that mesothelioma could be caused by environmental exposure (eleven of his thirty mesothelioma patients in the Cape had not worked with asbestos). But this proved only the start, as the conference presentations began dissolving the wall between the factory and the environment. The words 'non-occupational' and 'community' were heard at the conference and—perhaps for the first time—the emphasis was no longer on simply workers' cohorts, but on cities (London and Belfast), regions (Lombardy and Piedmont), and problems that affected everyone, such as atmospheric air pollution. Representatives from other industries noted the new emphasis. One DuPont man told his bosses:

> [T]he main interest of the meetings was in the drawing attention to the fact that asbestosis and complications of asbestosis have now been found in persons who would not in the ordinary way come to mind as being exposed to asbestos; that it to say, they do not work in asbestos mines, asbestos mills, or asbestos textile factories. It would appear from the consensus at this meeting that the risk of [such] persons . . . is greater than was originally thought. This fact should be borne in mind and embraces not only the person's occupation, but also his hobbies—for instance, a do-it-yourself worker at home can have material exposure to asbestos during home insulation.[9]

The newspapers, too, had little difficulty in picking out what would most interest their readers—namely, the fact that asbestos fibres were widely distributed in consumer products and also in the air of cities. One of

[7] P.E. Marchand, 'The Discovery of Mesothelioma in the Northwestern Cape Province in the Republic of South Africa', *AJIM* 19 (1991), pp. 241–6.

[8] Report of C.P. Visser, Assistant Inspector of Mines, Pneumoconiosis Research Section, Johannesburg, 4 March 1962.

[9] G.J. Stopps 'Trip Report' to J.A. Zapp Jr, 2 November 1964.

the first headlines on the conference was: 'Asbestos in Environment also Threatens City Dwellers', which reflected concern that one conference paper had reported that asbestos fibres had been found at *post mortem* in the lungs of city inhabitants as far afield as Cape Town, London, Miami, Pittsburgh, and Belfast.[10] Soon electron microscope studies would show that the lungs of nearly all adult urban dwellers contained asbestos fibres.

One paper presented at New York had a great impact. A London-based epidemiologist, Dr Molly Newhouse, and a social scientist, Hilda Thompson, had traced mesotheliomas at a major London hospital. This institution also happened to be the catchment area for the east London docklands and, more significantly, for the community around the Cape Asbestos factory.[11] Occupational exposure was a significant factor. But there were also cases of household exposure, where women had washed dusty overalls, and where individuals had never worked in the industry. Newhouse found 'evidence that neighbourhood exposures may be important', due to the fact that the individuals 'lived within half a mile of an asbestos factory'. Within days of the article appearing, a leading newspaper had dubbed asbestos the 'killer dust'. Not surprisingly, the Newhouse findings struck fear into the public, especially when other studies appeared.

In the US, a study in 1967 of forty-two mesothelioma cases in south-eastern Pennsylvania found that only ten had definite occupational exposure—many of the others were non-occupational or neighbourhood cases (including a clergyman, who lived for nineteen years across the road from an asbestos insulation plant).[12] In the same year, another study reported seventeen cases of mesothelioma in New Jersey. Two of those individuals had lived near the Johns-Manville asbestos factory, but they had no direct exposure to the mineral.[13] In Hamburg, too, there was the characteristic mix near an asbestos factory of workplace, neighbourhood, and domestic exposure leading to mesothelioma. The authors noted the 'remarkable fact that this exposure may be very short (minimum 14 days)

[10] J.G. Thomson, 'Asbestos and the Urban Dweller', *ANYAS* 132 (31 December 1965), i, pp. 196–214.

[11] M.L. Newhouse and H. Thompson, 'Mesothelioma of Pleura and Peritoneum Following Exposure to Asbestos in the London Area', *BJIM* 22 (1965), pp. 261–9.

[12] J. Lieben and H. Pistawka, 'Mesothelioma and Asbestos Exposure', *Archives of Environmental Health* 14 (April 1967), pp. 559–63.

[13] M. Borow, L.L. Livornese and N. Schalet, 'Mesothelioma and Its Association with Asbestosis', *JAMA* 201 (July–September 1967), pp. 587–91.

or so slight that it has not been enough to cause pulmonary asbestosis'.[14] Newhouse also continued to compile further tragic cases, such as: 'A patient suffering from a peritoneal mesothelioma... [who]... recalled playing with handfuls of asbestos on waste ground near a factory as a small boy, some forty years before he developed his tumour.'[15]

What Environmental Problem?

The public was disturbed to be told that asbestos filters contaminated their wine, beer, and water with millions of fibres; that city air contained fibres; that their houses, apartments, and office buildings contained millions of tonnes of asbestos; and that everything from cigarette filters to talcum powder could be a source of exposure. They were even more concerned when ARD cases began to appear in the newspapers not only among asbestos workers in the factories, but also amongst laggers, roofers, electricians, dockers, carpenters, hospital staff, bakers, teachers, and brake mechanics. Almost every category of worker discussed in Chapter 2—and many more—seemed to be at risk. For example, by the 1960s several workers who had assembled gas masks for Boots for only 5 months during the Second World War began to sicken with mesothelioma. With publicity like this, mesothelioma's linkage with environmental exposure was a catastrophe for the industry and posed its greatest challenge.

In the UK, the industry's main strategy was to launch its own investigations of the environmental problem through the Asbestos Research Council (ARC). Soon privately printed reassuring brochures appeared under the imprint of the industry's propaganda unit (the AIC). Wherever the industry looked—fibres in the air outside, asbestos from brake linings, the contamination of drinking water—fibres levels were invariably 'infinitesimal'. In 1970, for example, the AIC leaflet, *Asbestos in Buildings*, had a simple message: 'No Risk to Building's Occupants.' Less well publicized were the industry's confidential reports on the high dust counts encountered in demolition and in simple construction tasks, such as cutting boards, the increasing resistance of some customers to continued asbestos use,

[14] P. Dalquen, A.F. Dabbert, and I. Hinz, 'The Epidemiology of Pulmonary Mesothelioma: A Preliminary Report on 119 Cases from the Hamburg Area', *German Medical Monthly* 15 (February 1970), pp. 89–95.

[15] M.L. Newhouse, 'Asbestos in the Workplace and the Community', *Annals of Occupational Hygiene* 16 (1973), pp. 97–102.

and the stream of complaints from customers about the deterioration of insulation (caused, for example, by birds nesting in sprayed asbestos). The industry's standard reply was to downplay the mesothelioma hazard by arguing that exposure in environmental situations was negligible and usually involved 'safe' chrysotile. However, some knowledgeable British users in the late 1960s—the state railways, the electricity generating board, the Navy, universities, and health authorities—dropped the use of asbestos in buildings and ships. This was usually done discreetly, though as Turner & Newell noted: '[T]he fact that asbestos can be found almost everywhere is bound to be realised sometime.'[16]

Nevertheless, the industry's publicity machine, which constantly criticized substitute materials and emphasized the safety aspects of asbestos, ensured that public confidence in the material was not entirely destroyed. In the late 1970s, when mesothelioma was discovered amongst village communities in Greece and Turkey, and the exposure traced to naturally occurring rocks and soils containing tremolite or erionite, the asbestos industry was able to argue that such cancers were an ordinary risk of life and not always associated with industrial use.[17]

The UK government, too, was happy to see asbestos use in buildings continue, even though its Factory Inspectorate had warned in 1974 that future structural alterations could cause a danger to health. Asbestos use in construction materials surged ahead in the 1970s, despite disturbing evidence that carpenters had even more pleural plaques than laggers, beside a worrying incidence of lung cancers and mesothelioma.[18] Other studies showed that even household exposure (when cutting asbestos sheets) and hobbies (such as repairing brakes) were implicated in mesothelioma.[19] But the government was not interested in the environmental risk. A public enquiry into the asbestos scandal, the Simpson Committee, reported in 1979. But it was almost wholly concerned with factory problems, and even here the recommendations were studiously ignored by the government. Independent scientists despaired at the failure to react to 'one of the worst industrial health tragedies in history'.[20]

[16] T&N file re. Leeds University, 1967.

[17] P. Holt, 'Asbestos-Related Diseases: Neighbourhood and Domestic Hazard' (1987), pp. 157–66.

[18] D.E. Fletcher, 'Asbestos-Related Chest Diseases in Joiners', *Proceedings of Royal Society of Medicine* 64 (1971), pp. 837–8.

[19] M. Greenberg and T.A. Lloyd Davies, 'Mesothelioma Register 1967–68', *BJIM* 31 (1974), pp. 91–104.

[20] Paul Holt, discussing the Asbestos Advisory Committee Enquiry, at 'Asbestos in the Home and Environment', Lambeth Town Hall, 17 November 1982.

In Europe, the industry had even less to contend with as regards environmental matters. In 1971, when industry representatives gathered in London to discuss strategies for a global defence of the industry, the situation in most countries was quiescent.[21] In France, Finland, and Italy even the occupational hazards were only beginning to seep into the national consciousness. Only in Germany, where the industry representative stated that the environmental impact of asbestos was 'the greatest of our worries', and in Holland, where a study of mesothelioma in a shipyard community in 1968 had stirred up some unfavourable publicity,[22] was there any cause for immediate industry concern.

It was in America that the British industry found an equally beleaguered ally. Public and government reaction to the New York conference had initially been muted, but events gathered pace in the late 1960s. Selikoff was making his views increasingly known, the government was about to introduce environmental regulations, and journalists were on the case. In 1968, a series of articles entitled 'The Magic Mineral' appeared in the *New Yorker*, written by Paul Brodeur.[23] It sold so well it was later reprinted as a booklet. In a memorable passage, Brodeur described taking a Fifth Avenue bus from Mount Sinai Hospital. As he headed downtown, he noted demolition workers breaking up asbestos cement (a/c) sheets and asbestos spray blowing over pavements like thistledown.

Leading insurance companies, such as Travelers, took the situation seriously, too. They surveyed the growing environmental problems of asbestos and worried about being overwhelmed with claims—a situation made worse by the actions of leading companies, such as Johns-Manville. One Travelers' memo noted the words of a Johns-Manville attorney in 1969: 'Confidentially, Johns-Manville has been contaminating the "Hell" out of both the air and the water for quite some time.'[24] Apparently 'concerned and frightened over the implications', Johns-Manville responded by setting up an environmental task-force in 1967, with industrial hygienists such as Ed Fenner and Cliff Sheckler, and public relations adviser Matthew Swetonic. Legal issues were at the forefront of the company's concerns. The a/c wing of the industry (as the largest fibre user) realized

[21] International Conference of Asbestos Information Bodies, London, 24–25 November 1971. Typescript.

[22] Dutch physician J. Stumphius showed that a small quantity of asbestos could cause mesothelioma amongst shipyard workers in Walcheren and added: '[O]ne must fear an explosion of the same dimensions throughout the whole population'. Quoted in R.F. Ruers and N. Schouten, *The Tragedy of Asbestos* (2005), p. 13.

[23] P. Brodeur, *A Reporter at Large: The Magic Mineral* (1968).

[24] Travelers' memo from H.W. Rapp, 12 August 1969.

in the 1960s that it was vulnerable under product liability law to environmental claims from what it termed 'neighborhood plaintiffs', DIY enthusiasts, and members of the public.[25] Its reaction was to remind members that their first goal was 'to advise the public that asbestos cement products, in place and under normal use, present no known health risk to the user or to the general public'.[26] For product users, leaflets describing recommended codes of practice were to be distributed—but not too widely as one member, National Gypsum, felt that the contents would 'scare [the] hell out of everybody'.[27]

Nevertheless, Johns-Manville dutifully noted in its annual report for 1973 that it was working diligently to comply with environmental working conditions. Studies soon appeared that measured fibres in the ambient air and in potable water, apparently showing no risk from either. Like the ARC, Johns-Manville issued leaflets with comforting titles such as 'Asbestos—No Risk to Public'. In 1969, there was a much-publicized partnership with Selikoff, after Johns-Manville provided funding for an Industrial Insulation Health Research Program at Mount Sinai. It proved a disappointment for both parties, though internal documents show that the company did not become involved purely for medical reasons. A memo noted that the publicity was intended to downplay the asbestos risk, by 'making sure all statements for participants point out that the problem is *occupational* and does not involve a public risk' [emphasis in original].[28]

The Mount Sinai physicians were not so sure. In 1971, an episode occurred that typified the dilemmas that firms such as Johns-Manville faced and also explains why an *entente cordiale* between Mount Sinai physicians and the asbestos industry was doomed. Selikoff had become concerned at the import of women's coats containing 8 per cent chrysotile, which enabled the coats to be imported as asbestos products at a lower import duty. Dust counts taken after brushing the coats exceeded occupational safety levels. Swetonic told the importer that the industry would not support him; but Selikoff did more. He wrote warning letters to various federal agencies and then responded to the press interest. According to Swetonic, Selikoff cooked up a masterful piece of theatre for the media, with scientists in moon suits collecting dust as it was brushed

[25] Comments of H.M. Ball, Health & Safety Council/ACPA, Annual Meeting, 21 November 1969.

[26] Memo to the principals of Health & Safety Council/ACPA, 14 December 1970.

[27] Health & Safety Council/ACPA, 19 June 1968.

[28] Memo from C.M. Caton to W. Russell, re. 'Concealment Defense', 22 May 1986.

from a female mannequin. The event showed either Selikoff's sense of public duty or his taste for self-aggrandizement, depending on one's viewpoint. But there was no doubting the seriousness of such episodes for the industry. As Swetonic remarked: '[S]hould the American public come to accept the proposition that a certain coat in their closet is capable of giving them cancer, then our job of convincing people that asbestos-containing brake linings or roofing shingles are perfectly safe will become all that much harder, if not impossible'.[29]

After 1970, Johns-Manville tried to tackle environmental issues through its lobby group the AIA/NA. A watchful eye was kept on the EPA and OSHA plans to regulate asbestos; and support was provided wherever asbestos came under fire, as in the great set-piece battle over the Reserve Mining issue about waste discharges of taconite (and asbestos) into Lake Superior near Duluth, Minnesota.[30] Part of the funding that QAMA directed towards McGill University was to fund research into possible environmental ARDs in some of the Canadian asbestos mining communities. In Europe, the AIA monitored closely EEC directives for the environmental control of asbestos.

Johns-Manville's focus on these issues was sharpened by events at 'home'. According to Swetonic:

> We had all been dreading the inevitable discovery of the first 'neighborhood' or 'household' case of mesothelioma in the United States. When it did happen, the shock was incredible because... the victim was the wife of Cliff Sheckler.... [They] had never lived close to any Manville plant during his long career with the company... [and]... she had developed the disease from inhaling the asbestos dust he brought home on his clothes from work. More than any other development, the death of Sheckler's wife forced the entire industry to face up to the realities of the asbestos threat, something it had seemed reluctant to do previously.[31]

Ed Fenner was another mesothelioma casualty in 1984. This was not the only company to be affected. Several Cape Asbestos directors who were involved in the company's American operations also died of mesothelioma: they included Rudolf Wild in 1951, and also his daughter who was probably exposed to dust from her father's clothes or work samples. Tom Hale, a technical director, died of mesothelioma in 1961. So, too, did Cape's North American salesman, Bob Cryor, in 1970. Richard Gaze,

[29] Swetonic to AIA/NA, 18 June 1971.

[30] In 2007, these concerns were raised again when it was reported that thirty-five taconite miners in the region had died from mesothelioma between 1997 and 2005.

[31] M.M. Swetonic, 'Death of the Asbestos Industry', in J. Gottschalk (ed.), *Crisis Response* (1993), pp. 289–308.

who orchestrated Cape's North American operations, died from peritoneal mesothelioma in 1982.[32] Bystander victims in South Africa included Dr. Sluis-Cremer, who worked at the NCOH, and often visited the mines; a radiographer who also worked at the NCOH; and R.E.G. Rendall, who supervised baboon experiments. In Belgium, the Eternit director van der Rest and the company physician Dr Lepoutre both died of mesothelioma.[33] At James Hardie, a number of managers, safety officers, and personnel directors died from the disease.[34]

Whether these deaths, which were not publicized, really did force the industry to face up to reality is a moot point: Swetonic may be right when he says that they were a shock, but it is more likely that it was the launch in 1970 of the EPA, the Clean Air Act, and OSHA that forced the industry to react. In any case, the industry showed great reluctance to reform its ways.

Towards an Environmental Epidemic

In Britain, despite the Newhouse-Thompson study and the 1969 government regulations (which had accepted in principle a risk outside the factory), the asbestos industry and the government continued to focus on the industry itself and there was an evident reluctance to believe that asbestos could be a hazard to bystanders or that a new wave of environmental disease threatened. However, trade unionists and victim's action groups believed differently. By the mid-1970s, there was a groundswell of opposition to asbestos from the grassroots of the trade union movement, one result of which was Alan Dalton's activists' handbook, *Asbestos Killer Dust* (1979).[35] Dalton and his fellow trade unionists emphasized particularly the deteriorating asbestos within Britain's public buildings, such as hospitals—where blue asbestos lagging on pipes was crumbling—and on housing estates. This was dismissed as scaremongering by local authorities, but they and the asbestos industry soon had to contend with another critic, who in the early 1970s became active in championing the cause of asbestos victims.

[32] Information from death certificates, legal documents, personal information, and Paul Brodeur, 'Annals of Industry: Casualties of the Workplace. III Some Conflicts of Interest', *New Yorker*, 12 November 1973.

[33] Personal communication to G. Tweedale from Salvator Nay, 27 September 2003.

[34] G. Haigh, *Asbestos House: The Secret History of James Hardie Industries* (2006), p. 397.

[35] A. Dalton, *Asbestos Killer Dust* (1979).

Mrs Nancy Tait's husband had died of mesothelioma after exposure to asbestos as a telephone engineer. Not surprisingly, as her antagonism to asbestos began to grow, she showed great sensitivity to the dangers of transient exposure to asbestos and to the fact that, as she believed, almost anyone exposed was potentially at risk. Mrs Tait highlighted the fact that official statistics were now recording mesothelioma deaths among building workers, electricians, motor mechanics, tilers, and refuse workers—'occupations not normally investigated to ascertain the incidence of asbestos disease'.[36] She pondered whether the use of asbestos should be recorded in order to protect future workers, especially in the building trades which she regarded as a special risk. Tait felt that ARDs would become much more widespread than either the government or industry was suggesting.

In fact, T&N's managers exchanged letters in the late 1960s that warned of asbestosis and mesothelioma in a 'diverse character of occupations', such as porters, plumbers, joiners, and housewives.[37] But these files did not reach the public domain until the litigation of the 1990s. Tait had to find her own evidence. By 1981, local doctors helped her profile twenty-seven mesothelioma patients at a small East London hospital. The results were striking. Most of these patients had identifiable asbestos exposure, but far more disturbing was the fact that 'no patient has reported employment with an asbestos manufacturing company'.[38] Instead, mesothelioma appeared to be a particular risk for plumbers, joiners, and dockers.

The implications of this were worrying. It meant that the thrust of government regulation was based on a false premise that ARDs were only a problem for those in the asbestos industry. It also had profound implications for any projections of future ARDs—projections that were now being debated amongst epidemiologists. In 1982, Richard Peto (brother of Julian Peto, the asbestos epidemiologist) had entered into this debate with a projection of 50,000 ARD deaths over the next thirty years. This was bad enough, but Tait criticized the projection as too low, based as it was on industry-generated data. She emphasized the number of cases that she and her colleagues were seeing with no link to the asbestos industry—many with slight exposure to chrysotile.[39] David Gee, a leading health and safety officer for the trade unions, warned that the Peto projections could

[36] N. Tait, *Asbestos Kills* (1976), p. 11.

[37] A.B. Boath to H.C. Lewinsohn, 5 September 1967.

[38] N. Tait (with Dr V.J. Harvey), Presentation to Anglo/French Social Medicine Conference, Winchester, 1983. Typescript.

[39] N. Tait, 'The Dangers of Asbestos' (letter), *New Statesman*, 24 September 1982.

seriously underestimate the risk. Drawing upon the insights of Selikoff and his New York group, Gee predicted that over the next thirty years the dead would number nearer 70,000 and that the 'bulk of the deaths will occur in the shipyard and construction workers'.[40]

This was not what the industry wanted to hear, not least because it cast doubt on the work of their consultants, Sir Richard Doll and Julian Peto. In October 1982, T&N complained to them about Tait's 'uncompromising and extremist stance' and invited their views.[41] Doll and Peto were sceptical about a future asbestos epidemic and downplayed the 50,000 projection. Tait, they felt, was guilty of distortion. Peto believed Tait's information was 'biased and . . . worthless'. He felt that Tait and her organization had its uses, but that they also exaggerated the risks, were political, and were 'bound to be criticised by independent scientists'.[42] These private comments were to prove useful to the industry in attempting to discredit Tait, though the government was hardly inclined to listen to her anyway. It was already under a welter of public criticism precipitated by the television documentary, 'Alice—A Fight for Life', which had emphasized the environmental risks of asbestos and also criticized the government for its failure to implement the recommendations of a recent lengthy public enquiry. The government's reaction in September 1982 was to ask Doll and Peto to write yet another report on the asbestos problem.

The Doll and Peto review was published in 1985 in a sixty-one-page paperback under government imprint.[43] It was a sober academic evaluation of the epidemiological work conducted to date, most of which derived from industry studies. Mrs Tait had pleaded for some consideration of the risks outside the factory setting,[44] while David Gee had remarked that the need for an epidemiological study in the construction industry was 'overwhelming'.[45] But these areas were specifically excluded from the report. Doll simply refused to believe that there was any evidence that 'chance unsuspected' exposure existed—a strange argument to use with Nancy Tait, whose husband had died from exactly such exposure.[46] No alarming projections were advanced, though Doll and Peto suggested that the risk from indirect asbestos exposure in buildings amounted to no

[40] 'Asbestos Deaths' (letter), *New Statesman*, 1 October 1982.
[41] S. Marks to Doll, 13 October 1982.
[42] Doll to Marks, 15 October 1982; and Peto to S. Marks, 19 October 1982.
[43] R. Doll and J. Peto, *Asbestos: Effects on Health of Exposure to Asbestos* (1985).
[44] Tait to Doll, 30 April 1983.
[45] David Gee, 'Dangers of Asbestos' (letter), *Times*, 17 May 1984.
[46] R. Doll, 'Asbestos Dangers' (letter), *Times*, 25 May 1985, in reply to Tait letter, *Times*, 20 May 1985.

more than a death a year for the whole country (a reassuring calculation that was soon cited in the press). Nothing was said about building and maintenance workers; and only three paragraphs in the sixty-four-page report discussed 'other sectors of the asbestos industry', such as insulation. At a press launch, the HSE presented the report as something that would help everyone 'continue to live with asbestos'.[47]

Abategate: The Controversy over Asbestos in Buildings

Britain was not the only country where anxiety was expressed about environmental risks. In 1974, alarm about asbestos contamination in buildings had surfaced at Jussieu University in Paris. Built in the mid-1960s, on the Left Bank of Paris, Jussieu contained the biggest concentration of sprayed asbestos in Europe. As French journalist François Mayle has remarked, it was a 'veritable mine' of asbestos right in the heart of the city.[48] By the mid-1970s, a group of research staff led by Henri Pézerat, a toxicologist, and trade unionists had formed a syndicate to demand measures to deal with Jussieu's asbestos.[49] Soon they linked their efforts with disaffected workers at an asbestos textile factory (Amisol) in Clermont-Ferrand. This was an important step towards the 'social movement' that would eventually see asbestos banned in France.[50] By 1977, these groups had pushed the government towards a ban on sprayed asbestos and the recognition of mesothelioma as an occupational disease. But resistance was encountered from the industry's lobby group, the CPA. When a government-industry body was set up, its report on Jussieu and buildings contamination was reassuring. By 1980, government funds for decontaminating the site had dried up and, as in Britain, asbestos dropped from the headlines.

As the issue became buried in Britain and France, however, it flared up with renewed intensity in the US between the mid-1980s and early 1990s. The context was set by a number of factors: medical reports of ARDs

[47] HSE News Release, 24 April 1985. See also G. Tweedale, 'What You See Depends on Where You Sit: The Rochdale Asbestos Cancer Studies and the Politics of Epidemiology', *IJOEH* 13 (January/March 2007), pp. 70–9.

[48] F. Malye, *Amiante: Le Dossier de L'Air Contaminé* (1996), p. 36. See also R. Lenglet, *L'Affaire de L'Amiante* (1996); E. Henry, *Amiante: Un Scandal Improbable. Sociologie d'un Problème Public* (2007).

[49] Collectif Intersyndical Securité des Universites-Jussieu CFDT, CGT, FEN, *Danger! Amiante* (1977).

[50] A. Thébaud-Mony, 'Justice for Asbestos Victims and the Politics of Compensation: The French Experience', *IJOEH* 9 (July/September 2003), pp. 280–6.

(including mesotheliomas) from environmental exposures[51]; the finding of asbestos bodies in the lungs of city dwellers; and an awareness of the crumbling asbestos in older buildings. Particularly significant was a growing environmental movement and increasing concern over the impact of industrial pollutants and chemicals. This followed the publication of Rachel Carson's *Silent Spring* (1962), which targeted the problems caused by pesticides and heightened public fears about cancer. In America, these concerns were translated into a growing number of government agencies and environmental regulations. Chief amongst these were the EPA, the Clean Air Act, and the OSHA—all launched in 1970.

A sign of the way the wind was blowing was the EPA's decision in 1973 to ban the spraying of asbestos in buildings. Selikoff had been a prominent critic of this notoriously dusty process, especially its use in the construction of the World Trade Center. In 1969, Selikoff had highlighted the severe asbestos contamination experienced not only by the sprayers but also by those in the surrounding environment.[52] Even the industry had to agree. Swetonic has left us with a striking account of the occasion when a ban became inevitable. He had attended a meeting in New York in an office which overlooked asbestos sprayers at work on the construction of the W.R. Grace building. The 'virtual snow shower' of fibre that was hitting the sidewalk convinced Swetonic that Johns-Manville would have to 'take a walk on those people' and abandon its support for spraying.[53]

But what about asbestos already inside buildings? According to Johns-Manville, there was 'no evidence of danger in the use of asbestos in schools or other public buildings'.[54] But by the end of the 1970s, the government was forming a different view. The EPA was particularly concerned with the airborne asbestos risk to children, after reports of flaking asbestos inside schoolrooms and other buildings. At Yale, sprayed asbestos in one faculty building had to be removed because dust levels

[51] N.J. Vianna and A.K. Polan, 'Non-Occupational Exposure to Asbestos and Malignant Mesothelioma in Females', *The Lancet*, i, 20 May 1978, pp. 1061–3. This New York study showed that women run the risk of mesothelioma 'if they are indirectly exposed to environmental asbestos'. See also K.H. Kilburn, R. Warshaw, and J.C. Thornton, 'Asbestos Diseases and Pulmonary Symptoms and Signs in Shipyard Workers and Their Families in Los Angeles', *Archives of Internal Medicine* 146 (November 1986), pp. 2213–20, which found asbestosis rates of 11% in wives, 8% in sons, and 2% in daughters (64% of the male sample also had the disease).

[52] W.B. Reitze et al., 'Application of Sprayed Inorganic Fiber Containing Asbestos: Occupational Health Hazards', *American Industrial Hygiene Association Journal* 33 (March 1972), pp. 178–91.

[53] Swetonic, 'Death of Asbestos', pp. 301–2.

[54] J-M, *Annual Report* (1978), p. 28.

were above industry levels.[55] The EPA's privately commissioned reports showed that 'asbestos levels in certain buildings [sprayed with asbestos] are as high as those measured in the neighborhood of asbestos plants where mesotheliomas have been reported'.[56] In 1980, the EPA stated that 8,500 schools contained friable asbestos, placing three million pupils at risk (aside from staff), and raising the spectre of thousands of premature deaths. Between 1984 and 1986, President Reagan had signed an Abatement Act and the Asbestos Hazard Emergency Response Act (AHERA). The aim was to inspect school properties for friable asbestos and introduce management plans for problem areas, with the option to remove ('abate') asbestos if it was judged a health hazard.[57] As the programme was rolled out, up to $100 million a year was appropriated and a new industry was born—asbestos removal—which rapidly flourished as abatement gathered pace.

The implications of this were enormous, not only for schools (where thousands of abatements might be needed), but also for hundreds of thousands of commercial buildings throughout America, many of which contained friable asbestos. If EPA plans extended into the commercial sector, as seemed likely, then costs would be astronomic. One industry analyst estimated that property owners could spend $100 billion for asbestos removal over the next twenty-five years. In 1989, for example, New York City estimated that 68 per cent of all its buildings contained asbestos and that 81 per cent of that asbestos was in such poor condition that it was a public health risk. Since AHERA did not provide funds for abatement, some schools launched legal actions against the asbestos companies, thus ensuring that during the late 1980s asbestos-in-buildings litigation became another growth industry.

Not surprisingly, EPA actions were soon shadowed by those with a vested interest in the properties under discussion—notably a group known as the Safe Buildings Alliance (SBA), which had been formed in 1984. Internal asbestos industry documents show that the SBA was organized by Hill & Knowlton for those companies—Celotex, US Gypsum, and W.R. Grace—with asbestos liabilities in the construction industry. Ostensibly set up to advise property owners on how to deal with asbestos,

[55] R.N. Sawyer, 'Asbestos Exposure in a Yale Building: Analysis and Resolution', *Environmental Research* 13 (1977), pp. 146–69.

[56] P. Sebastien et al., 'Measurement of Asbestos Air Pollution Inside Buildings Sprayed with Asbestos' (EPA 560/13-80-026, August 1980), p. 29.

[57] C. Harvey and M. Rollinson, *Asbestos in Schools: A Guide for School Administrators, Teachers and Parents* (1987).

the main interest of the SBA was in public relations and finding ways of discrediting and opposing government legislation.[58] This was done effectively, when in 1989 the SBA and several real estate associations sponsored a Symposium at Harvard University to discuss the health impact of asbestos in buildings.[59] It was a select audience and many of those attending the conference were representatives of realtors, asbestos companies, or from university departments with links to the Canadian mining industry. They judged the EPA's strategy ill-advised. Instead, the Harvard Symposium publicized the view that the danger of asbestos in buildings was negligible, made much of the fact that most of the asbestos used in construction was chrysotile (and therefore innocuous), and vigorously opposed government plans to remove the material in schools, and by implication, commercial properties.

In America, this debate—centred on the highly sensitive issue of children's health—suddenly became national news. Similar 'scares' about asbestos in schools hit the headlines in Japan and Canada, but the scale of media reporting in America was far greater. As public concern mounted, so too did opposition to government measures. Michael Fumento's *The Asbestos Rip-Off* attempted to debunk the idea that 'abategate' was necessary and asserted that extortionate profits were being made by the asbestos removal business.[60] *USA Today* likened asbestos to a sleeping dog: best left alone, because if it was hammered or sawed it could get up and bite.[61]

The debate hit the academic press, where the Harvard group got their blow in first. At the heart of the opposition to abatement were pathologist Brooke Mossman, physician Bernard Gee, and engineer Morton Corn.[62] Corbett McDonald and his colleagues at McGill University were supportive, as were many academics in the UK. In 1989, an article duly appeared in the *New England Journal of Medicine* (*NEJM*) and another followed shortly afterwards in the popular journal *Science*, both warning against any rash removal policy. The 'asbestos panic must be curtailed', opined the authors: it would release unnecessary fibre and expose more workers and bystanders. This was a valid concern, but others were more questionable, such as the suggestion that only amphiboles were a problem,

[58] Safe Buildings Alliance, *Asbestos in Buildings: What Owners and Managers Should Know* (1989).

[59] Harvard University Energy & Environmental Policy Center, 'Summary of Symposium on Health Aspects of Exposure to Asbestos in Buildings', August 1989.

[60] M. Fumento, 'The Asbestos Rip-Off', *American Spectator* (October 1989).

[61] Quoted in Fumento, 'Rip-Off'. [62] Castleman, *Asbestos*, pp. 822–7.

not chrysotile.[63] Nevertheless, Philip Abelson, the editor of *Science*, wrote an editorial supporting the authors. Abelson lamented how 'the asbestos removal fiasco' could cost as much as $150 billion and registered dismay that chrysotile had become 'tarred' by a mistaken association with crocidolite.[64]

Both articles generated enormous publicity and, in turn, a spate of magazine and newspaper comment. The tabloid *The National Examiner* announced its own interpretation of the findings under the headline: 'Asbestos Is Safe'. The *NEJM* and *Science* articles also drew heavy criticism, not only because of the views expressed, but because it transpired that the authors had undisclosed asbestos industry consultancies or had served as expert defence witnesses. One critic of the *NEJM* article argued that it subtly presented the views of that industry and he disclosed that an attorney for an asbestos company had presented him with an advance copy! The critic asked that the authors disclose their interests.[65] Mossman and Gee took offence at this suggestion, asserting that they had disclosed their asbestos affiliations prior to publication in line with *NEJM* policy that 'conflicts of interest should always be disclosed'. This was confirmed by the *NEJM* editors, who then mysteriously told the journal's readers: 'We kept [the authors'] disclosure [of asbestos consultancies] on file, but chose not to publish it.'[66] One consequence of this debacle was a major review of the *NEJM*'s policy towards the publication of papers by authors who had 'interests'.

Critics of the Harvard Symposium and the SBA eventually coalesced around Selikoff and his Mount Sinai Hospital colleagues. In 1990, under the auspices of the Collegium Ramazzini and with the support of organized labour and plaintiffs' lawyers, they organized a three-day conference in New York as a counterpoint to the Harvard meeting.[67] It was a more ambitious affair than its predecessor: the number of speakers was greater and the proceedings, like the great 1964 New York conference, filled a complete volume of the *Annals* of the New York Academy of Sciences. The scientists who attended had a different view on the risk of asbestos in buildings, arguing that it was all too real and that the

[63] B.T. Mossman and J.B.L. Gee, 'Asbestos-Related Diseases', *NEJM* 320 (29 June 1989), pp. 1721–30; B.T. Mossman et al., 'Asbestos: Scientific Developments and Implications for Public Policy' *Science* 247 (19 January 1990), pp. 294–301.

[64] P.H. Abelson, 'The Asbestos Removal Fiasco,' *Science* 247 (2 March 1990), p. 1017.

[65] O. Eliasson, letter to editor, *NEJM* 322 (11 January 1990), p. 130.

[66] Editor's note, *NEJM* 322 (11 January 1990), p. 131.

[67] P.J. Landrigan and H. Kazemi (eds.), 'The Third Wave of Asbestos Disease: Exposure to Asbestos in Place. Public Health Control', *ANYAS 643* (1991), pp. 1–628.

exposure of schoolchildren, teachers, maintenance staff, and construction workers could result in a 'Third Wave' of ARDs.[68] Although the conference concluded that asbestos should only be removed when necessary and with care, on almost every other point they took issue with the Harvard group. The New York conference presented evidence that low-level exposures were a potential source of ARDs; that asbestos in buildings could be a constant source of uncontrolled exposures; and that chrysotile, although less dangerous than amphiboles, was a significant factor in ARDs, including mesothelioma.

The two conferences increased animosity between the two camps. A McGill colleague of McDonald, Dr Bruce Case, who attended the Third Wave conference, wrote a furious letter to Bernard Gee that lambasted Selikoff and described the proceedings as a piece of Broadway theatre.[69] A copy of this letter stirred up Julian Peto, who felt moved to write to the chief of Mount Sinai Hospital complaining that most of the papers had been scientifically 'dubious' and would tip the debate in what Peto described as the 'wrong way'.[70] With such influential friends, it appeared that the Harvard group had the upper hand. In 1991, a Washington journalist Michael J. Bennett, who had links with the Center for the Defense of Free Enterprise, launched a full-scale attack on Selikoff in his book *The Asbestos Racket.*[71] Labelling the whole environmental campaign as 'asbestos hysteria', Bennett blamed it for the Challenger Space Shuttle disaster and characterized Selikoff as Chicken Little (the fairy tale character, who runs round warning everyone that the sky is falling). Morton Corn endorsed the book as 'required reading'. In an attempt to settle the debate, the EPA set up an expert panel under the auspices of the Health Effects Institute (HEI) (a body funded by the government and industrial interests). The HEI panel—many of whom, as Selikoff pointed out, were defence witnesses in asbestos property damage litigation—largely confirmed the view that the risk from non-occupational exposure to asbestos in buildings was slight and that removal posed a greater risk.[72]

[68] The first wave of ARDs occurred amongst workers in the primary manufacturing and mining of asbestos; the second, in those engaged in the use of asbestos products, such as insulation. See M. Huncharek, 'Changing Risk Groups for Malignant Mesothelioma', *Cancer* 69 (1 June 1992), pp. 2704–11.

[69] Case to Gee, 14 June 1990.

[70] Peto to Thomas Chalmers, President and Dean, Mt Sinai Medical Center, 28 June 1990.

[71] M.J. Bennett, *The Asbestos Racket: An Environmental Parable* (1991).

[72] Health Effects Institute, *Asbestos in Commercial Buildings: A Literature Review and Synthesis of Current Knowledge* (1991).

Some members of the HEI panel pressed their case further, when in the fall of 1993 New York City authorities ordered the closure of schools until an emergency asbestos inspection had been conducted. The scientists (which included Mossman, Corn, and Gee, besides, *inter alia*, Robert Murray, Molly Newhouse, and Christopher Wagner from the UK) wrote to the *New York Times*, deploring the asbestos 'scare' in the city. Their letter concluded thus:

> Rational voices must be heard. Interviews with individuals who pioneered this field should be broadcast to the public, not only in New York, but nationwide, to understand the conditions under which asbestos-related cancers arose historically. If this were done, the public would appreciate that those conditions of the past could never exist in classrooms today. More importantly, calm would prevail. Science, and not unreasonable emotion, should guide both the administrative and public response.[73]

The tide began to turn. In 1990, the EPA suddenly agreed that millions had been wasted on unnecessary asbestos removal and assured property owners that all that might be needed was an active in-place management programme. Maintenance and containment became the watchwords, not removal, and Congressional asbestos laws for commercial and public buildings were quietly shelved. The spectre of property litigation receded, too. Hundreds of actions had been launched by school authorities against manufacturers, such as US Gypsum and W.R. Grace, but although there were a few settlements the plaintiffs found the cases expensive and difficult to win. The number of property damage actions outside the schools arena soon waned, especially after 1995 when even the mighty Chase Manhattan Bank was humbled in the courts when it tried to pin the costs for removing sprayed asbestos from its Wall Street skyscraper on T&N. As 'abategate' abated, the asbestos companies, insurers, and real estate owners avoided severe financial damage. Academics who regarded asbestos removal as a 'nefarious activity' were now able to write off the whole episode as a public policy debacle.[74] The SBA felt its campaign had been a great success. As one of its representatives stated: 'Simply put, the Safe Buildings Alliance has altered the way in which the public, news media, legislators, and regulators view the issue of asbestos in buildings

[73] A. Churg et al., 'Call on the Asbestos Scare', *New York Times*, 23 September 1993.

[74] M. Ross, 'The Schoolroom Asbestos Abatement Program: A Public Policy Debacle', *Environmental Geology* 26 (October 1995), pp. 182–8.

today; and that view is a far cry from the views that were held just five short years ago.'[75] Calm prevailed.

Asbestos in Buildings: The Problem that Refused to Go Away

The asbestos-in-schools controversy shows the involvement of a familiar range of economic and political 'interests'. In the final shake-out, the message for the public was reassuring. The asbestos dog was best left alone. This view has recently been restated at length by Jacqueline Corn, in the first historical account of what she terms the asbestos-in-schools 'fiasco'.[76] Whilst accepting no blame for the actions of the asbestos companies involved (one of whom, W.R. Grace, sponsored her study), Corn criticizes the media, the EPA, and the 'rhetoric' of Mount Sinai physicians for drowning out the voice of those sober scientists (amongst whom was her husband, Morton Corn) who warned against asbestos removal.

The EPA, however, is not so easily dismissed. Certainly it was legitimate to question the wisdom of badly executed asbestos removals, but the arguments of the EPA's critics were always contradictory: they stated that most asbestos in buildings was harmless chrysotile, yet simultaneously argued that it was dangerous to remove it. Amidst the publicity, the risks associated with maintenance work and the consequent disturbance of old asbestos had been glossed over. Most scientists had little appreciation of the dirty realities of building work and maintenance. Doll argued that there was 'a lot of loose thinking about the removal of existing asbestos which is often safest left where it is—but marked so that people know about it when the building is finally pulled down'.[77] Aside from the problem of marking all asbestos, this simplistic approach took no account of the fact that buildings are not static until they are demolished: owners change, constant maintenance is required, and a range of emergencies can occur (such as burst pipes and fires) that can expose workers to invisible and uncontrolled risk. It was known by the 1980s from several studies that, for example, sheet-metal workers involved in installing air ducts and heating systems suffered significant ARDs.[78] Some of this was likely

[75] R.J. Day to C.F.N. Hope (T&N), 26 October 1989.

[76] J.K. Corn, *Environmental Public Health Policy for Asbestos in Schools: Unintended Consequences* (1999).

[77] Doll to David Gee (General & Municipal Workers' Union), 23 July 1980.

[78] E. Drucker et al., 'Exposure of Sheet-Metal Workers to Asbestos During the Construction and Renovation of Commercial Buildings in New York City', *ANYAS* 502 (1987), pp. 230–44.

to have been from construction work before spray was banned in 1972; but exposure *continued* as these workers maintained and renovated the same buildings. Office plenums (the false ceilings often used as air ducts) were identified as especially hazardous, not only for construction workers, but also for office workers who inhaled the dust as it spread through the building. US Gypsum was well aware of this problem. In 1984, when it was organizing the SBA to oppose asbestos removal, it began relocating its offices because of the problems caused by asbestos removal in its own Chicago headquarters! USG was worried particularly about the spread of dust in the plenums.[79] The type of regulation proposed by the EPA was probably the only answer to this type of exposure.

There was another risk, too, that was hardly mentioned—except perhaps by Tait—the problem of asbestos in the home. This was a multi-layered problem, ranging from dust brought home on workers' clothes to asbestos in numerous household products and fittings. This represented a risk to trades people and also to occupiers of domestic properties. Dust exposure in such situations would obviously be slight, but could be endlessly repeated, so that those at home all day could be perpetually inhaling asbestos particles shed from walls and carpets.

The problem of old asbestos was to emerge again in both Britain and the US in episodes (not mentioned by Corn) that demonstrated that the situation was more complex than vested interests allowed. In Britain, evidence was accumulating about bystander risk. In 1982, the death from mesothelioma was reported of a London school buildings inspector (aged 41).[80] In the same year, Fred Lodge, who grew up in a house next to the Cape factory in Barking, and had never worked with asbestos, died of mesothelioma aged 39.[81] Tait also highlighted the case of Mrs Kuflewski, a 54-year-old office worker, who had died in 1983 from mesothelioma. Electron microscope analysis of the lungs showed significant amounts of amosite—the same material that had been sprayed onto the ceilings of her office. There was no other significant asbestos exposure.[82] Cases of asbestosis and mesothelioma were by now regularly being reported in the press. Typical was the fate of Winifred Wilson, 73, whose death

The authors presciently suggested a registry to document asbestos in buildings—an idea later taken up in Britain (see below).

[79] USG Memo, 26 September 1984.

[80] Paul Holt, discussing Asbestos Advisory Committee Enquiry.

[81] London Hazards Centre, *Rising from the Dust: Building a Support Network on Asbestos and Environmental Disease in Barking and Dagenham* (2004).

[82] R.C. Stein et al., 'Pleural Mesothelioma Resulting from Exposure to Amosite Asbestos in a Building', *Respiratory Medicine* 83 (1989), pp. 237–9.

from mesothelioma was the subject of an inquest in 1985. She had been exposed at home to asbestos on her husband's overalls, after he had finished work each day at the London docks. He had died of mesothelioma four years previously.[83] Tait remarked of the case: 'We are beginning to establish a pattern, I am sure, in which indirect exposure to asbestos is causing mesothelioma and therefore we should be worried about the effects of asbestos in the environment.'[84] More ominous in 1993 was the mesothelioma death in south London of a schoolteacher, 37-year-old Sarah Gibson, who the inquest decided had been exposed to asbestos in the classroom.[85] The school was closed and the local authority spent £50,000 surveying other schools.

Belatedly, some scientists began to change their perceptions. Julian Peto and his colleagues switched from looking specifically at factory workers to analysing mesothelioma death rates in the whole of the UK. With asbestos production in decline since the 1970s and with exposure in buildings supposedly negligible, there was an assumption amongst scientists and the government that mesothelioma deaths would soon peak and decline. Peto and his colleagues, in a classic study, found the opposite.[86] Not only was mesothelioma mortality rising (and projected to do so until at least 2020, when a peak was expected of about 3,000 men a year), but a striking percentage of workers were either plumbers, electricians, construction workers, or painters—even schoolteachers. In fact, a quarter of the deaths in the study were from building and maintenance work. Peto had been at the Harvard Symposium, one of the experts on the HEI panel, and, as his criticisms of Tait and Selikoff had shown, viewed Third Wave risks as negligible. The new research, published in 1995, gave a fresh perspective. Whereas Peto had once thought estimates of an epidemic of 50,000 deaths greatly overdrawn, his latest projections were at least three or four times higher. It was now clear that concentrating on factory cohorts had led epidemiologists to underestimate greatly the wider risks.

Not surprisingly, this article created widespread anxiety and resurrected the whole problem of asbestos in buildings. Asbestos was now dubbed the 'building workers' plague'.[87] The asbestos industry's critics no longer seemed so extreme, especially given the steady stream of bystander mesothelioma cases reported almost weekly in the press. This was when

[83] 'Asbestos in Clothes Causes Widow's Death', *The Guardian*, 23 February 1985.

[84] M. Horsnell, 'Wife Killed by Dust in Husband's Clothing', *Times*, 4 March 1985.

[85] S. Weale, 'School Asbestos Death Claim', *Guardian*, 12 February 1994.

[86] J. Peto et al., 'Continuing Increase in Mesothelioma Mortality in Britain,' *The Lancet* 345 (4 March 1995), pp. 535–9.

[87] London Hazards Centre, *The Asbestos Hazards Handbook* (1995).

the scandal of T&N's Roberts' factory made national news. This small factory in Leeds had closed in 1957, yet a local newspaper had linked numerous mesothelioma deaths in the 1980s and early 1990s with its activities. Victims included former workers, but also their families—contaminated by dust brought home on clothes—and many individuals who simply lived nearby. John Kennally did not work with asbestos or live near Roberts' factory: but his mother worked there for four years before its closure and used to place her coat over her son's bed in cold winter weather. After he died from mesothelioma in 1988 (aged 43), a solicitor John Pickering discovered ten similar cases linked to Roberts'. One victim, June Hancock (1936–1997), was exposed to asbestos as a child in the streets around the factory. She developed mesothelioma, which had also killed her mother. In 1995, June Hancock won the first environmental case in the courts against T&N.[88] Since then, there have been over fifty bystander cases from this factory alone, though there are many others from other T&N sites.

It was in France, however, where Peto's paper had the most dramatic impact. A decade of reassurances from the CPA had not made anxiety about the environmental risk disappear. In 1990, the IARC building in Lyon was evacuated for removal of asbestos and the problem of asbestos in buildings began attracting national attention. One catalyst was the cancer deaths of six teachers at a school sprayed with asbestos in Gerardmer (Vosges), which resulted in a group of widows launching legal actions for manslaughter. Another was a newly constituted Comité Anti-Amiante de Jussieu (CAAJ), which under Henri Pézerat became increasingly vocal in its warnings over 'l'air contaminé'. Already at Jussieu, there were twenty cases of 'maladies professionelles', including two mesotheliomas.[89] When the CAAJ organized a conference in Paris on 21 March 1995, it was conveniently timed: Peto's publication was only two weeks old and he agreed to address the conference.[90] Though Peto was sceptical about asbestos removal (he felt it was motivated by fear rather than evidence), his findings helped destroy the doctrine of 'controlled use' in France and supported the CAAJ's claim that asbestos could not be controlled in an environmental setting. The resultant national publicity—combined with fresh government reports and the formation of national victims' action groups, such as Association Nationale de Défence des Victimes de L'Amiante (ANDEVA) (see Chapter 6)—led the French to reject asbestos

[88] G. Tweedale, 'Management Strategies for Health: J.W. Roberts and the Armley Asbestos Tragedy, 1920–1958', *Journal of Industrial History* 2 (1999), pp. 72–95.

[89] Malye, *Amiante*, p. 130. [90] Lenglet, *L'Affaire*, pp. 112–13.

in 1997. Meanwhile, the government pledged £141 million to launch the world's biggest asbestos decontamination project at Jussieu.

British government reaction was tardier. Chrysotile was not banned until 1999 and not until 2002 did the government update its workplace regulations to compel property owners to manage asbestos *in situ* and provide workers with an asbestos register for their property. This was alongside a publicity campaign that emphasized that some one and a half million British workplaces contained asbestos and that 5,000 lives might be saved by the regulations. As in America, however, these modest proposals, which made no mention of the millions of households contaminated with asbestos, provoked a vigorous response from British property interests. In 2002, the Conservative Party opposed the measures in Parliament, while government regulations were also denounced by the conservative national newspaper, the *Daily Telegraph*. The journalist who led the attack was Christopher Booker, who wrote a steady stream of articles that alleged that most asbestos in buildings was chrysotile—a mineral Booker believed was as safe as talcum powder.[91] Booker claimed the backing of experts. These included a/c businessman John Bridle, whose standing was severely damaged after a BBC Radio exposé cast serious doubt on his professorial qualifications.

The Disaster in Libby

One of the most graphic illustrations of the problems caused by environmental exposure to asbestos is the Libby scandal. This affair, not mentioned by Corn, has all the hallmarks of the genre: corporate culpability, medical collusion (and occasional self-sacrifice), ineffectual government officials, victims aplenty, and the media spotlight.[92] However, it was unusual in one respect; one does not expect a full-blown asbestos scandal and America's greatest man-made environmental disaster to erupt a century after asbestosis was first identified.

The story began in the 1920s, when an obscure Zonolite mine began production in the picturesque Cabinet Mountains at a town named Libby

[91] See, for example, C. Booker, 'Unnecessary Asbestos Bill Will Top £8bn', *Daily Telegraph*, 27 January 2002. The analogy was unfortunate, given that talc can contain asbestos.

[92] A. Schneider and D. McCumber, *An Air that Kills* (2004); A. Peacock, *Libby, Montana: Asbestos and the Deadly Silence of an American Corporation* (2003); M. Bowker, *Fatal Deception* (2003).

in Montana. Zonolite is a trade name for vermiculite, a clay mineral that because of its water content can be heated and 'popped' into a feather light material that can be used to insulate, fireproof, expand, and aerate products, both in industry and the home. It was widely used in lofts and for potting soil and garden fertilizers. By the 1940s, Libby was the main source of American vermiculite. The mine was small, with only a few score workers originally, but it was profitable enough to attract one of America's largest companies, W.R. Grace, which acquired the mine in 1963. Grace was soon supplying nearly 70 per cent of the western world's vermiculite.

Two factors paved the way to a health disaster. First, although vermiculite itself is relatively harmless, Libby deposits contained a dangerous type of asbestos—tremolite. Libby's vermiculite, both at the mine and when it was bagged for sale, was contaminated with this deadly fibre (as high as 26 per cent in the ore). Second, this situation was made worse by the nature of the product and its manufacture. As an unwanted 'tramp' material, most of the tremolite was removed in the milling process and expelled in tons of tremolite-laden dust over the town. In the mid-1960s, Grace noted how it coated downtown Libby. The mill and mine conditions were so bad that workers took asbestos-laden dirt home on their clothes and boots.

But the chain of environmental contamination did not stop there, as not all the tremolite was removed. To avoid shipping bulky bags of Zonolite, Grace preferred to ship the semi-processed ore for 'popping' at about fifty US expansion plants (mostly part Zonolite owned). Thus the ore was shipped to thirty-eight states and five Canadian provinces; then billions of pounds of bagged vermiculite were shipped around the world. Grace alone is known to have shipped to 750 locations. It then found its way into numerous industrial applications, such as Grace's sprayed Monokote. It was Libby asbestos that was released into the Manhattan air on 9/11, since the product had been used in the World Trade Center. More ominously, vermiculite and its deadly cargo were taken right into the home, when the bags were emptied into loft spaces, walls, greenhouses, and even cat litters. The work was done by unsuspecting tradespeople and householders.

The state government in Montana had inspected the mill since the late 1940s and was well aware of its severe dust problems and consequent dangers. State reports were copied to the Zonolite company, but they remained confidential—as did a company medical survey in 1961 that showed that 82 out of 130 workers had lung disease. That included a death from asbestosis. Soon after acquiring Libby in 1963, Grace knew the hazards of tremolite and had even been warned by its insurer. In 1968, state hygienists urged the Libby plant manager to exercise 'continued vigilance', after they had found that 'whatever tremolite is present in the ore is ending up in the air'.[93] Grace's attitude to such suggestions has been characterized as one of 'mind-boggling cynicism'.[94] In 1972, a Grace 'asbestos update' recognized that tremolite was identical to that 'bad actor' crocidolite and recommended 'building up a bank of knowledge'. But this was not to protect the workers, but 'to show that Libby has no significant problems'.[95] As a start, Grace funded experiments at a New Jersey university on hamsters, which by the late 1970s would confirm that tremolite induced fibrosis and mesothelioma. Significantly, Grace recognized that 'small amounts [of tremolite] are carried to expanding plants and ultimately into finished products'.[96] Tremolite was also carried far afield in sprayed Monokote, though communications to outsiders denied this.[97]

From its own medical examinations, which had begun in Libby in 1959, Grace knew that lung disease was inevitable for long-term workers at its plant. For example, a list of abnormal X-rays compiled in 1976 by Harry Eschenbach, Grace's medical director in Boston, showed about a fifth of the staff with X-ray changes consistent with asbestos exposure (with some already dead from asbestosis and cancer).[98] When workers' wives began to develop asbestosis, a local doctor, Richard Irons, publicized the cases and wrote to Selikoff. But his attempts to interest Grace in taking more vigorous action were unfruitful and Irons soon left Libby to its fate. Bad news was also being posted from Canada, from where one ore processor told Grace in 1968 of a case of asbestosis in Winnipeg. He warned: '[I]t won't take many more biopsy reports before we get fingered.'[99]

By the mid-1980s, alongside the first bystander cases, litigation began.[100] A corporate presentation summarized the bad news for Grace:

[93] J.R. Lynch (Montana State Dept of Health) to B.F. Wake, 17 October 1968; B.F. Wake to E.D. Lovick (Zonolite Co), 27 November 1968.

[94] Schneider and McCumber, *Air that Kills*, p. 183.

[95] H.A. Eschenbach to R.M. Vining, 5 June 1972.

[96] E.S. Wood to C. Brookes and C.N. Graf, 24 May 1977.

[97] B.R. Williams to E.S. Wood, 31 March 1977.

[98] H. Eschenbach, 'Libby: 1976 Abnormal Chest X-Rays'.

[99] www.cbc.ca/national/news/deadly_dust

[100] The first state compensation case had been quietly settled by Grace's insurers in 1965 to avoid publicity. The insurer had warned the company that it might be guilty of 'wilful and wanton conduct'. See Peacock, *Libby*, pp. 81–2.

3,604 personal-injury lawsuits (both from Libby and vermiculite processors and users), over a hundred property damage suits, and disputes with insurers. The cost by the end of 1986 was $21 million. For Grace the main aspect of the 'asbestos management problem' was not dealing with the suffering of individuals, but 'the handling of litigation and related settlement deliberations... [and]... attempting to convince property owners from removing asbestos-containing materials, including influencing regulatory and legislative actions at the Federal, State and Municipal Level'.[101] Grace was not yet ready to get out of the business, especially while much of the situation remained hidden. Workers were not informed of the results of medical examinations, and Grace sent no warnings about the risks to either workers or users of vermiculite. Zonolite trade leaflets famously proclaimed that the product was 'non-irritant'. Grace's company documents and medical experiments were confidential. A memo about the hamster experiments to J. Peter Grace (the company's CEO) contained a familiar industry proviso that, 'any publication of the findings will be made only by mutual agreement'.[102] Not surprisingly, these results of the carcinogenicity of Libby fibre were never reported. Another idea was that Grace's lawyers should commission research, so that any 'erroneous' or 'misleading' results could be kept confidential within the attorney/client relationship.[103]

To be sure, epidemiological studies investigating the problems at the Libby mine appeared in the medical literature after the mid-1980s.[104] Some of the studies were commissioned by Grace, who recruited Corbett McDonald and the McGill scientists. Their findings showed excess death rates and mesotheliomas (though a 'Vermiculite Operations Fact Book', distributed by Grace in 1985, insisted that this was partly due to smoking and that current dust levels could not possibly harm workers). Such reports would not be read by workers, but they should have prompted official action. So too should events in 1978 at the O.M. Scott plant at Marysville, Ohio, where lawn-product workers developed such serious pulmonary problems that NIOSH was asked to investigate. Scott was Libby's best customer for vermiculite. Government agencies monitored the situation at Libby at the start of the 1980s and privately commissioned EPA reports flagged, not only the dangers at the mine, but also nationwide

[101] RCW/BLD Review, 5 December 1986.

[102] C.N. Graf to J. Peter Grace Jr, 4 March 1976.

[103] R.H. Locke to H.C. Duecker, 30 June 1977.

[104] F. Moatamed, J.E. Lockey, and W.T. Parry, 'Fiber Contamination of Vermiculites: A Potential Occupational and Environmental Health Hazard', *Environmental Research* 41 (1986), pp. 207–18.

environmental exposure to tremolite. But these agencies rapidly lost interest after 1980, some drawing a connection between this and the ascendancy of J. Peter Grace as President Reagan's choice to head a commission to deregulate business.

The Libby mine was allowed to operate until 1990 and it was yet another decade before Libby was headline news. By 2001, Grace's construction division had filed for bankruptcy, saddled by billions of dollars of legal claims, clean up costs, and government fines. In the aftermath, a screening of over 6,000 Libby residents revealed that nearly a fifth had lung abnormalities consistent with asbestos exposure, some of whom had never worked for Grace. Although most Libby residents have not been affected (and may never be), the latest figures show that well over 200 people have been killed by the mine, with at least another 375 with potentially fatal ARDs.

Libby now ranks with Acre Mill and Armley in Britain, Penge in South Africa, and Wittenoom in Australia, as among the world's worst industrial disasters. Libby has been well documented by journalists, particularly Andrew Schneider at *Seattle Post-Intelligencer*, who broke the story in 1999.[105] More worrying are the effects of vermiculite outside Libby. Grace estimated in 1985 that 30,000 additional asbestos lung cancers could result among those 'involved in the application of our products'.[106] These figures are speculative. What is more certain is that millions of American homes contain Libby vermiculite/tremolite (estimates ranging between fifteen to thirty-five million). Even at a local or regional level the amount of exposure in staggering. For example, 300 million pounds of Libby ore were processed at a Grace plant at Dearborn (whose plant manager, Dayton Prouty, died of mesothelioma in 1999); and it has been estimated that about 800,000 homes in Michigan contain Zonolite. In Minnesota, 2,600 north-east Minneapolis residents reported exposure to Libby vermiculite, with cases of bystander asbestosis and mesothelioma around the plant in Osseo.[107] This is aside from vermiculite sent to other countries, such as Canada, where Zonolite-induced mesotheliomas have now appeared. In 2004, the Royal Brisbane Hospital in Australia settled a mesothelioma case involving a 49-year-old former nurse, Vicki

[105] 'A Town Left to Die', *Seattle P-I* Special On-line Report: http://seattlepi.nwsource.com/uncivilaction/

[106] R.C. Walsh to K.T. O'Reilly, 1 November 1985.

[107] G. Gordon, 'Asbestos Bill Gets New Life in Congress', *Star Tribune*, 27 November 2004. Posted at: www.startribune.com/stories/587/5106776.html

Benson, after documents were found showing that Monokote had been extensively sprayed on a hospital block in the 1970s.[108]

In 2000, the US Public Health Service warned that 'even minimal handling by workers or residents poses a substantial health risk'.[109] But not until 2003 did the EPA issue a tepid press release containing advice about how to deal with vermiculite attic insulation. This advice boiled down to a simple instruction: Stay out of attics! This assumed that residents and maintenance workers knew about vermiculite's dangers: even if they did, vermiculite could sift through ceilings and it was unlikely that people could stay out of their attics forever. Meanwhile, the core Grace company remains a profitable business, with sales of about $2 billion and 6,000 staff in nearly forty countries.

Environmental Pollution: A Global Problem

In Libby, the EPA found nearly 1,500 properties with dangerous fibre levels; about half of these had been cleaned up by 2006 at a cost of nearly $130 million. In 2008, Grace agreed to pay the government $250 million towards the clean-up. The EPA estimates that decontamination could take another seven years. Medical costs for the sick Libby residents, not all of whom can claim from Grace, are expected to be at least $32 million by the end of the decade. There have even been suggestions that the town should be evacuated. But whatever Libby's problems, they pale in comparison with the continued environmental contamination in the South African mining regions. The hundreds of unreclaimed mines in South Africa have left large areas of the Northern Cape contaminated. Fibre is picked up by the prevailing winds and moved ever closer to the population centres of Johannesburg and Pretoria. The Department of the Environment has a budget of R100 million to clean up the mines, but far more will be needed before the landscape is made safe. South Africa is a poor country with glaring disparities of wealth and poverty. The money being spent on reclaiming mines abandoned by foreign companies is sorely needed to build schools and provide health care. The Swazi state is even poorer, and it is unlikely the dumps at Havelock will ever be reclaimed.

However, as this chapter has shown, the South African problems are only unique because of their degree. Asbestos in the environment is a

[108] H. Thomas, 'Government Accepts Hospital Housed Killer Asbestos', *Courier-Mail* [Queensland], 11 February 2005.

[109] Schneider and McCumber, *Air that Kills*, p. 368.

continuing problem wherever it has been mined, milled, and used. Key 'asbestos' into an Internet news search-engine, and the thousands of hits attest to a problem that has not disappeared and will be with us for some time. In the US, recent research has confirmed community risk, even in areas producing mostly chrysotile. Residents of Manville have been found to have an elevated risk of mesothelioma, probably due to ambient air and household contact with workers, and because 'emissions regularly coated cars, homes, and yards like a fresh snowfall in much of the immediate community'.[110] This contamination could persist due to the stability of the fibres. Meanwhile, the asbestos-in-buildings issue is far from moribund, despite the claims that the problem could somehow be encapsulated and deferred. However low the dust levels in most buildings and schools, ARDs have nevertheless been reported in building custodians and maintenance personnel. One study highlighted thirteen American schoolteachers with malignant mesothelioma, whose only identifiable asbestos exposure was in schools.[111] In 2004, another source highlighted that 'elementary schoolteacher' was the eighth most frequently cited occupation on mesothelioma death certificates (hospital jobs ranked fourth).[112] A quarter of all US government-reported mesothelioma mortality occurred in people who did not work directly with asbestos.

In Canada, public health authorities reported in 2004 on the surprisingly high rates of mesothelioma in Quebec women, which pointed to an environmental cause.[113] Thirty tailing dumps at Thetford Mines are one possible source of exposure; another could be the 300 school buildings that the Quebec government has identified as needing immediate remedial action for decaying sprayed asbestos. In 2004, Canadian university professors called for action on asbestos in buildings, after the mesothelioma deaths of two professors at the University of Manitoba. By 2005, one faculty member and a retired maintenance engineer had also been affected. Ironically, for a country that remained committed to exporting asbestos, in 2005 there were calls for the Parliament Building in Ottawa to be evacuated, because it was contaminated with sprayed amosite.[114]

[110] M. Berry, 'Mesothelioma Incidence and Community Asbestos Exposure', *Environmental Research* 75 (1997), pp. 34–40. See also A. Miller, 'Mesothelioma in Household Members of Asbestos-Exposed Workers: 32 United States Cases Since 1990', *AJIM* 47 (2005), pp. 458–62.

[111] L.C. Oliver, 'Asbestos in Buildings: Management and Related Health Effects' (1994), pp. 174–88.

[112] Environmental Working Group, *Asbestos Think Again* (2004), pp. 24–5. Posted at: www.ewg.org/reports/asbestos

[113] Institut National de Santé du Quebec, *The Epidemiology of Asbestos-Related Diseases in Quebec* (July 2004), p. 53.

[114] T. Naimetz, 'Report Warns of West Block Asbestos Risk', *Ottawa Gazette*, 13 March 2005.

In November 2005, Ontario introduced tighter regulations to manage asbestos *in situ*.

In Britain, the government is attempting to implement its new regulations, but early indications are that trying to compel property owners to inventory their asbestos is going to be difficult. Indeed, it seems clear already that the government has not earmarked sufficient resources to monitor compliance. It has also sent out mixed messages. On the one hand, it has exempted certain asbestos decorative building materials from its regulations. On the other, it has warned schoolteachers of the dangers of classroom asbestos, after the government published figures showing that 147 people in education had died from mesothelioma between 1991 and 2000. Similar building decrees now exist in France, but, again, the government has no way of knowing if they are applied.[115] Meanwhile, the decontamination of Jussieu will continue until at least 2008: seven teaching staff are now reported dead from mesothelioma, amongst over 100 cases of ARDs associated with the site.[116] Environmental cases have occurred elsewhere in France: for example, 49-year-old Pierre Leonard developed mesothelioma in 1995, probably from childhood exposure at his school, which was only a few metres from a factory processing asbestos in Aulnay-sous-Bois, near Paris.

In Brussels, the Berlaymont building, commissioned as the headquarters of the EU, has provided member states with a decade of experience in dealing with asbestos-in-place. Built in 1968 with liberal quantities of asbestos, the 'Berlaymonster', as it became known, was evacuated in 1991 for decontamination.[117] By the time it reopened in 2004, nearly £1 billion had been spent. Elsewhere in Belgium, the manufacture of asbestos cement (a/c) has left a dismal legacy of environmental cases. These have included members of the Jonckheere family. Francoise Jonckheere died of mesothelioma in 2000, after living for thirty years in an old farmhouse near an Eternit plant in Kappelle-op-den-Bos. Her husband, Pierre, had worked at the factory and died of mesothelioma in 1987. That was not the end of the tragedy. In 2003, one of the couple's sons, Pierre Paul, also died

[115] C. Prieur, 'L'État est Impuissant a Dresser L'Inventaire de L'Amiante en France', *Le Monde*, 22 April 2005.

[116] F. Malye, *Amiante: 100,000 Morts à Venir* (2004), p. 135; *Senate Rapport (No 37) D'Information... Sur Le Bilan et Les Conséquences de La Contamination par L'Amiante* (Paris, 2005), pp. 319–21. The exact source of previous exposure is after problematic in mesothelioma cases; however, the five most recent cases are believed to have occurred from purely Jussieu exposure.

[117] S. Nay, *Mortel Amiante* (1997), pp. 110–11.

of the disease.[118] In the Netherlands, mesothelioma cases (five so far) have been identified among women without occupational or household exposure to asbestos. Some had lived during childhood along contaminated roads close to an Eternit a/c facility in the community of Hof van Twente in north-east Netherlands (the factory had for over fifty years made the waste available for paving roads).[119]

In Italy, Casale Monferrato has been badly contaminated by the activities of an Eternit a/c factory that operated between 1905 and 1985. A local oncologist (and town councillor) reported in 2000 that asbestos cancer cases had reached thirty each year, 'not only among people exposed professionally, but above all in the normal population'.[120] The Liguria region, with its tradition of shipbuilding and heavy industry, has left a legacy of mesothelioma—including many female cases that have no known exposure (pointing to an environmental cause).[121] Elsewhere in Italy, indirect deaths from mesothelioma have been amply documented in association with sugar refineries and railroads. In the latter industry, mesothelioma cases have been logged around the world and have included clerical workers and those who have experienced prolonged exposure to asbestos objects in their home.[122]

Mining in Wittenoom has caused ARDs among the children of senior management; and there have also been high rates of disease among those who transported Wittenoom fibre to the coast.[123] In Australia generally the country is coming to terms with the ubiquity of James Hardie's products. Mesothelioma in that country has been referred to as the 'home renovator's' disease.[124] In Western Australia, it was reported in 2005 that building workers and DIY enthusiasts account for more new cases of

[118] S. Nay, 'Asbestos in Belgium: Use and Abuse', *IJOEH* 9 (July/September 2003), pp. 287–93. Another son was diagnosed with mesothelioma in 2007.

[119] A. Burdorf, M. Dahhan and P. Swuste, 'Occupational Characteristics of Cases with Asbestos-Related Diseases in the Netherlands', *Annals of Occupational Hygiene* 47 (2003), pp. 485–92; A. Burdorf, M. Dahhan and P. Swuste, 'Pleural Mesothelioma in Women Associated with Environmental Exposure to Asbestos', *Ned Tijdschr Geneeskd* 148 (28 August 2004), pp. 1727–31; S.P. McGiffen (ed.), *The Polluter Pays: Notes from the International Conference on Asbestos Held in Amsterdam in May 2004* (2005).

[120] D. Degiovanni, 'History of the Asbestos Ban in Casale Monferrato' (17–20 September 2000).

[121] V. Gennaro et al., 'Incidence of Pleural Mesothelioma in Liguria Region, Italy (1996–2002)', *European Journal of Cancer* 47 (November 2005), pp. 2709–14.

[122] C. Maltoni, 'The Long-Lasting Legacy of Industrial Carcinogens: The Lesson of Asbestos', *ANYAS* 837 (1997), pp. 570–86.

[123] E. Merler et al., 'On the Italian Migrants to Australia Who Worked at the Crocidolite Mine at Wittenoom Gorge, Western Australia' (1999), pp. 295–6.

[124] J. Leigh and T. Driscoll, 'Malignant Mesothelioma in Australia, 1945–2002', *IJOEH* 9 (July/September 2003), pp. 206–17, 213.

mesothelioma (30 per cent) than former workers at Wittenoom (16 per cent).[125]

In 2005, a series of major Japanese industrial companies suddenly disclosed for the first time that several hundred of their employees had died from asbestos exposure. Amongst those deaths were several non-occupational cases. For example, in Hyogo Prefecture (which contained Amagasaki, a district in Osaka, that was home to several asbestos-using companies), 143 people died of ARDs between 2002 and 2004. Of those, twenty had never worked with asbestos.[126] The machinery manufacturer Kubota Corporation, which operated an a/c factory in Amagasaki, revealed—in a pronouncement that became known as the 'Kubota shock'—that seventy-nine of its employees had died from ARDs between 1978 and 2004. Kubota also said that three residents, who lived near the defunct factory in Amagasaki had mesothelioma and that the wife of a factory employee had died of mesothelioma.[127] Other cases reported included a 70-year-old Osaka stationery shop owner who died of mesothelioma in 2004, after he had inhaled asbestos from the linings of the walls of his shop, beneath an elevated railway.[128] In 2005, a survey showed that many public buildings had a potential asbestos risk, including nearly 500 schools and well over 300 hospitals.

At a more anecdotal level, we are all familiar with that increasingly frequent and depressing news item: the painter or lagger, the carpenter or fitter, the schoolteacher or fireman, the housewife or hairdresser dead or dying from mesothelioma, and often struck down in middle age or about to enjoy a hard-won retirement. The fact that exposure was indirect, environmental, or even unknown, adds a particularly bitter twist to what is a cruel illness. Social status, wealth, a blameless lifestyle, youth, and no connection with the asbestos industry, are no protection against mesothelioma. In Britain in 2004, one of the youngest ever mesothelioma sufferers, Barry Welch, died aged only 32. He had contracted the disease while playing on his stepfather's knee (he was a scaffolder in a power station). The injustice can be devastating. John Kennally lamented, after physicians told him that he had mesothelioma from the dust at a factory where he had never worked: 'You think to yourself, Jesus, it's a death warrant, isn't it? For doin' nowt.'[129]

[125] Western Australian Cancer Registry, *Cancer in Western Australia* (2005).

[126] *Yomiuri Shimbun*, 11 February 2006.

[127] E. Arita, 'Asbestos Deaths Just Tip of Iceberg', *Japan Times*, 22 July 2005. Kubota confirmed thirty-three more ARD deaths in 2006.

[128] *Japan Times*, 23 August 2005.

[129] 'Too Close to Home', First Tuesday documentary, Yorkshire TV, 6 December 1988.

Better known individuals than Kennally have been struck down. Exposure to asbestos cannot be proven in every case of mesothelioma (though non-asbestos induced mesothelioma is believed to be as rare as one case per million of the population per year) but, taken together, the following cases—most of which have proven exposure to asbestos—provide a stark illustration of the environmental and bystander risks. The film star Steve McQueen died of mesothelioma in 1980, aged 50: an attempt by Mexican doctors to remove the tumour simply hastened his end. Mesothelioma claimed Hollywood actor and *Die Hard* star Paul Gleason in 2006. Musical celebrities who have died from mesothelioma include Mickie Most [Michael Hayes] and Warren Zevon. The palaeontologist Stephen Jay Gould was diagnosed with peritoneal mesothelioma in 1982 and succumbed to cancer twenty years later.

In North America, politicians and public figures have died from mesothelioma: notably, Democrats Mark Hannaford and Bruce Vento, besides Bill Clinton's former presidential campaign chief, Eli J. Segal. In 2005, Chuck Strahl, a Conservative MP in Ottawa, revealed that he was suffering from mesothelioma. Toronto police chief Jack Marks died from it in 2007. In Australia, Sir David Martin, the Governor of New South Wales, died of the disease. Judges with mesothelioma have included Judge William L. Forbes—a retired Chesapeake Circuit Court Judge—Bob Bellear, the first indigenous Australian judge, and Ann Ebsworth, an English High Court judge. In America, the malignancy has killed Admirals (Elmo R. Zumwalt Jr), distinguished physicians (Joseph Sagura), computer pioneers (Jerre D. Noe, Bob Miner), distinguished architects (Paul Rudolph[130]), and Olympic gold medal wrestlers (Terry McCann). In Britain, mesothelioma killed consultant plastic surgeon James Emerson, university chief Sir Gordon Beveridge, Swindon footballer George Hunt, writers Michael Coney and Rob Dawber, and wealthy businessman Anthony Farmer.

The environmental hazards of asbestos can no longer be doubted.[131] Most scientists now accept that low-dose exposure in the home or in the general environment can, in certain circumstances, carry a measurable risk of malignant pleural mesothelioma.[132] This chapter has presented a multifaceted picture of the uncertainties and conflicts surrounding the

[130] The interior of Rudolph's Yale Art and Architecture Building was finished with sprayed asbestos. In September 1974, the building had to be closed because of asbestos particles falling from the ceiling.

[131] P. Boffetta and F. Nyberg, 'Contribution of Environmental Factors to Cancer Risk', *British Medical Bulletin* 68 (2003), pp. 71–94.

[132] C. Magnani et al., 'Multicentric Study on Malignant Pleural Mesothelioma and Non-Occupational Exposure to Asbestos', *British Journal of Cancer* 83 (2000), pp. 104–11.

connection between the environment and health, which demonstrates that science is open to political manipulation and that epidemiological data are riddled with uncertainty.[133] In asbestos, much of that uncertainty was created by the attitude of the asbestos companies, which for so long tried to dismiss environmental concerns. For over forty years the industry, in alliance with governments and the medical profession, was a significant drag on the recognition of the problem. Not all their arguments were sinister or ridiculous. The health impacts of environmental exposure can be slight and difficult to quantify; the hazards of asbestos in many environmental situations are sometimes small, compared to other risks in life; and many of the problems, such as asbestos in buildings, have never been susceptible to an easy fix. Moreover, not every mesothelioma can be tied with asbestos. Nevertheless, the resistance of the industry to the idea that asbestos was an environmental hazard ensured that the mineral's use in construction materials was untrammelled in the 1960s and 1970s—something that has made a major contribution to the current health disaster. This also has profound importance for countries still using asbestos, especially in the developing world, which run the risk of experiencing many of the problems described here.

[133] See generally S. Kroll-Smith, P. Brown, and V.J. Gunter (eds.), *Illness and the Environment: A Reader in Contested Medicine* (2000).

8

Pushing Asbestos in the Developing World

Now that we have invested so much time, money, effort and human suffering learning to use asbestos safely what should we do? Should we ban it world-wide, thereby losing the use of an abundant and useful natural resource, or should we continue to use it wisely and carefully to gather some return on what we have invested into it. The safe use of asbestos could be our memorial to those who have died before the danger was recognised.

F.J. Wicks
Royal Ontario Museum
Canadian Society of Exploration Geophysics Conference
Abstract, 2000

We worked hard for Cape [Asbestos] and now we hear the dust was poisoning us. The dust was everywhere, in our houses, in our clothes, and when you coughed the phlegm was full of dust. Now my health is poor, I cough and my chest pains. The dust has eaten holes in my lungs. Must I just accept it?

Ragel Olyn (66)
South African asbestosis sufferer
Quoted in Hein Du Plessis
'Asbestos's Sorrowful Legacy: A Photo-Essay', *IJOEH* 9
(July/September 2003), p. 237

During most of the twentieth century, asbestos mined in economically depressed regions was used to manufacture products in the industrial heartlands of the North. Although the Canadian mines were to some extent an exception, Quebec is one of Canada's poorest provinces and the men and women employed at Thetford were amongst the most vulnerable of Canadian workers. Over the past three decades that pattern has changed. Today asbestos manufacture is concentrated in Africa, Asia, and Latin America, where a wide range of cheap and durable products are made. The pandemic of asbestos disease in the US and the UK should have served as a barrier to the continued use of chrysotile in the developing world. Unfortunately, that has not happened and the industry has been so successful in promoting the view that asbestos can be used safely that in 2006 more than 2 million tonnes were consumed. Hazardous work conditions combined with a lack of medical services in countries like Brazil, China, and Thailand will only worsen the suffering of those who contract ARDs during the industry's renaissance.

World markets for asbestos have been influenced by a number of factors, including changes in technology, consumer fashion, and global conflicts. No factor has been more important than the health scares about mesothelioma. In 1973, US demand peaked at 800,000 tonnes. In response to pressure from trade unions, social movements, and medical specialists, by the year 2000 US consumption had fallen to 15,000 tonnes.[1] The same decline occurred in Western Europe as the major companies moved out of asbestos and diversified. The British firm Cape plc went into asbestos removal, while Johns-Manville turned to the manufacture of PVC pipes, industrial lighting, and fibreglass. W.R. Grace shifted into non-asbestos construction materials. As the major corporations were reinventing themselves, the same obsolete and dangerous technologies that those companies had used so profitably reappeared in the hands of other producers in Asia, Southeast Asia, and Latin America. The industry has enjoyed such a revival—as we have emphasized—that around half the world's consumption of asbestos during the twentieth century occurred after 1976.[2]

The history of asbestos manufacture and mining is complex, and in terms of public health there have, over the past three decades, been losses and gains. There have, however, been two constants: the struggle by labour in developing countries for safer work conditions and the malevolent role played by Canada in promoting asbestos use in the developing world. Canada is a member of the G8 and it carries some influence on the global stage. Its industry and government, backed by a sophisticated scientific community, have used their access to elite forums, including the WHO and the WTO, to promote asbestos. They have also

[1] R.L. Virta, *Worldwide Asbestos Supply and Consumption Trends from 1900 through 2003* (2006), pp. 5–6.

[2] Virta, *Worldwide Asbestos*, p. 1.

used their influence on smaller stages. In 1997, for example, the Canadian Embassy in Seoul persuaded the South Korean government to withdraw labelling legislation about the dangers of chrysotile.[3] In February 2006, the Canadian High Commissioner in South Africa complained to the ruling ANC government that a proposed South African ban on asbestos would damage Canada's $5 million trade to Africa.[4]

The regulation of asbestos use, like the levels of public awareness of its dangers, has varied from country to country. Regulation has also varied between the individual parts of corporations. The introduction of legislation in Britain in 1931 saw conditions in T&N's domestic factories improve, but they remained unchanged in its Canadian and southern African mines. T&N enjoyed the support of regional governments in Quebec and British Columbia, which were keen to encourage investment. In southern Africa, T&N and Cape Asbestos took advantage of the harsh labour conditions characteristic of minority rule. When challenged about their labour policies, T&N and Cape always claimed to be using uniform safety standards across their global holdings. The evidence suggests otherwise. British manufacturers voluntarily ceased importing blue asbestos in 1970 because it is so dangerous, yet mining of crocidolite continued in South Africa until 1996.

In developing countries, the links between industry and governments are often intimate. That has been particularly true of South Africa where the Departments of Mines and Health supported a largely foreign-owned industry. In Swaziland, the Havelock mine was for a period jointly owned by T&N and the Swazi government, while in Zimbabwe the Shabanie and Gaths mines were expropriated by the Mugabe regime in November 2004.[5] The lack of effective regulation means the disease burden has gone largely unrecognized. As late as 1995, for example, the Department of Health in Kimberly was unaware of a pandemic of mesothelioma in the nearby mining communities of the Northern Cape.[6]

Governments in developing countries tend to be hostile to trade unions, which they accuse of discouraging investment. They are certainly fearful of organized labour as a source of political dissent. Less than 10 per cent of workers in developing countries are protected by occupational health and safety legislation, yet globally they make up the

[3] J. Brophy, 'The Public Health Disaster Canada Chose to Ignore', *Chrysotile Asbestos: Hazardous to Humans, Deadly to the Rotterdam Convention* (2006), pp. 17–20.

[4] C. Smith, *'Don't Ask, Don't Tell'*, in *Chrysotile... Hazardous to Humans*, pp. 11–13.

[5] 'Zimbabwe SMM Asbestos Seized', *The Herald* (Harare), 16 November 2004.

[6] See J. McCulloch, 'Beating the Odds: The Quest for Justice by South African Asbestos Mining Communities', *Review of African Political Economy* 32 (March 2005), pp. 63–77.

majority of those disabled in the workplace.[7] In parts of Asia and Africa, children account for over 10 per cent of industrial workers.[8] The most dangerous jobs are in small factories, which are rarely if ever visited by a state inspector.

The UN and the ILO are the only public agencies with influence over global labour practices. The UN has limited access to workplaces and little or no capacity to impose any form of regulation. The ILO Convention on Occupational Safety & Health carries some moral weight but it has been ratified by only thirty-seven of the ILO's 175 member states. A mere twenty-three states have ratified the ILO's Employment Injury Benefits Convention that lists diseases for which compensation should be paid. The ILO's Asbestos Convention, which is fifteen years old, has been ratified by only twenty-five countries. The provisions are feeble and proscribe only crocidolite and certain manufacturing processes, thereby leaving the mainstream asbestos industry intact. The ILO and the WHO have limited budgets and lack trained personnel. In 2004 the WHO Program for Occupational Health had a staff of four.[9]

The reasons for the rebirth of the asbestos industry have tended to vary from region to region, but there are a number of common features. They include the lack of state capacity and the rampant corruption which are typical of the developing world. There is the industry's influence over medical knowledge, and the commercial appeal of asbestos as an ingredient in simply-made building materials. The new industry is also part of the global transfer of technologies that have become politically unsustainable in North America and Western Europe. Those transfers include the manufacture and use of pesticides and PCBs, and the dumping of toxic waste.[10] When things go wrong, as in Bhopal (1984) and the Ivory Coast (2006), the costs of corporate misconduct can be minimized because of the corporate veil. Finally, there is the role played by the Canadian government in promoting the use of chrysotile. In October 2006, Canada successfully opposed the listing of chrysotile as a hazardous substance under the Rotterdam Convention thereby denying importing countries access to knowledge about the risks (see Chapter 9).

[7] J. LaDou, 'International Occupational Health', *International Journal of Hygiene and Environmental Health* 206 (August 2003), pp. 303–13.

[8] LaDou, 'International Occupational Health', p. 305.

[9] LaDou, 'International Occupational Health', p. 303.

[10] B.I. Castleman, 'The Export of Hazardous Factories to Developing Nations', *IJHS* 9 (1979), pp. 569–606; J. Clapp, *Toxic Exports: The Transfer of Hazardous Wastes from Rich to Poor Countries* (2001). See also Blacksmith Institute, *The Hidden Tragedy: Pollution in the Developing World* (2000).

Table 8.1. World Asbestos Production by Country, 2006

Country	Tonnes (estimated)
Russia	925,000
Kazakhstan	355,000
China	350,000
Canada	243,000
Brazil	236,000
Zimbabwe	100,000
Columbia	60,000

World total, including other nations: 2,300,000 tonnes.
Source: US Geological Survey, *2006 Minerals Yearbook: Asbestos* (June 2007). http://minerals.usgs.gov

The countries that now use asbestos need to provide cheap housing, and improved access to clean water and sanitation for their growing populations. Brazil, China, Russia, India, Japan, and Thailand currently account for 90 per cent of asbestos consumption. Brazil and the China have mines but most developing countries rely upon imports from the countries shown in Table 8.1. It is ironic that Canada, which has been the major exporter of chrysotile, consumed only 500 grams per capita in the year 2000. The figure for the US was even lower at 100 grams.[11] In 2004 Thailand had a per capita consumption of 1.9 kilogram which was among the highest in the world.[12]

The industry has long been aware of the potential markets in the South. The International Conference of Asbestos Information Bodies held in London in November 1971 was attended by delegates from the UK, the US, France, Belgium, the Netherlands, Italy, Germany, and Scandinavia. Following the open sessions, Bill Raines from Johns-Manville and T&N's chairman, Ralph Bateman, discussed the marketing of asbestos-cement (a/c) products in developing countries. According to Raines, Bateman has this to say:

His point is that in many of these countries the life expectancy is so low, as a result of deaths from diseases from impure drinking water, for example, as well as starvation, and inadequate housing, that the question of the very, very, small risk of mesothelioma that may exist in exposure to asbestos in some situations, is totally outweighed by the contribution that asbestos pipe and other products

[11] L. Scavone, F. Giannasi, and A. Thébaud-Mony, 'Asbestos Diseases in Brazil and the Building of Counter-Powers: A Study in Health, Work, and Gender', *Annals of the Global Asbestos Congress—Past, Present and Future* (17–20 September 2000).
[12] S. Pandita, 'Banning Asbestos in Asia', *IJOEH* 12 (2006), pp. 248–53.

can make to improving the living standards and, indeed, the life expectancy of people in these countries. This is an interesting philosophy, although it has to be expressed rather carefully.[13]

The New Industry

The slow emergence of asbestos manufacture in the Third World was due to a number of factors. They included delayed industrial development and the lower profits on offer to the companies which controlled the lucrative North American market. In 1920, only 7,000 tonnes of fibre was used in the whole of Asia, where manufacture was confined to three countries. It was not until the 1950s that consumption began to rise and by 1960 222,000 tonnes were used in the region, with China and Japan as the major customers.[14] Twenty years later, Asian consumption had risen to over 1 million tonnes with significant manufacture of a/c sheeting, pipes, and roof tiles in China, India, Indonesia, Iran, Japan, South Korea, Malaysia, and Thailand. The pattern has been much the same in Latin America. During the 1960s total regional consumption was less than 100,000 tonnes with almost all fibre being imported. Over the past three decades, regional manufacture has become significant.

At first glance, it appears that after 1975 there was a dramatic shift to new economies as the major mining companies sought new customers. The change was not that simple and there were in effect a series of shifts to regional markets which in some instances already had a small manufacturing capacity. It is also true that every increase in government regulation in Britain, France, and the US encouraged the industry to relocate.

The initial shift saw the manufacture of particularly hazardous products, such as asbestos textiles moved off-shore. From the late 1960s there was a fall in asbestos textile production in the US, while imports of those same products from Mexico and Venezuela rose.[15] The companies involved were subsidiaries of US corporations anxious to avoid litigation in American courts. Asbestos textiles imports also came from affiliates of a Japanese company, which operated plants in Taiwan and South Korea. Johns-Manville stopped domestic production of Thermobestos at the end of 1973, two years before the EPA banned the use of asbestos in thermal

[13] Raines's Report on International Conference of Asbestos Information Bodies, 1971, to F.J. Solon, 10 December 1971.

[14] Virta, *Worldwide Asbestos*, p. 14.

[15] B.I. Castleman, 'Controlled Use of Asbestos', *IJOEH* 9 (July/September 2003), pp. 294–8.

insulation and pipe covering. Yet in 1976, the company was making the same products in Brazil, without warning labels, and continued to do so into the early 1980s.[16] The next part of the industry to migrate was the a/c plants. From 1970 data on exposure levels for those working in a/c manufacture showed a high risk of lung disease. In response to growing public concern, US a/c plants began to close.[17]

Johns-Manville had a shareholding in a number of Mexican plants, and in June 1977 its director of environmental safety, William B. Reitze, was sent from head office in Denver to review their operation. Reitze first visited the Asbestos de Mexico factory which used both blue and white fibre and employed around 900 workers. A recently completed Social Security Administration survey had found that almost a third of the workforce had ARDs. Conditions throughout the plant were hazardous and according to Reitze the local manager appeared unaware of the risks. At the Almeria factory the problems of pollution extended to the adjacent community. 'The plant manager informed me', wrote Reitze, 'that over two tons of fines [short fibre] go out of the stack every day. The plume from the plant is visible for many miles and creates not only a problem for the surrounding village, but also an-in-plant problem'.[18] Although Johns-Manville held a minority share in those factories, Reitze warned his Denver colleagues: 'I believe that our continued participation in this venture, without insisting upon a general overall cleanup is immoral, if not criminal.'[19]

The development of new chrysotile mines in Brazil after 1971 has seen that country become an important producer and consumer of asbestos. The national government invested US$9 million in the development of mines, which saw production rise to 237,000 tonnes in 1991.[20] Most asbestos has gone into local a/c manufacture with exports to Argentina, India, and Mexico accounting for around a third of sales. Subsidiaries of the leading French asbestos and glass manufacturer Saint Gobain hold a majority share in Canabrava, the country's largest mine. Following the 1997 French ban on chrysotile, Saint Gobain converted its domestic plants to non-asbestos technology while in Brazil it continued to invest in chrysotile. The primary industry in Brazil employs 10,000 workers in mining and manufacture, while a further 200,000 are engaged in

[16] B.I. Castleman, *Asbestos: Medical and Legal Aspects* (2005), p. 790.

[17] Castleman, 'Controlled Use', p. 294.

[18] Memo: W.B. Reitze to H. Moreno, 10 June 1977. Re. Visit to Almeria, Meisa, and Asbestos de Mexico Johns-Manville/JM Mexico.

[19] Memo: Reitze to H. Moreno. [20] Virta, *Worldwide Asbestos*, p. 7.

distribution, sales, construction, and repair.[21] Asbestos is used in building materials, brake linings, cloth, paints, and floor tiles, bringing the annual per capita consumption to 1,400 grams.[22]

Work conditions in Brazil are typical of developing countries. The industry is poorly regulated and the labour turnover is high. More than half of those employed in Brazil's construction industry, for example, work 'off the books' and there is little data on disease rates. In 1992, with fifteen million registered workers in Brazil only 8,299 cases of occupational disease were reported by the National Institute of Social Security.[23] In 2000 fewer than a hundred cases of asbestos disease were cited in the Brazilian literature. There were only four mesotheliomas for a population in excess of 160 million.[24] Legislation covering asbestos was first introduced in 1991, but in the official statistics mesothelioma was not differentiated from other cancers until four years later. Since 1997 the industry has been regulated by the Permanent National Commission on Asbestos, a body composed of representatives from government, labour, and management. Its role is to oversee the mining, manufacture, sale, and transport of chrysotile. The Commission's approach is based on the assumption, common to such bodies, that work is inherently dangerous and it is more important to maintain jobs than reduce risk in the workplace. Its attempts at regulation have been feeble and only since 1999 have asbestos companies been obliged to provide industrial laundries for workers' clothes.[25]

Asbestos was first mined in Korea during the 1930s, but only small amounts of fibre were produced and for two decades the industry was minuscule.[26] Rapid economic development in South Korea during the 1960s saw consumption increase with a/c construction materials playing a major role. The industry also expanded into asbestos textiles and brake linings for the new automobile industry. By the early 1970s, the tightening of regulations on asbestos textile manufacture in Western Europe saw manufacture move to countries like South Korea where there were no specific regulations governing the mineral's use.[27] Initially, the asbestos

[21] Scavone, et al., 'Asbestos Diseases in Brazil', p. 9.
[22] Scavone, et al., 'Asbestos Diseases in Brazil', p. 1.
[23] Scavone, et al., 'Asbestos Diseases in Brazil', p. 2.
[24] Scavone, et al., 'Asbestos Diseases in Brazil', p. 2.
[25] Scavone, et al., 'Asbestos Diseases in Brazil', p. 4.
[26] D. Paek, 'Asbestos Problems in Korea: History and Current Situation', *Annals of the Global Asbestos Congress*.
[27] Paek, 'Asbestos Problems in Korea', p. 3.

used in textiles came from insulation recovered from dismantled ships.[28] With the introduction of the Occupational Safety and Health Act in Japan in 1974 a number of local asbestos textile manufactures moved their operations to Seoul. So too did the German company Rex-Asbest.

The introduction of monitoring programs in South Korean factories in the mid-1980s revealed high fibre levels in a/c and textile plants. That resulted in stricter workplace regulations, higher costs of production, and a fall in output. Following tightened regulations in Western Europe, the car industry was forced to switch to asbestos substitutes in order to retain access to those markets. To offset the fall in domestic production, cheaper asbestos textiles were imported into South Korea from China and Indonesia. Korea now imports asbestos friction materials and textiles and exports fabricated asbestos products.

In South Korea, the Industrial Health and Safety Act regulates the use of asbestos as a raw material. But small factories and workshops which use asbestos-based products are exempt and those commodities are still sold at hardware stores without any warning labels. Asbestosis screening for textile workers was only introduced in 1993. In that same year, the first case of mesothelioma was diagnosed in a former asbestos textile worker. From the year 2000, around forty to fifty mesothelioma cases have been reported annually. South Korean manufacturers cannot claim, as have their OECD predecessors, that they have been victims of the deficiencies of medical science.

The pattern has been similar in South Asia, and during 2004 India consumed more than 60,000 tonnes of asbestos. While most of that fibre was imported from Canada and Brazil about a fifth of domestic needs was met by tremolite from local mines. A part of domestic needs are also met by the recycling of asbestos insulation. Around 6,000 workers are employed in the primary industry while another 100,000 work in asbestos manufacture. Asbestos mining and milling are carried out by small firms. Manufacture is regulated under the Factories Act (1948) and the Environment Protection Act (1986). The exposure limits are 2 f/cc: elsewhere in the world they vary between 0.1 f/cc and 0.5 f/cc. A number of studies of asbestosis have been conducted by the National Institute of Occupational Health (NIOH) in Ahmedabad. There have been no studies of lung cancer or mesothelioma.[29]

[28] Paek, 'Asbestos Problems in Korea', p. 4.

[29] Toxicslink, 'Asbestos: Fibre of Subterfuge', No. 11 (2001); G. Krishna, 'Asbestos Kills Europeans, Australians and Japanese But Not Indians', *Toxicslink* (6 January 2006).

Half of the 700 ships dismantled each year in the world are processed in India. The shipbreaking industry is centred at Alang in Gujarat state and employs around 35,000 workers. Large ships contain tonnes of lagging which are stripped by hand and recycled for insulation and roofing material.[30] Most workers are migrants from the poor states of Orissa and Bihar who are paid less than US$3 a day. There are no trade unions and work conditions are hazardous. There is no protective equipment and the nearby houses are full of dust. The industry falls under various acts including the Workmen's Compensation Act of 1923, the Factories Act of 1948, and the Environment Protection Act of 1986.[31] But the regulatory systems are lax and the Ministry of Labour, for example, keeps no data on morbidity or mortality rates. In fact rather than discourage the use of asbestos, between 1995 and 2000 the Indian government lowered import duties on raw fibre and asbestos-based products thereby giving the industry an advantage over competitors using PVA.[32]

In August 2003, the Indian government joined Canada in frustrating attempts to have chrysotile listed as a toxic material under the Rotterdam Convention (see Chapter 9).[33] Four months later the International Conference on Chrysotile Asbestos Cement Products was held in New Delhi. The conference, sponsored by the Asbestos Cement Product Manufacturers Association and the Asbestos Institute of Canada, had all the earmarks of government endorsement. Delegates were welcomed by the Ministers for Environment and Forests and Urban Development and those attending included the Zimbabwe Minister for Labour, July Moyo, and the Zimbabwe Minister for Mines, Mr. Brian Dickson; the Canadian Deputy High Commissioner was also present.[34] While demands for a global ban were being made at the World Social Forum at Mumbai in January 2004, the asbestos industry flooded the national dailies with advertisements promoting the safety of chrysotile. The industry also appealed successfully to the national government to allow the Ministry for Rural Development to use asbestos sheets in housing projects.[35]

The impact of asbestos on health is complex and the definitions of risk are as varied as are the settings in which fibre is used. In the US,

[30] L. Braun et al., 'Scientific Controversy and Asbestos: Making Disease Invisible', *IJOEH* 9 (July/September 2003), pp. 194–205, 197.

[31] Braun et al., 'Scientific Controversy'.

[32] 'T.K. Joshi and Asbestos in India: A Message from the Collegium Ramazzini', *AJIM* 45 (2004), pp. 125–8.

[33] G. Krishna, 'Say No to White Asbestos', *Business Standard*, 12 February 2004.

[34] 'Joshi and India', p. 126. [35] Krishna, 'Say No'.

for example, it is all but impossible to construct an inventory of the individuals at risk.[36] The problem is compounded by the fact that less than 5 per cent of mesotheliomas occur among those employed in the primary industry.[37] The widespread use of asbestos in domestic dwellings combined with the recycling of building materials makes the threat in poor countries even greater.

Effective knowledge about risk requires that employees are aware of the presence of asbestos and understand the hazard. Communities need to know that asbestos is being used in a nearby factory or that it is dangerous for a worker to bring fibre home on his or her clothes. Such forms of knowledge are imperfect even in the OECD where there is a massive state capacity, a free press, trade unions, and universal literacy. In developing countries, state regulation is often non-existent and the impact of asbestos on community health goes unrecorded. It is also likely to be obscured by more immediate threats to family survival such as unemployment, infantile diarrhoea, HIV/AIDS, and malaria. The most rudimentary knowledge of all, that when diseases such as mesothelioma occur the cause is asbestos, is often out of reach.

Southern Africa, Mexico, and Brazil

Under apartheid, South Africa had the most regulated economy on the continent, yet there are few data on asbestos fabrication and even less on occupational disease. Most of T&N's South African records have disappeared; we do not have access to Department of Labour inspection reports after 1962; and out-of-court settlements in 2003 in London and Johannesburg by Cape and Gefco were predicated upon the destruction of all documents disclosed during the case (see Sources).[38] Nevertheless, the surviving evidence is disturbing.

In 1962, the Department of Labour conducted a study of ARDs in Johannesburg factories. Asbestos was at that time used in cladding for the South African Railways, as insulation for steam pipes by the state electricity commission (Escom), and in the acoustic spraying of public

[36] B. Terracini, 'A Precautionary Programme for Workers Who Have Been Exposed to Asbestos?' *Annals of the Global Asbestos Congress*.

[37] A. Tossavainen, 'Epidemiological Trends for Asbestos-Related Cancers', *Annals of the Global Asbestos Congress*.

[38] 'Summary: Settlement between The Claimants, Cape Plc, and Leigh Day & Co, London May 2002'. See also K. Maguire, 'Law Firm Agreed to Shred Vital Data', *The Guardian*, 27 September 2004.

buildings.[39] The Department identified almost 2,000 workers along the Reef who were 'materially exposed'. It also found that because of inadequate records the number of exposed workers could not be estimated. The Department surveyed 714 men employed in six factories. In addition to the sixteen men with tuberculosis, 145 were found to have asbestosis, a rate of lung disease six times higher than among gold miners.[40] The Chief Inspector viewed the results as alarming. He noted that workers had no information about the risks they faced and that unlike miners they were not given regular medical examinations. In addition, migratory labour and a lack of coordination between factory inspectors and local authorities further obscured the disease burden. The 1962 survey appears to be the only South Africa review of its kind.

T&N operated a factory near Johannesburg, but its investments in mining were far more important. Inside South Africa the British company was shielded by apartheid and the poverty of mining communities. But T&N's management was well aware that publicity at home about conditions on the mines would have damaged its standing with shareholders and the British public. Rather than investing in work safety, T&N set out to ensure that information about work conditions and disease rates did not reach the press. In the early 1970s, however, conscientious shareholders in London began questioning the labour policies of Cape Asbestos.[41] The resulting publicity and a wider national concern about apartheid led in 1974 to a public inquiry into the conduct of British companies in South Africa. T&N told the committee that it paid above average wages and provided good working conditions for its black employees.[42] The government committee disagreed. It found that the average wage for the lowest paid whites was more than double the wages received by the highest paid African workers. The wage for an African crane driver of R116 per month, for example, was just above the poverty level of R94.[43] In its final report the committee was critical of both T&N and Cape.[44]

[39] Memo: Asbestosis: from the Workmen's Compensation Commissioner to the Secretary for Labour, Pretoria, 23 June 1961. The Secretary for Labour: ARD 3609.

[40] Memo: Disease of the Lung Caused by the Inhalation of Asbestos Dust. Chief Inspector of Factories (Machinery) Pretoria, 1 July 1962. The Secretary for Labour: ARD 3609.

[41] One shareholder claimed that at the Cape mines five black South Africans a week were being incapacitated by asbestosis. See A. Batty, 'South African Asbestos Workers', Letter to Editor, *Times*, 26 July 1971.

[42] See T&N, *Annual Report* (1973).

[43] House of Commons Expenditure Committee (Trade & Industry Sub-Committee), Session 1973/4, *Fifth Report... on Wages and Conditions of African Workers Employed by British Firms in South Africa. HC 116; Minutes of Evidence and Memoranda.* HC 21 I-IV, pp. 78–9.

[44] *Wages and Conditions of African Workers*, pp. 754–63, 1973/4.

The next wave of bad publicity came from two television documentaries screened in the UK. In 1981 *Dust to Dust* scandalized British viewers by showing black children playing on dumps of blue asbestos.[45] The following year, the *Alice: A Fight for Life* documentary (described in Chapter 4) sent T&N stocks plummeting. T&N sought to prevent the programme being broadcast in Swaziland, where it owned a major share in the Havelock mine. The T&N officer sent to control the damage reported back to head office: 'You will be pleased to know that during my visit to Swaziland last week I was able to prevail on Mr. Bill Mummery who heads up the Swazi TV Company, not to show the Alice film.'[46]

Following criticism of work conditions at a T&N factory in Bombay, published in *New Scientist* in 1981, T&N compiled a report on dust levels at a number of its overseas subsidiaries. At the Havelock mine most of the mill complex was found to be well above the then current UK standard of 2 f/cc. So too were the mines at Shabanie and Mashaba, in Southern Rhodesia.[47] The report noted that it would be difficult to achieve better results. In the case of Havelock, the only solution was to build a new mill which would have cost £1 million. At Shabanie and Mashaba, T&N's advisers doubted there was any technology which could make the mills safe. Zimbabwe achieved majority rule in 1980 and T&N was soon in conflict with the new government. Stephen Gibbs, one of T&N's senior officers, wrote to the company's general manager Mr. C. Newton in August 1982: 'Public access to the information that we have regarding [work conditions] at Shabanie and Mashaba Mines could explode a bomb that would wreck T&N without any assistance from the Zimbabwe government and that we may find ourselves under considerable pressure from our shareholders to withdraw from Zimbabwe.'[48] In 1982, a new mill at Shabanie was opened but it made little difference. Dr Peter Elmes, a former head of the PRU in the UK and now T&N's consultant, visited the mine in September 1985 and noted: 'Under existing conditions, and as the [Zimbabwean] authorities adopt international hygiene standards, this mine is likely to be closed because of the continued dusty conditions in the mill and the manifest X-rays changes in the older workers.'[49] T&N

[45] ITV World in Action, 'Dust to Dust', 16 November 1981, Transcript. See also L. Flynn, *Studded with Diamonds and Paved with Gold: Miners, Mining Companies and Human Rights in Southern Africa* (1992), pp. 189–202.

[46] K. Nixon memo to H.D.S. Hardie re. Alice film, 21 June 1983.

[47] 'Asbestos Dust Levels in Overseas Companies', T&N Executive Committee Meeting, March 1981.

[48] Memo from S. Gibbs to C. Newton, 10 August 1982.

[49] Dr P. Elmes, 'Report on the visit to Shabanie Mine', 19 September 1985.

ran those mines until 1995, when it sold out to local investors. At no time were work conditions safe. In the past decade they have deteriorated.

In Mexico, as in Zimbabwe, there is little effective state regulation. There is also a lack of information about environmental exposure. In their late 1990s study of working class communities living near asbestos plants in the Valle de Mexico, G. Aguilar-Madrid and her co-researcher examined the knowledge of risk.[50] The twenty-nine factories in the Valle de Mexico employed over 4,000 workers. The average age of the workers was thirty and the majority had worked with asbestos for less than six years. That is surprising as the factories offered steady employment and paid three times the basic wage. The high turnover of labour may reflect a management policy of shedding labour before disease becomes apparent. Less than one in four thought their employment dangerous and almost 80 per cent were satisfied with their jobs. Less than half were aware that they worked with asbestos.[51] Over two-thirds of those in the nearby communities were married women with low levels of formal education. The majority knew that the factories made brake pads and 'car parts' and they perceived the factories as posing a health risk to their families. But they had no access to reliable information about the risks and no avenues for protest.

A study of mesothelioma in the state of São Paulo, Brazil, conducted between 1996 and 1997, again illustrates the problem of knowledge. Lucila Scavone and Jovana de Melo interviewed surviving relatives about the disease and their understanding of its cause. None had received help or information from physicians or state agencies but they could identify the disease and understood its symptoms. They had gained that knowledge by caring for a sick relative. Few, however, were aware that mesothelioma is caused by asbestos.[52]

The Other Empire

It is tempting to view the histories of asbestos in southern Africa and even Latin America through a post-Colonial lens and to attribute the legacy of disease and environmental pollution to the darker side of capitalism. However, many of the problems that characterize the industry in Asia

[50] G. Aguilar-Madrid et al., 'Occupational and Environmental Rights Violation in the Asbestos Industry in Mexico', *Annals of the Global Asbestos Congress*.

[51] Aguilar-Madrid et al., 'Occupational and Environmental Rights'.

[52] L. Scavone and J. de Melo, 'Institutional Knowledge, Common Knowledge: Occupational Asbestos Diseases and Gender', *Annals of the Global Asbestos Congress*.

and Africa are found also within what Ronald Reagan once referred to as the *evil empire*. Western capitalism and Soviet Marxism may have been founded upon contrasting political philosophies, yet in generating, preserving, and transferring heat, their very different industrial systems relied upon asbestos. As a result, the states which have recently emerged from the break-up of the Soviet Union share a common history with other developing economies.

Large chrysotile deposits were discovered in the Russian Urals in 1884 and mining began two years later. With the exception of the Second World War, output rose steadily. At its peak in 1975/6 the Uralasbest complex produced 1.5 million tonnes of fibre.[53] During the twentieth century, over 40 million tonnes of asbestos were extracted from 4 billion tonnes of processed rock. Mining began in the Soviet territory of Kazakhstan in 1965. By 1975 the combined outputs of Kazakhstan and Russia had surpassed Canada as the world's leading producer. Production in the Urals accounted for 73 per cent, followed by Kazakhstan with 24 per cent and the Tuva Republic with 4 per cent.[54] In 2006, over a third of the world's asbestos output comes from Russia's mines (see Table 8.1).

The initial mining at Bazhenovskoye, in the Urals, was done in the warm months from May to October by labour drawn from nearby villages and the excess ore was stored for processing during the winter. All the work including the crushing of the host rock was manual. The final product was packed into boxes and sent to consumers. By the turn of the century the workforce numbered 12,000.[55] On the eve of the Revolution in 1917 there were twenty open pit mines. None other than Armand Hammer was one of those who enjoyed a concession to mine asbestos in the Ural Mountains, courtesy of Lenin. During the 1920s mechanization was gradually introduced with pneumatic drilling first used in 1925. There was no mechanical ventilation in the mills and anecdotal evidence suggests that the dust levels were very high. The completion of a branch line to the trans-Siberian railway in 1927 enhanced access to markets and soon Russia accounted for 15 per cent of world output. The mills constructed in the 1950s incorporated dust extraction systems and work safety improved. But the ore continued to be sorted by hand until the early 1970s.

[53] S. Shcherbakov et al., 'The Health Effects of Mining and Milling Chrysotile: The Russian Experience', in R.P. Nolan et al., *The Health Effects of Chrysotile Asbestos* (2001), pp. 187–98.

[54] Virta, *Worldwide Asbestos*, pp. 11–12.

[55] For a history of the early period, see S.V. Kashansky, S.V. Shcherbakov, and F. Kogan, 'Dust Levels in Workplace Air (A Retrospective View of "Uralasbest")', in G.A. Peters and B. Peters (eds.), *Current Asbestos Issues* (1994), pp. 337–54.

In the 1930s, the USSR began to develop an asbestos manufacturing industry. A five-year plan, which commenced in 1937, saw the manufacture of a range of a/c products, which used most of the 200,000 tonnes of chrysotile mined each year. In 1950, new plants were constructed to make brakes for trains, trucks, cars, and motorbikes. By the early 1960s, more than 80 per cent of Bazhenovskoye's chrysotile went into building materials such as roof tiles, wall panels, pipes, and moulded products. Lower grades of fibre were used in insulation. Treated mill waste went into reinforced asphalt, and roofing paper; untreated waste found its way into railway ballast and road surfacing. Despite a fall in demand, Bazhenovskoye's factories continue to produce 7,000 tonnes of insulation, 6,000 tonnes of textiles, and forty million units of friction materials annually. They also make paper which is 96 per cent asbestos.[56] The factories of Asbest City are all located within the town itself.

The Uralasbest region is unique in terms of the industry's longevity, the large number of workers and bystanders exposed, and the diversity of asbestos manufacture. Over the past two decades, conditions in the mills have improved. Even so, by Western standards dust levels remain high and injuries from noise and vibration are common.[57] To reduce exposures in the worst areas, shifts have been shortened to six hours and 'hazard pay' is given as compensation. Respirators are used routinely, a practice long frowned upon in the West.

The first research into occupational disease at Bazhenovskoye was carried out by the Ekaterinburg Medical Research Centre in 1932. At that time Asbest City had the highest rate of tuberculosis in the Urals and the project was aimed at reducing infectious disease rather than addressing the problem of asbestosis. Until the late 1940s there were no routine medical examinations of miners or mill workers. Officially such a programme was deemed unnecessary as only 'brawny men with robust health' were employed in the industry.[58] A review of miners in 1947 found asbestosis in almost 30 per cent of employees. The majority worked in the crushing plant or the finishing shops. Among those with more than twenty years' exposure the rate was 90 per cent.[59] There were several cases of asbestosis in men aged under 20 who had less than three years in the mills.[60] Over

[56] Shcherbakov et al., 'Health Effects of Mining', p. 194.

[57] S. Kashansky et al., 'Retrospective View of Airborne Dust Levels in Workplace of a Chrysotile Mine in Ural, Russia', *Industrial Health* 39 (2001), pp. 51–6.

[58] Kashansky, 'Retrospective View', p. 54.

[59] Shcherbakov et al., 'Health Effects of Mining', p. 191.

[60] Shcherbakov et al., 'Health Effects of Mining', pp. 190–1.

the next decade the rate of new cases declined sharply due to the transfer of workers with evidence of fibrosis, improved ventilation, and the use of respirators.

In Russia, as elsewhere, the highest risk of lung cancer occurs in those employed in asbestos textile manufacture. In 1964, when a programme to reduce fibre emissions was begun in Asbest City, the death rate from lung cancer was three times higher than the average for the highly industrialized Urals region as a whole.[61] In the decade up to 1982, 181 cases of mesothelioma were reported for the Urals region which has a population of five million.[62] Those returns suggest an incidence of mesothelioma in the occupationally exposed miners and millers similar to that reported in Canada. It is difficult, however, to accept the veracity of such data. During the Soviet period, industrial development had priority over occupational safety, while since 1991 the decline felt throughout the national economy has seen work conditions deteriorate. The Russian attitude towards asbestos disease has also been influenced by the work of Corbett McDonald and his Canadian colleagues.

The Central Asian republic of Kazakhstan, which is the second largest producer of asbestos in the world behind Russia (see Table 8.1), is four times the size of Texas. It was conquered by Russia in the eighteenth century and became a Soviet Republic in 1936. Kazakhstan is blessed with mineral resources, including vast deposits of chrysotile. The collapse of the Soviet empire created a political vacuum and when Kazakhstan gained independence in 1991, most public utilities and capital works were privatized to the benefit of a few individuals. The current President Nursultan Nazarbayev is reported to be among the half dozen richest people on earth.

Soviet industrialization marks every part of Kazakhstan's urban landscape. The large cities are dominated by massive factories, the endless tower blocks of public housing, and the Tets power generation plants. The coal mine at Irikatrust in the northeast is one of the few constructed objects visible from space. The scale of the industrial system created by the Soviets means that the problems of occupational safety and environmental health tend to converge. That legacy includes the zone known as the Polygone, where over 500 nuclear tests were carried out. Kazakhstan is also blighted by numerous toxic chemical dumps and much pollution. The glorification of physical labour was a central feature of

[61] Shcherbakov et al., 'Health Effects of Mining', p. 196.
[62] Shcherbakov et al., 'Health Effects of Mining', p. 193.

Soviet ideology, yet that sentiment did not translate into a commitment to worker safety.

Semipalatinsk is a two-hour flight from Almaty over a largely barren landscape. The city of 350,000 lay at the centre of the Gulag system, which built the region's industrial base. Prisoners built the Tets One power station, which supplies the city with heat in winter. The construction of Tets One began in 1931. The technicians were from Moscow and the plant is a monument to high-quality engineering and political fanaticism. Many of the bricks used were taken from churches demolished as part of the Soviet policy of destroying religious faith. The boilers were insulated with asbestos, as were the heating pipes that line the city's streets.

Coal is trucked by rail to Tets One and then moved by overhead cranes into a crusher where it is pulverized into a fine dust. It is then fed into the boilers and the heated water piped throughout the city. That part of the Tets system is well managed, but the plant itself has many structural flaws; the expansion joints on the boilers are inadequate and the constant vibration breaks down insulation releasing asbestos fibre into the air. There is massive heat loss from the boilers and even more heat is lost in the piping system. The overground pipes are covered in a slurry of asbestos, which soon flakes off and lies about in heaps.

The power plants in Semipalatinsk were sold off in 1997 at a nominal figure to the US-based AES corporation. The Kazakh government wanted foreign exchange and it lacked the resources to run, let alone maintain, the facilities. By any standards the plants are hazardous. At Tets One Jock McCulloch saw bare wiring hanging from fuse boxes, furnace doors which open onto walkways, and buckled and brittle railings a hundred feet above the factory floor. The power stations should by rights have been torn down, but Kazakh consumers can neither support the high costs nor survive a winter without heat when temperatures fall to −40°C. When ATEC took over, every surface subject to heat was covered in asbestos. Insulation was made on the spot by maintenance teams, who mixed a slurry of chrysolite and water, which was then applied by hand to the open surfaces. The covering would soon break apart from vibration or friction and a new coating of paste then applied. The habit of 'washing down' in which loose fibre was hosed onto the ground floor merely shifted pollution from one part of the plant to another. Once the slurry dried it was circulated as dust throughout the building by the movement of workers and the cross-draft from the ceiling vents. During

McCulloch's visit dust was clearly visible in the beams of daylight which stream into the work space from the skylights above. No one knows how many cases of asbestosis there have been in the Tets facilities. ARDs have been hidden by the Soviet orthodoxy that asbestos is harmless, the high rates of tuberculous within the general community, the lack of adequate X-ray equipment, and by a lack of medical knowledge about mesothelioma.

One of the major problems in the Tets plants is the work culture and in particular the attitude to risk. There are no effective trade unions, but there is a deeply ingrained tolerance of hazardous work. The men and women employed in those plants are used to wading through asbestos and breathing in dust. Life is hard in Kazakhstan and any job is valuable.

Once it sold the Tets plants, the national government has taken no interest in their operation. There are no TLVs in Kazakhstan and AES could simply have ignored the asbestos problem. But the AES is also aware that sooner or later there will be litigation. All that is required is a case of asbestosis or mesothelioma to find its way into the hands of a keen lawyer. Perhaps partly for that reason, AES has contracted Atec, a company headed by John Gilbert, who once worked for Cape plc, to remove the insulation. In the first six months of 2002, ATEC removed 1,500 bags of asbestos each weighing 30 kilograms from Tets One. The material is packed into attractive red plastic bags, sealed with string, and dumped at the tip. Without authorization the locals recycle the bags for domestic use.

Two hundred and fifty kilometres northwest of Semipalatinsk is the mining and industrial city of Karaganda. It features a copper-processing plant that bellows a permanent cloud of smog over the town, blurring the horizon for a distance of 30 kilometres. In the late 1940s, Alexander Solzhenitsyn worked as a Gulag prisoner on the local Tets station. Karaganda is the home of the National Centre for Labour Hygiene and Occupational Diseases. The Centre is part of the Ministry of Care of Public Health and its director is Professor Cabdulla Abdikozhaevich. The Centre treats occupational disease and it also conducts research. The emphasis is on prevention and cooperation with industry, but a lack of commitment by private industry, and the absence of trade unions, are barriers to enhanced safety. The presence of foreign investment is both helpful and a problem. Overseas companies have resources lacking in the local economy, but they have little incentive to improve work conditions.[63]

[63] J. McCulloch interview with Professor C. Abdikozhaevich, Karaganda, 2 June 2002.

The most common problem in Karaganda is chronic bronchitis due mainly to the copper smelter. There is silicosis and asbestosis, and obstructive pulmonary disease from gas and oil contaminants. There are also the elevated cancer rates from the Polygone, which affect the entire Semipalatinsk region. The sources of occupational exposure to asbestos include the mines in the northwest (which are currently owned by a Russian consortium), the numerous a/c sheet factories, and the power plants. Extensive exposure in domestic dwellings also occurs from insulated water pipes and boilers, and a/c roofing tiles. According to Soviet ideology, asbestos was harmless and Soviet scientists carried out no studies of occupational disease in Kazakhstan. When the asbestos industry held a World Symposium in Montreal in 1982 the USSR and the eastern European bloc countries declined an invitation saying there was no asbestos disease in their factories or mines.[64] Since independence, sixty cases of asbestosis have been identified at the Centre, and by June 2002 there were a further fifty possible cases as miners have become more aware of the risk.[65] Three cases of mesothelioma were also diagnosed at autopsy. The figures on ARDs are unreliable and Professor Abdikozhaevich believes that a massive problem exists. A national program, which enjoys presidential support, has been set up to improve diagnosis and prevention.

Kazakhstan shares with other developing countries a history on the edge of a major imperial system. That system may have been driven by a very different political ideology, but in regard to asbestos the outcome has been similar to capitalism. The widespread use of asbestos in Kazakhstan has been due to a number of factors. There are large local deposits of the mineral and a severe climate that make insulation essential; and there was the role played by Soviet scientists, who denied the mineral's toxicity. Added to that was the absence of a free and critical press, the general lack of concern with occupational disease, and the lack of independent trade unions or social movements to question work conditions. Underlying all of those elements was the state control of knowledge, which deprived labour of information about the risk. Soviet scientists found their double in Canada, where industry-funded researchers at McGill agreed that chrysotile is a safe and modern material.

[64] Yvon Turcot, Secretariat of the Symposium, in Canadian Asbestos Information Centre, *Asbestos, Health and Society Proceedings of the World Symposium on Asbestos* (1982), p. 533.

[65] McCulloch interview with Abdikozhaevich.

Canada and the World Trade Organization

Canadian governments have played a progressive role on a range of issues including land rights for indigenous peoples and the resettlement of refugees. Unfortunately, the determination of national and provincial governments to create jobs for Quebecois has seen them support the continued export of chrysotile to the developing world. It has also given the industry remarkable leverage in domestic politics.[66] During an asbestos summit held at Parliament House, Ottawa, in September 2003, delegates were bemused to learn that for the safety of MPs asbestos insulation was being removed from the building. That work was undertaken while Canada was still promoting the export of asbestos to developing countries.

Opposition to asbestos use in the OECD states threw the Canadian industry into a prolonged crisis. The QAMA repeatedly warned the federal government that without new markets the mines would have to close. During the 1976 provincial election, Le Parti Quebecois (PQ) attacked Ottawa for its failure to support the industry. When elected, the PQ under René Levesque nationalized the Asbestos Corporation and created the Société National de L'Amiante. Public ownership of the mines was part of the 'Maitres Chez-Nous' programme designed to increase the independence of the Canadian economy from its powerful southern neighbour. As the crisis over mesothelioma deepened, the Asbestos Institute (later to be renamed the Chrysotile Institute) was formed in Quebec in 1984.[67] From its inception, the Institute has always been heavily subsidized by Canadian tax payers. Between 1984 and 2001, the AI, which has always described itself as a 'not-for-profit' body, received C$54 million from government sources. The money was largely spent on promoting the 'safe use of chrysotile' through conferences, public relations initiatives, and scientific publications.[68] To counter consumer resistance, the industry continued to use phrases such as 'locked in' and 'encapsulated' to describe products in which fibre is supposedly rendered safe by being bonded with cement or vinyl. In particular, the AI and Canada relied heavily on the principle of *controlled use* to market chrysotile. Supposedly, once *controlled*

[66] See L. Tataryn, *Dying for a Living: The Politics of Industrial Death* (1979), pp. 15–60.

[67] L. Kazan-Allen, 'The Asbestos War', *IJOEH* 9 (July/September 2003), pp. 173–93.

[68] See J. Kuyek, 'Asbestos Mining in Canada' paper presented to the International Ban Asbestos Conference, Ottawa, 13 September 2003. In that year, the Canadian government expressed its continued support for the industry with a $775,000 grant for the Asbestos Institute spread over three years.

use is established, workers and the users of asbestos products face no significant risk.

The Institute had some success in finding markets in the developing world, but a number of Canada's new customers lacked foreign currency. To overcome that problem the Canadian government working with the World Bank tied specific aid projects in Asia, Africa, and South America to the use of asbestos. Simultaneously, the government in Ottawa began to remove asbestos from public buildings in Canada.[69] In addition to government funding for the Chrysotile Institute, the industry has also been the beneficiary of generous tax concessions. Mining taxes in Quebec are the lowest in the country and the annual royalties paid by industry usually total as little as a fifth of the subsidies it has received. In 2002, although only 6,992 people worked in all types of mines in Quebec, total public expenditure on the mines was C$108 million. The subsidy per job amounted to C$10,120 from the Quebec government and over C$13,000 from federal sources.[70] That scale of support has enabled the industry to shape the public perception of the asbestos hazard both at home and abroad.

During the 1970s and 1980s, France was a major user of Canadian chrysotile and opposed any suggestion of a ban in the EC. The French attitude was at odds with that of its neighbours as Iceland (1983), Norway (1984), Denmark (1986,) Sweden (1986), Austria (1990), the Netherlands (1991), Italy (1992), and Germany (1993) introduced unilateral bans. Those decisions increased the pressure on France, as did the activities of an informal coalition of French workers, trade unions, scientists, and environmentalists who campaigned for an end to asbestos imports. The Association Nationale de Defence des Victimes de L'Amiante (ANDEVA) also lobbied strongly. In 1993, the EC introduced a draft for a total ban on asbestos. The initiative was opposed by France, Spain, Portugal, and Greece and the plan was shelved. But that decision only increased the domestic opposition to imports. The French Medical Research Council (INSERM) was asked in 1996 to review the scientific literature. The day after the release of its finding that all types of asbestos cause cancer, France announced its intention to ban chrysotile.[71]

[69] K. Sentes, 'Asbestos and the World Trade Organization: A Case Study Challenging the Legitimacy of the WTO as a Forum for the Adjudication of Public Health and Environmental Issues' (University of Alberta MA 2002), p. 24.

[70] M. Winfield et al., *Looking Beneath the Surface: an Assessment of the Value of Public Support for the Metal Mining Industry in Canada* (2002).

[71] L. Kazan-Allen, 'The WTO Speaks: Chrysotile is Bad for You', *British Asbestos Newsletter* Issue 39 (Summer 2000), p. 2.

The Canadian government was hit by protests at home that it was not doing enough to protect jobs. More worrying were the claims by its critics in Quebec that federalism had failed, thereby once again raising the spectre of separatism. The national government set aside C$500,000 to promote 'the responsible use of chrysotile' through a series of missions, consultations, conferences, and international workshops.[72] Consular officials lobbied scientists, politicians, trade unions, and journalists in Belgium, France, the UK, Brazil, South Africa, and Russia.[73]

By the late 1990s, France was using minuscule amounts of chrysotile and the loss of that market would have had little impact upon Canadian exports. What the Quebec mining companies feared was a domino effect in the developing world, particularly in the former French colonies of Morocco, Tunisia, and Algeria. It was for that reason, with the backing of Brazil and Zimbabwe, that in October 1998 Canada complained to the WTO that France was illegally resticting trade. The basis of the Canadian case was that chrysotile presents no risk to health; compliance to current exposure limits means chrysotile can be used safely; and substitute fibres are hazardous.[74] France was supported by the EU and the US.[75] Under the WTO mandate a member country deemed to have 'excessively restricted trade' can face penalties. A failure to comply with a ruling can then lead to trade sanctions. Given the nature of the WTO it is not surprising that its appeals panels have invariably supported trade; by 1999, of the twenty-two cases heard only three were decided in favour of public health or the environment.[76] The principle of *controlled use* formed the basis of the Canadian case before the WTO.

According to the Chrysotile Institute, *controlled use* is multifaceted. It requires that all distributors or manufacturers hold an import permit which can be withdrawn if a company fails to meet strict requirements; all users should be licensed and workers trained to use the material safely; users are forbidden from selling to third parties; and unused material has to be returned to the manufacturer, who in turn must provide products cut strictly to specifications, thereby avoiding the dumping of waste. In addition, downstream use of a/c materials should be carefully policed.[77] Canada has never suggested it should be part of such a regulatory process or that it should bear any responsibility for the end use of asbestos

[72] Sentes, 'Asbestos and WTO', pp. 27, 36. [73] Kazan-Allen, 'The WTO Speaks'.

[74] Sentes, 'Asbestos and the WTO', p. 36.

[75] For an inside account of the proceedings, see B.I. Castleman 'WTO Confidential: The Case of Asbestos', *IJHS* 32 (2002), pp. 489–510.

[76] Sentes, 'Asbestos and WTO', p. 33. [77] Castleman, 'Controlled Use', p. 294.

products. The burden of *controlled use* and surveillance falls upon the host country. That leaves unanswered the question as to why a manufacturer in Thailand or Bangladesh or India would carry out voluntary industrial hygiene surveillance of their customers. Alternatively, the operator could wear a respirator. On closer examination the principle of *controlled use* dissolves into contradictions. If chrysotile is harmless, why is *controlled* use necessary? Alternatively, why would any manufacturer use a material that requires such scrupulous regulation?

In OECD states *controlled use* would be both expensive and difficult to police; in poor countries, where a/c building materials are constantly recycled for domestic or commercial use, it is fanciful. Asbestos may be 'locked in' to a/c products and therefore present far less of an immediate hazard to labour, but it is also true that as early as the 1970s there was evidence showing high levels of exposure for those working with such materials. A Johns-Manville report in 1976 showed that the use of power tools on a/c sheets could cause exposures over 250 f/cc.[78] To reduce that exposure to less than 1 f/cc would require a metal hood fixed over the saw blade, connected to an exhaust hose, which was in turn connected to an efficient fan. Even then, it would pollute the environment.

In September 2000, the WTO's Dispute Resolution Panel endorsed France's right to impose a ban. Although the WTO procedures are confidential, it is known that the scientists commissioned to review the medical literature were unanimous on two issues: chrysotile is a carcinogen and controlled use is impractical.[79] The Panel also endorsed the use of substitutes. On the surface the result showed the WTO willing to put health and the environment ahead of commercial gain. On closer inspection the WTO had little to lose in siding with France. Chrysotile is acknowledged as a carcinogen by the ILO, the WHO, the IARC, the IPCS, the World Bank, the International Commission on Occupational Health (ICOH), and the Collegium Ramazzini. Since 1989, the Building & Woodworkers' International, which represents 350 trade unions with a total membership of thirteen million workers in 135 countries, has campaigned against asbestos. In summary, asbestos is so widely regarded as a carcinogen it was unlikely the WTO ruling would set a precedent for other hazardous materials. The WTO benefited from the decision, because it enhanced its standing at a time it was beset by anti-globalization

[78] Castleman, 'Controlled Use', p. 296.

[79] Kazan-Allen, 'The WTO Speaks', p. 4. See also WTO, 'European Community—Measures Affecting Asbestos and Asbestos-Containing Products'. Report of the Appellate Body, WT/DS135/AB/R, 12 March 2001; Report of the Panel, WT/DS135/R, 18 September 2000.

demonstrations.[80] Its decision ensured that markets for white asbestos in the developing world would be more fiercely contested.[81]

The Invisible Costs

The social impact of asbestos is usually measured in terms of the numbers who have died. Such calculations are always approximate and in poor countries official returns on occupational disease are notoriously unreliable. Between 1994 and 1997, for example, only sixty-six cases of mesothelioma were compensated in South Africa: of those over a third were white males, who made up less than 5 per cent of mine labour.[82] In contrast, independent audits of blacks and coloured people from the Northern Cape suggest a disease rate of between 25 and 45 per cent.[83] Even more disturbing is J.C.A. Davies' study of women cobbers in Limpopo Province, which shows the incidence of asbestosis to be over 90 per cent.[84] Access to compensation in South Africa has never been easy for black or coloured workers. Between 1977 and 1998, only 10,520 people with ARDs were compensated by the Medical Bureau of Occupational Diseases (MBOD) in Johannesburg.[85] The South African compensation system changed with majority rule in 1994, but many of the same problems remain. Most people in the Northern Cape lack access to transport and in many cases they lack proof of employment. Illiterate men and women find the application process intimidating.

The impact of disability depends very much upon the kind of work required for a decent life. Among the poor, an individual's capacity to perform hard labour is often their greatest asset. For that reason ARDs are a major contributing factor to poverty in Limpopo and the Northern Cape. Sixty per cent of households in Limpopo are poor and within that

[80] Sentes, 'Asbestos and WTO', pp. 92–5, argues that the anti-globalization movement, or as she terms it the 'Seattle factor'—a fear of further political fallout against the organization—explains the WTO decision.

[81] See B.I. Castleman, 'Implications of the World Trade Organisation Verdict for Public Health and Global Trade', in *Annals of the Global Asbestos Congress*.

[82] J.E. Roberts, 'What is the Price of 80 KG's? The Failure of the Detection of, and Compensation for, Asbestos-Related Disease: Social Exclusion in Sekhukhuneland' (University of Natal, MA thesis, December 2000), p. 7.

[83] S. Kisting, 'Health Implications' (Industrial Health Research Group, University of Cape Town) (1998), p. 44.

[84] See J.C.A. Davies et al., 'Asbestos-Related Lung Disease Among Women in the Northern Province of South Africa', *The South African Journal of Science*, 97 (March/April 2001), pp. 87–92.

[85] Kisting, 'Health Implications', p. 43.

population those with ARDs have the lowest incomes. They also have the lowest levels of education.[86] Mesothelioma is a hideous disease for a person who has the benefit of advanced medical care. For a man or woman living at the end of a dirt road with no access to medication, it is even worse.

Under South African legislation, first-degree disability means that a man or woman is deemed still capable of performing some types of work. But work for the poor requires the kind of physical labour which is beyond the reach of someone for whom even minimal exertion results in breathlessness. Those with second-degree disability are breathless at rest. To receive compensation claimants have to be aware of their rights and be able to access a medical examination. They also need proof of employment. The process is so drawn out that by the time a payout arrives families are often deep in debt. There is no compensation for those who were employed under the age of 16 and none for those with bystander exposure. Compensation does not allow for the fact that a person with asbestosis will never work again. A survey from the year 2000 found the average total payout for second-degree disability was R19,137, or less than US$ 2,000.[87]

Lost income is only the most obvious consequence of occupational disease. Chronic illness increases dependency within families and time lost from labour to care for a sick relative further reduces incomes and opportunities. That burden falls mostly upon women, who as carers are taken out of wage labour or prevented from participating in the informal economy. The ancillary costs of medical care are often high and even the journey to a hospital for treatment may ruin a family's budget for a month. Those costs impact upon children who suffer lost nutrition and access to schooling. Occupational disease ensures that poverty is passed on generationally.

The story of Ezekial Mavuso, who worked at Penge, is an all too common one in mining communities.[88] As a 12-year-old in 1960, Mavuso began work in the mill, which was full of dust. Fifteen years later, when he became ill, he consulted traditional healers. As his health deteriorated, he finally had an X-ray, which revealed asbestosis. In 1998, he received a lump sum of R10,340 (about US$1,000).[89] Mavuso spends his days and nights coughing and in pain. Any physical activity leaves him breathless

[86] Roberts, 'What is the Price?', p. 84.
[87] Roberts, 'What is the Price?', p. 71.
[88] This case study is taken from J. Roberts, 'What is the Price?', pp. 85–6. The name of the former miner is fictitious.
[89] Roberts, 'What is the Price?', p. 86.

and he is unable to carry firewood, which in the early years of his illness he gathered to sell. His wife and seven children now do that work. On bad days he cannot dress himself. Mavuso seeks medical care only when the pain is severe, but the clinic where he can get paracetamol is 17 kilometres away. The bus fare costs R11 (US$1) and so he rarely makes that journey.[90]

Men like Ezekial Mavuso enjoy better prospects than miners who live across the border in Swaziland. The Havelock/Bulembu mine, which operated from 1939 until 2001, was the main source of wage labour at Pigg's Peak, sustaining both individual families and the regional economy. The labour policies of T&N, the absence of trade unions, and an ineffective regulatory authority meant that the mill was full of dust. Internal correspondence shows that four years after the mine opened, T&N's management was aware that as many as half of those employed in the mill would eventually develop asbestosis. The mine's management made no effort to inform the workforce of the risks they faced. In effect, T&N decided that the health of its Swazi workforce was expendable.[91]

The town of Bulembu is encircled by tailings dumps which are two or three hundred feet high. The milling process was crude and so the dumps are saturated with chrysotile. We know that during T&N's tenure pollution from the tailings was such that a number of women who lived at Pigg's Peak developed the occupational disease asbestosis from environmental exposure.[92] Today fibre from the dumps is blown over the town and the nearby primary school. The only evidence on disease rates is anecdotal. On an interview list of twenty former miners that Jock McCulloch was given in June 2004 by the Nelspruit legal firm of Ntuli, Noble & Spoor Inc., eight had died in the previous eighteen months. In addition, most of the miners McCulloch spoke with at Pigg's Peak have asbestosis. They receive no medical care and they face an uncertain future. The men spent their lives doing heavy work and now they are breathless after walking a hundred metres. They say their chests are rotten and they cannot serve their families. Joel Dlamini, who worked in the mine for more than twenty years, has an asbestos corn on his hand and he can feel the fibre burying itself deeper and deeper into his skin. He fears that is what is happening to his lungs.[93]

[90] Roberts, 'What is the Price?', p. 86.

[91] J. McCulloch, 'Dust, Disease and Labour at Havelock Mine', *Journal of Southern African Studies* 31 (June 2005), pp. 251–66.

[92] See Dr J. Allardice, Memo re. Asbestos Workers Survey at Havelock, 17 December 1979.

[93] J. McCulloch interview with J. Dlamini at Pigg's Peak, Swaziland, 13 June 2004.

In 1997, T&N was taken over by the US company Federal Mogul. Four years later Federal Mogul filed for Chapter 11 bankruptcy, thereby avoiding litigation in the US. That manoeuvre also brought to an abrupt halt the legal action begun in a British court in September 2000 on behalf of Swazi miners. Swaziland is one of the poorest countries in Africa and its government can neither afford to clean up the waste dumps nor provide medical care for miners and their families.

A New Politics

The second wave of asbestos industrialization has generally been welcomed by governments in the developing world. That is true of India where, until the formation of the Ban Asbestos Network in 2002, manufacturers faced little opposition. However, in other countries such as Brazil it has encountered strong resistance from trade unions, scientists, social movements, and health workers. Loose confederations of those groups were responsible for the asbestos summits held in Johannesburg (1998), Osasco (2000), Tokyo (2004), and Bangkok (2006). They were novel events in the sense of bringing together the industry's opponents. They were also a powerful reminder of the importance of knowledge. Since the 1930s, asbestos producers have held their own closed conferences to share information and design ways of keeping medical knowledge from employees and governments. Until the Johannesburg Summit the industry's victims had no such forums.

In opening the National Asbestos Summit in November 1998, the Deputy Minister of Environmental Affairs Peter Mokaba made reference to the Bhopal tragedy and the Rio Declaration, thereby placing the South African experience within a global setting.[94] The Summit's aims were to share information, hear the voices of those affected, arrive at a picture of the industry's history, identify the extent of the health problem, and formulate practical solutions.[95] The Summit began with presentations by provincial delegates, who described the impact of mining. Mr. Zac Mabileji, from Limpopo, spoke of how his community was wracked with illness and of children who spend eight hours a day in classrooms made from asbestos bricks. Like other delegates he wanted the mine owners

[94] The Deputy Minister of Environment Affairs and Tourism Peter Mokaba, in *National Asbestos Summit*, pp. 15–18.

[95] Hon. Jerry Ndou, Chair Asbestos Subcommittee, Asbestos Subcommittee Report, in *National Asbestos Summit*, pp. 20–1.

tracked down: 'We feel that they are the people that we need to confront at some stage and make them account for what they have done to our communities and to our people.'[96] The Declaration called for a review of compensation and other remedial systems, a strengthening of health care, the promotion of community development and rehabilitation, and an immediate ban on amphiboles and a phasing out of chrysotile.[97]

The Summit was a political forum and besides its stated objectives other agendas were at play. The industry, which was well-represented by delegates from Zimbabwe and Canada, viewed the proceedings as a threat to the continued use of asbestos in southern Africa. An Australian industry representative, Mr. Don Monro, spoke passionately in favour of chrysotile. Aided by Canada, the Zimbabwe delegation was able to fend off an immediate South African ban. It would be a further eight years before a ban was enforced.

The Brazilian Association of Workers Exposed to Asbestos (ABREA) was formed in the late 1990s to fight for the rights of labour and work for the banning of asbestos. The ABREA has almost a thousand members and, like other social movements in Latin America whose membership is based upon a shared deficit, it has self-consciously set out to challenge the idea that politics is owned by the state. The ABREA has successfully moved the problem of occupational disease into a wider forum drawing upon a loose coalition of women, the young, and those unfit for work, who in Brazil as elsewhere are usually excluded from the public arena.[98] It was largely through the efforts of ABREA that the Osasco Conference was held in September 2000. The conference which attracted delegates from more than two dozen countries involved a sharing of mutual concerns and information across national boundaries. A series of workshops involved, for example, Brazilian construction labourers and Canadian auto-workers, who share common interests. The Osasco declaration was written around a set of broad objectives. They include promoting solidarity among anti-asbestos campaigners, campaigning for bans, assisting asbestos victims, opposing industry efforts to transfer discredited technologies to the developing world, lobbying politicians and trade unions, and placing asbestos issues into a global context.[99] The conference saw immediate

[96] Mr Z. Mabileji, the Northern Province, in *National Asbestos Summit*, p. 23.

[97] 'Declaration of the National Asbestos Summit, November 1998', in *National Asbestos Summit*, p. 6.

[98] Scavone et al., 'Asbestos Diseases in Brazil'.

[99] See 'The Osasco Declaration', *Annals of the Global Asbestos Congress*.

bans introduced in the cities of Sao Paulo and Rio de Janeiro. National bans soon followed in Chile and Argentina. Before the end of that year the major Brazilian producer Saint-Gobain announced that it would stop using asbestos in Brazil by 2005. Osasco was followed by the Global Asbestos Congress (GAC) held in Tokyo in November 2004. It attracted more than 800 delegates from forty countries including representatives from Asian trade unions and the ILO.

The Johannesburg, Osasco, and Tokyo summits were democratic forums at which non-expert opinion was treated with respect. In that sense they were the opposite of the litigation driven hearings found in OECD states. Courts of law and public inquires are owned by experts who treat the experience of those with ARDs as mere anecdote. It is one thing to have a dreadful disease like asbestosis; it is another to be denied the right to express an opinion as to why one's health has been ruined. The Ontario Royal Commission of 1984, for example, was dominated by legal and medical experts, which left little room for the voices of those who had worked in the industry.[100] When they did speak, their testimony carried no weight. The most important legacy of the Summits was neither the regional bans in Brazil, Argentina, and Chile nor the eventual ban on asbestos six years later in South Africa, but the ethos they embodied. The success of those coalitions in South Africa and Brazil could well be a forerunner of other environmental movements in the developing world.

The Tide Turns

Against the background of the first two summits an important legal case was being played out. In 1997, almost 8,000 asbestos miners and their families began an action in a London court against their former employer Cape Asbestos (now known as Cape plc). At that time it appeared that the claimants had no chance of success. The plaintiffs were South Africans, and their injuries were sustained in South Africa while they were employed by fully owned subsidiaries of a British company. The corporate veil, which since 1948 Cape had utilized to separate the parent company from its South African subsidiaries, appeared to offer an insurmountable barrier to litigants. Cape plc had no assets in South Africa and it knew that there were no precedents for such a case to be

[100] *Report of the Royal Commission on Matters of Health and Safety Arising from the Use of Asbestos in Ontario* (1984), vol. 1.

heard in Britain. Probably Cape assumed that no British government, even a Labour government, would have welcomed a British company with an apartheid-era history being scrutinized in a London court. The company also had time on its side for the longer the hearings were drawn out the fewer plaintiffs would be left alive. Cape was so successful in delaying proceedings that 776 claimants died before the final settlement. Finally, a loss by Cape plc would have carried serious implications for multinationals based in the EU.

The case became bogged down in legal argument for six years. Eventually Cape was co-joined as a defendant by the South African corporation, Gencor, which had taken over Cape's mines in 1980. The case was finally settled on 13 March 2003.[101] Cape plc agreed to pay £7.5 million in compensation to 7,500 workers, while Gencor agreed to set up a trust fund for its workers worth R448 million (£37.5 million). In addition, R40 million was set aside for the rehabilitation of mining sites under a scheme to be administered by the Department of Mineral and Energy Affairs.[102]

The result was remarkable. The Cape claimants won the right to run a class action in a British court for injuries sustained in South Africa, while the Gencor settlement is the first time in South Africa's history that black miners have received compensation under common law for workplace injuries.[103] Besides bringing relief to miners and their families, the settlement may change the way in which multinational companies based in the EU behave in the developing world. Such cases are expensive to defend and generate bad publicity. In future multinational corporations will either have to step back from the operations of their subsidiaries so that they have genuine autonomy or take a hands-on approach to ensure that breaches of occupational safety do not occur.[104] The second strategy will force multinationals to close the gap between their occupational health and safety practices at home and abroad, thereby ending

[101] 'Black Workers to Receive £45 million Asbestos Settlement', *The Guardian*, 14 March 2003. See also R. Meeran, 'Cape Plc: South African Mineworkers' Quest for Justice', *IJOEH* 9 (September 2003), pp. 218–27; P. Muchlinski, 'Corporations in International Litigation: Problems of Jurisdiction and the United Kingdom Asbestos Cases', *International and Comparative Law Quarterly* 50 (January 2001), pp. 1–25.

[102] 'Gencor pays R460m to Settle Asbestos Claims', *Mail & Guardian* [Johannesburg], 30 June 2003.

[103] Summary Gencor Settlement between the Claimants, Gencor and Ntuli, Noble & Spoor, Johannesburg, 13 March 2003.

[104] See, for example, 'South Africa Asbestos Ruling May Bring Flood of Claims to London', *Lloyds List*, 30 August 2000; and 'Litigation that Looms from Lands Far Away', *Financial Times*, 22 August 2000.

the double standards which make developing countries so attractive to unscrupulous investors.

The community groups which brought the claims against Cape and Gencor had few natural allies and several determined opponents. But they did have some notable advantages. Work conditions on the mines were appalling and, as the interest group Action for Southern Africa pointed out, the case arose because a British company took advantage of the racist labour practices typical of apartheid. Once the issue of a forum was decided it would have been relatively easy to convince a British court that Cape plc was negligent. The Concerned People Against Asbestos had the imagination and flexibility to fight the case on the local, national, and international fronts. There was a high level of political skill in the community and its members and leadership had the endurance and persistence, which were essential for survival under minority rule. It had the support of the National Union of Mineworkers (NUM) and the International Chemical, Energy & Mineworkers Federation, and NGOs such as One World Action, the World Development Movement, and Amnesty International. A number of concerned individuals provided much needed expertise, as did some officers in the Department of Health at Kimberly. The provision of legal aid in the UK and the commitment of the lawyers Ntuli, Noble, and Spoor were vital to the case being successful. The South African government had good reason to support the claimants and to make representations on their behalf. If the claims had been lost it would have been left to pay the massive costs for rehabilitation and medical care.

Despite its support of the Cape claimants, the South African government was reluctant to ban chrysotile. The complex regional relationship between South Africa and Zimbabwe gave the industry a reprieve and it took more than eighteen months after the Cape/Gencor settlement and six years after the Johannesburg Summit for the Mbeki government to finally agree to ban all forms of asbestos. That decision was achieved in the context of a dying industry. The last of the amphibole mines in South Africa closed in 1996, as the markets for blue and brown fibre in the Middle East dried up. They were followed by the closure of chrysotile mines at Msauli (2000) and then by the Havelock/Malembu mine in Swaziland (2002), leaving Zimbabwe as the sole regional producer.

The Zimbabwe regime fought hard to prevent the ban. It argued that the Western obsession with environmentalism and occupational health are luxuries that Africa cannot afford. The Zimbabwe press described the campaign as 'vicious'. It pointed out that over 90 per cent of asbestos fibre mined in Zimbabwe is exported and earns around US$65 million per annum in foreign exchange for a crumbling economy.[105] Conflict between the South African NUM and its Zimbabwean counterparts intensified in the wake of the World Summit on Sustainable Development in September 2002. Despite the efforts of Zimbabwe's Chamber of Mines and trade unions, which argued that a ban would cost thousands of jobs, in March 2003 the Southern Africa Development Community passed a resolution for a regional phasing out of asbestos.

In June 2004, the South African Minister for the Environment announced that asbestos products would be banned.[106] While the ban is a compromise in that certain products will be given a five-year phasing-out period, it is an important step in enacting the resolutions taken six years earlier at the Summit. The delay has come at a cost for some South Africans. For more than a decade, Zimbabwe has shipped 90 per cent of its fibre through Durban and has stored large quantities of chrysotile in Durban warehouses. Until 1995, when Shabanie was a subsidiary of T&N, fibre left the mines in bags marked with the warning labels, 'Contains Asbestos Fibres', 'Avoid Creating Dust', and 'Breathing Asbestos Dust May Cause Cancer and Other Fatal Diseases'.[107] After the mines were bought by indigenous capital, that practice was abandoned. Supposedly at the insistence of customers, the bags are now marked with 'Product of Zimbabwe' and a code number.[108] Once the bags leave the mine railway workers, stevedores, and truckers who handle the material are unaware that it is asbestos. Even heavy plastic bags can split when loaded by cranes onto railway carriages or dropped onto wharves. There has also been a wider cost. If South Africa had banned asbestos in 1998 the number of its citizens exposed in mining, processing, transporting, and manufacture would have been greatly reduced. Such a ban by the region's most important economy would also have encouraged other countries in Southern Africa to do the same.

[105] 'Proposed Asbestos Ban Causes Anxiety in Zimbabwe', *The Herald* (Harare) 27 August 2002.

[106] 'Asbestos Products to Become Illegal, says Van Schalkwyk', *The Cape Times*, 22 June 2004.

[107] Correspondence: Shabanie & Mashaba-Asbestos and Health, March 1985

[108] J. McCulloch interview with L. Thompson, manager of Shabanie mine, Zvishavane, Zimbabwe, 11 March 1998.

Conclusion

The first cycle of the asbestos story (1890–1972) was played out with frightful consequences in Western Europe and North America. When faced with litigation or public censure, the industry has always protested that if it had known of the dangers no one would have died of asbestos disease. Within the second cycle, where work conditions are at least as hazardous as those found in US factories in the 1920s and health care is poor or non-existent, there are no such excuses.

The industry's survival into the twenty-first century has been abetted by a number of factors. In countries like India, Kazakhstan, Brazil, and South Korea, the industry has been welcomed by compliant governments and encouraged by the absence of effective trade unions or a critical press. Its greatest advantage has been the lack of an informed public opposed to the use of a known carcinogen. The Canadian industry, strongly supported by its national government, has played a key role in promoting the fiction that white fibre is harmless. To that end it has manipulated the medical record and corrupted public debate. It has also influenced forums such as the WHO.[109]

The industry has never welcomed public scrutiny, and for that reason it has shifted ever further away from effective regulation. First, the manufacture of particularly hazardous products such as asbestos textiles was moved to new economies. When local opposition arose production was moved again. As countries become successful industrially they tend to export their hazardous industries to poorer neighbours. In the past decade, the wealthier economies in Asia have reduced their use of asbestos, while the poorer countries in that region now consume more fibre.

The new generation asbestos companies are very different to those corporations which once were so dominant. The majority are small indigenous firms that tend to be insensitive to bad publicity. They are opportunistic in seeking the short-term profits that can be made from a/c products and they work in the kind of unregulated environments enjoyed eighty years ago by US and British corporations. The industry in the developing world is too new and the epidemiology on disease too limited for the impact of manufacturing to be fully understood. There is, for example, no reliable data on mesothelioma for Latin America, Asia,

[109] See J. LaDou, 'An International Review of Occupational and Environmental Asbestos Issues: Funding of International Agencies', Paper presented to the International Ban Asbestos Conference, Ottawa, 13 September 2003.

or Africa. The only thing we know for certain is that inadequate regulation has resulted in hazardous work conditions for which labour will eventually pay.[110] In twenty years, when claims for occupational disease and environmental exposure surface in India, China, Russia, Brazil, and Thailand, few of the new companies will still be in business. Even fewer will have assets available to plaintiffs.

It is easy to conjure up the ideal setting for the industry simply by identifying those features which have rendered it obsolete in North America. To flourish, the industry should have strategic importance in an economy in which it provides much needed employment. A compliant government and a lack of scrupulous or competent regulatory authorities is certainly an advantage. So too is the absence of a critical press, an environmental lobby, or social movements to highlight the dangers. Weak trade unions and preferably migrant workers who will become ill out of sight are also an advantage. So too is a low life expectancy. If life expectancy is sufficiently low there will be no mesothelioma, because asbestos workers will not live long enough for the disease to become manifest.

Zimbabwe presents such an environment. It has an anti-democratic government intolerant of criticism and brutal in its suppression of opponents. Zimbabwe also has a political culture of misinformation and deceit. There is a minuscule scientific community dependent for employment upon the state. The impact of HIV/AIDS and government mismanagement has seen life expectancy for a male plummet below thirty years. In the future that is certain to obscure the impact of asbestos. The industry is also sheltered by a political discourse about imperialism. In defending Zimbabwe's industry the Mugabe regime has been tireless in its attacks upon what it terms 'the self-indulgent environmentalism of the West'.[111]

Variations on the theme of *controlled use* have been adopted by mining interests in Zimbabwe and Brazil. The Brazilian industry has argued that its chrysotile is safer than foreign chrysotile, local industry uses asbestos in a responsible manner, substitutes are too expensive, ARDs are rare, and a ban would lead to unemployment.[112] Zimbabwe has used the same arguments to justify the continued mining at Shabanie. It is ironic that in Brazil and Zimbabwe the idea of asbestos as a quintessentially modern material, an idea which for decades lay at the heart of the industry's

[110] See 'Worldwide Proliferation of Asbestos Disease', Castleman, *Asbestos*, pp. 780–94.

[111] See 'NUM Lashes Out at Zimbabwe Objection to SAADC Resolution on Asbestos', *Financial Gazette* [Harare], 6 March 2003.

[112] F. Giannasi, 'Globalisation from Below: Building an Anti-Asbestos Movement in Brazil', *Annals of the Global Asbestos Congress*.

advertising, has been superseded. In the twenty-first century asbestos has been transformed into a product ideally suited to poor countries, because it is so cheap. Behind that facade lies the assumption that, although the occupational and environmental disease which goes with asbestos manufacture is no longer tolerable in the North, it is acceptable in the South. That in turn involves a discounting of lives across the geographical divide between the rich and poor.

If asbestos disease was a tragedy the first time around, what word can be used to describe the injuries suffered by men, women, and children during the industry's reincarnation?

9

Defending Asbestos: Twenty-First-Century Perspectives

Dear Sir, I am one of the aftermaths of asbestos as my husband was one of the victims of dreaded asbestosis. It is something one never forgets, the horror of seeing one's loved one die so cruelly. In my judgment, my husband was murdered just as surely as if he'd been beaten to death with a stick. He was not aware of the dangers of asbestos materials when he went to work in 1942.

Letter to Professor Irving J. Selikoff MD from Mrs Eunice Bailey
28 January 1986

Charges of cover-up and conspiracy are unfounded Johns-Manville Corporation

Annual Report (1978), p. 26

In 1973, Matt Swetonic addressed the Asbestos Textile Institute in Arlington, Virginia. Despite the industry's travails, he was his usual eloquent self, and he posed his audience a question: why had asbestos been singled out as the prime target for so many assaults? His answer: '[I]t was vulnerable to attack, because there was someone willing to lead that attack [Selikoff], and because Congress had graciously provided the vehicles for such an assault.'[1] This refrain—that the industry had itself been a victim of asbestos—was to become a familiar one after the 1970s. In June 1977, the managing director of Cape Asbestos, Geoffrey Higham, appeared before a government committee in London and told its members that asbestos had been misunderstood and that his company had nothing to conceal: 'The main feature of the criticisms against the industry has been

[1] M.M. Swetonic, 'Why Asbestos?', Presentation before ATI, 7 June 1973.

the lack of appreciation of what the industry has done and how it has changed over the years.'[2]

Supposedly the industry, like its employees and the users of asbestos products, had been trapped by imperfect science and the unique properties of the magic mineral, which meant that ARDs only appeared many years after workplaces had been made safe. That particular view of history was endorsed by the Ontario Royal Commission in 1984 when it concluded: 'To learn from the asbestos experience is to learn from the harshest of schoolmasters—human tragedy.'[3] It is also a view that has been endorsed by some of the medical community, and recently by Peter Bartrip, Rachel Maines, and Robert Virta.[4] Given that tragedies are unavoidable events, nothing could be further from the truth. The industry leaders were well aware of the hazards of asbestos long before the US and European public, and they spent large sums of money hiding medical evidence about those hazards. Miners in Canada and South African were never told of the true extent of the risks they faced, and neither were regulatory authorities, nor shareholders, nor the wider medical community.

The history of occupational health in the asbestos industry has been punctuated by three crises: asbestosis (1920s), asbestos-related lung cancer (1940s), and mesothelioma (1960s). In that sense, the industry was a victim, but its reaction was never that of an injured and vulnerable party. As this book has shown, during each crisis the industry was able to organize strategies and defences that grew more sophisticated as the evidence against asbestos mounted. The story has certain similarities to Big Tobacco: except that asbestos industrialists had been defending asbestos far longer than the tobacco barons.[5]

The ability of companies to defend asbestos was based, in the first instance, on their own knowledge. Employers knew far more than regulatory authorities in individual states about disease in the workplace. They knew, for example, which parts of the productive process were the dustiest, and they had access to the health records of employees. Unlike regulatory authorities they were well informed about disease in mines *and* factories. The industry leaders entered into commercial alliances,

[2] G. Higham, Cape Asbestos Company. Testimony before the Simpson Committee, 29 June 1977, transcript, p. 146.

[3] *Report of the Royal Commission on Matters of Health and Safety Arising from the Use of Asbestos in Ontario* (1984), Vol. 1, p. 818.

[4] P. Bartrip, *The Way from Dusty Death* (2001); Bartrip, *Beyond the Factory Gates* (2005); R. Maines, *Asbestos and Fire* (2005); R.L. Virta, *Worldwide Asbestos Supply and Consumption Trends from 1900 through 2003* (2006).

[5] S. Glantz et al. (eds.), *The Cigarette Papers* (1996).

they exchanged technical knowledge, and they jointly funded medical research. The minutes of an Asbestos Cement Producers' Association (ACPA) meeting held at Johns-Manville's offices, New York, in November 1969 illustrate how well informed were the leading manufacturers. At that meeting, a discussion of product liability identified four classes of potential litigants: those who worked daily with a/c products; those who lived near a plant or mine and as result were exposed to airborne fibre; those who periodically used asbestos products; and lastly the general public. The meeting was told that if mesothelioma resulted from trivial exposure then a large number of claims could result from all four categories.[6] It was a perfect assessment of what was to happen in US courts more than a decade later.[7]

The defence of the product also reflected the industry's characteristics. Its product line spanned a huge variety of commodities and manufacturing processes. In its heyday, the industry was a major employer in Western economies and penetrated into many key industries, such as construction and defence. It was an industry with political friends. Moreover, it spanned the globe. Over fifty years before the term 'globalization' came into vogue among economists and academics, the asbestos industrialists were carving up the planet into spheres of commercial influence. While these men searched for profits and markets, they also networked and cooperated through cartels and trading associations that eventually numbered over a hundred. After the 1960s, these became the springboard for the industry's propaganda and scientific bodies—such as the ARC and AIA—that almost from the outset were international in scope.

As we have shown, a number of key strategies were important in protecting knowledge and defending asbestos. The first was suppression of information. There are some who would deny this and argue that information about the health hazards was always easily available in the classic literatures of Merewether, Doll, and Wagner. This is the view of Rachel Maines, who has argued that the evidence of an asbestos conspiracy rests only on a small group of documents and a few mouse experiments conducted in the 1940s and 1950s. This is false. The major

[6] Memo to Principals, Representatives and Guests Attending Health & Safety Council/ Asbestos Cement Producers Association, First Annual Meeting, 21 November 1969.

[7] This did not stop members of ACPA expanding production, or even opening new facilities. For example, Capco, an a/c pipemaker in Ragland, Alabama, opened in 1965 and operated throughout the 1970s, using both chrysotile and amphiboles, with disastrous consequences for the workforce. See J. Centers and M. Korade, 'Special Report: Cement Asbestos Products Company (Capco)', *The Anniston Star*, 27–30 March 2005. Posted at: http://www.annistonstar.com/showcase/2005/as-specialreport-0327-0-5c28k0245.htm

companies were so adept at controlling the flow of knowledge that a host of important studies—one might term them lost literatures—never reached the public domain and would only be read decades later in courtrooms. In South Africa, they include Slade's study of mine workers in the Transvaal (1930) and the PRU Survey from the Northern Cape (1962). In addition, twenty years of scientific sabotage at the NCOH in Johannesburg rendered most South African asbestos science irrelevant. South Africa was an anti-democratic state where a small and fragile research community was dependent for its funding upon industry and the public purse. Under apartheid the mining companies were able to secure the boundaries to knowledge thereby ensuring trade unions and regulatory authorities outside South Africa were kept ignorant of the true extent of the dangers. No fiction was unthinkable and in addressing a national audience the industry linked criticism of asbestos to wider political agendas. In particular, the SAAPAC, abetted by the Department of Mines, spun the fantasy that the critics of apartheid lay behind the opposition to South Africa's crocidolite. The industry was so successful in hiding the evidence of risk that conditions on the mines barely changed after 1960, and South Africa continued to export crocidolite until 1996.

In America, the lost literatures included the Johns-Manville funded research by Pedley and Stevenson (1930); Leroy Gardner's industry-sponsored work on lung cancer at Saranac (1940); and the ATI's Hemeon Report (1944). These studies, which could have directly benefited workers involved in fatal occupations such as insulation, were simply never made available. In other countries, such as the UK, Canada, and Italy, even studies that saw the light of day had their texts altered, were subject to industry pressure, or the findings were ignored. Vigliani's work on manufacturing in Turin (1942) falls into that category; so does Doll's 1955 T&N study, and the Canadian research of Braun and Truan in 1957. Most of the asbestos used in the US came from Canada, which put the mines of Quebec at centre stage in any discussion of risk. Industry claims from the early 1930s that there was no asbestosis at the Thetford or Asbestos mines set a precedent. Soon it would be argued that not only was there no asbestosis, but also that there was no lung cancer and then no mesothelioma. In the short term, those deceptions saved the mining companies from having to make workplaces safer. More importantly, it gave the industry a method, which it slowly refined, of how to use science for commercial advantage. By the time the mesothelioma crisis arrived the industry was well versed in the myth that Canadian fibre is benign. The scene was set for the coping stone of that deception: the chrysotile defence.

Another key strategy was to exploit the ambiguity of medical knowledge. That ambiguity stemmed from the singular position of occupational medicine, which was faced with the problem of reconciling medical and commercial imperatives.[8] In the careers of company physicians such as Paul Cartier, R.H. Stevenson, Kenneth Smith, John Knox, and Walter Smither, that dilemma was usually resolved in favour of the latter. When Smither was asked in the 1970s, whether his duty to patients conflicted with his duty to the company, he replied smoothly: 'I don't see any conflict there at all—anymore than I see a conflict in sewer workers who are exposed to horrible diseases. These men give a service in a hazardous occupation. My . . . duty . . . [is] to the community—as is everyone's.'[9] At Johns-Manville, Kenneth Smith (like Cartier and Knox) had hidden diagnoses of ARDs from patients and notoriously asserted that Johns-Manville had no cancer problem. Smith shared data collected by MetLife with business associations such as the ATI and the Asbestos Cement Products Group, but he did not share it with employees, trade unions, or regulatory authorities. It is a dismal picture for those who believe in the impartiality of science and professional integrity, but it stemmed partly from powerful social and institutional constraints. Smith as a company physician probably had little capacity to alter things. Even when he made recommendations, as he made clear, 'The decision was the company's, not mine'.[10] Eventually, Smith escaped those conflicts by first leaving the company (according to Johns-Manville, he was fired because of his heavy drinking) and then in 1976 agreeing to give damning depositions against his former employer in two important court cases. Two months later he was killed in a hit-and-run accident in his home town of Windsor, Ontario.

The silence of the medical community has an even darker side. Too many scientists served the industry's cause by arguing for the mineral's indispensability, warning about lost jobs, and testifying against workers in compensation cases. Like the tobacco industry, asbestos manufacturers never had any problems in finding physicians and epidemiologists ready to argue that the risk of working with asbestos was 'acceptable'. From the malleable clay of the medical profession, the industry moulded a key

[8] T.H. Murray, 'Regulating Asbestos: Ethics, Politics, and the Values of Science' in R. Bayer (ed.), *The Health and Safety of Workers* (1988), pp. 271–92. See also D. Ozonoff, 'Failed Warnings: Asbestos-Related Disease and Industrial Medicine' in R. Bayer (ed.), *The Health and Safety of Workers* (1988), pp. 139–218.

[9] BBC TV Horizon, 'Killer Dust', 20 January 1975.

[10] Deposition of Dr K.W. Smith in *Estate of William Virgil Sampson v. Johns-Manville Corp.*, Jefferson Circuit Court 7th Division No. 164–922, 21 April 1976, p. 22.

part of its defence: uncertainty. This involved all the paraphernalia of industry funded/commissioned studies, front organizations, and attorney and media-driven disinformation. As David Egilman and his colleagues have remarked: 'The success of these strategies over many years is apparent in their continued impact on standard setting, limitation of tort liability, and delay and weakening of regulatory oversight.'[11] That success also contains a paradox: asbestos production did not fall when Wagner's research became known. On the contrary, world production rose to record levels. Mesothelioma killed workers, but not the industry.

In America and Europe, however, rising morbidity and mortality from ARDs meant that eventually the dam had to break. The fateful day would arrive when it was no longer possible for the industry to maintain its public posture of innocence in the face of growing public knowledge. For the world's biggest asbestos company that day dawned in 1978. Johns-Manville issued a combative annual report (much influenced by Hill & Knowlton) that dismissed all the media stories as fantasy and sensationalism. The report tried to cast doubt on whether asbestos could cause lung cancer, before chairman John McKinney told shareholders: 'Claims that Johns-Manville did not inform its employees and the public of the hazards of asbestos are unfounded.'[12] Within months, McKinney received a confidential letter from one of his executives, which told him bluntly: 'We must face reality and ask ourselves whether or not we are are playing semantic games in constructing our defense, which may be appropriate for a court of law, but fade into incredulity under the state of public opinion.' The letter cited many examples that were intended to show McKinney, 'why we have members of Congress calling us liars'.

- JM position: Prior to 1964, we lacked scientific knowledge about cancer hazards of asbestos.

  Fact: In the 1950s, JM officials knew of scientific studies showing a relationship between cancer and asbestos.
- JM position: There will be no future problems with asbestos workers.

  Fact: Workers in Mexico are not being protected.
- JM position: Since 1964 JM has 'come clean' with employees about asbestos hazards.

[11] S.R. Bohme, J. Zorabadeian, and D.S. Egilman, 'Maximising Profit and Endangering Health: Corporate Strategies to Avoid Litigation and Regulation', *IJOEH* 11 (2005), pp. 338–48.

[12] Johns-Manville Corporation, *Annual Report* (1978), p. 26.

Fact: A current 41-year employee of JM has called me to say that he has a spot on his lung... but no one at JM has ever communicated with him about it or warned him of the dangers.

- JM position: There was no cover-up.

  Fact: A current employee (same as above) says his father worked for JM for 35 years at Manville and was never told about dangers. Father was dismissed because of 'miner's asthma' and later (in 60s) he had to take JM to court to get compensation for an asbestos-related disease.

- JM position: We have communicated openly with employees about illnesses.

  Fact: A current JM employee tells me it was company practice into the early 1970s not to tell a person about his illness.[13]

A decade later, David Austern, the attorney who had assumed charge of the Manville Trust after the company's bankruptcy, reviewed the Manville archive. He noted that Manville's long-standing defence had been that it was unaware until 1964 that exposure to asbestos could cause injury and that it had no relevant documentation. Austern discovered so many documents that he estimated it would take one person twenty years to read them. He was shocked not only to find that the documents demonstrated corporate knowledge of the dangers dating back to 1934, but also that the documents were so damaging. He noted that 'there are so many embarrassing documents that people disagree as to which group of any ten documents is the worst'.[14] So damning, that Manville told one of its employees that he ought to hire his own lawyer, because the documents showed that he could be indicted for manslaughter. Austern noted that most of Johns-Manville's defences were false and that the plaintiffs' bar would have a field day using such evidence of a corporate conspiracy. Austern told his Trustees:

> While it is not my intention to be an alarmist, I believe the documents evidence corporate irresponsibility of a magnitude, which is understated in [Brodeur's] Outrageous Misconduct. The content and tone of the documents demonstrate that Manville officers, directors and employees... held secret information that had it been revealed, would have prevented the deaths of thousands of people.

These memoranda reveal the truth behind the façade.

[13] C.G. Linke to J.A. McKinney, 12 January 1979.
[14] D.T. Austern memo, Memo re. Manville Documents, 8 February 1988.

By then, the major players in the industry were in flight from asbestos: mines were being closed, assets hastily sold, and the old names shed—all against a backdrop of increasing litigation. It was an unedifying spectacle. For many, the James Hardie story in Australia (detailed in Chapter 6) serves as a microcosm of the global industry, featuring as it does hazardous products, inadequate regulation, and corporate misconduct. It is also a story in which the NSW state has played an ambiguous role, having bought James Hardie products but failed to regulate mining or manufacture. In 2007, however, Australia's financial corporate regulator began legal proceedings against the Hardie board, on the basis that every director of Hardie in 2001 failed to meet basic standards of boardroom behaviour. W.R. Grace, former owners of the Libby mine, also faced a raft of state and federal actions. In 2005, Grace and seven of its current and former executives were indicted on federal charges that they knowingly placed their workers and the public in danger. Meanwhile, embarrassing documents on the industry's past continued to spew forth, though not from every direction. Despite Geoffrey Higham's assertion that Cape Asbestos had nothing to hide, as part of its modest £7 million out-of-court settlement with South African miners, Cape insisted that payment was dependent on the destruction of all evidence submitted during the hearings.[15]

Nonetheless, bankruptcy or diversification strategies saw the most reviled companies—even Manville—reincarnated as respectable and profitable members of the business community, often engaged in ripping out the asbestos they had installed. In America, the much-vaunted fund for compensating victims stalled in the US Senate in early 2006, after some conservative senators feared that costs would be passed to taxpayers. Another factor was that by then it was apparent that the number of new asbestos filings was falling dramatically—the result of state tort reforms and judicial rulings aimed at restricting unimpaired claims. Some believed, too, that there was evidence that mesothelioma in America was on the wane. In 2007, a US Senate bill was at last drawn up to ban asbestos.

This softening picture is not being duplicated elsewhere, particularly on the other side of the world. Australia has the world's highest recorded incidence of mesothelioma—a legacy of the three decades after 1945 when Australia was the world's highest per capita user of asbestos.[16] In

[15] B. Clement, 'Lawyers in Asbestos Case Agreed to Destroy Evidence', *Independent*, 27 September 2004.

[16] NSW Government: Cabinet Office, *Report of the Special Commission of Inquiry into the Medical Research and Compensation Foundation* (September 2004). 'Asbestos and James Hardie', Annexure J, p. 117.

Japan, mesothelioma cases are also still rising in the aftermath of the 'Kubota Shock'. As to other countries in Asia and the developing world, their problems are only beginning. According to ILO estimates, the global workforce suffers over 250 million accidents each year resulting in as many as a million fatalities.[17] The majority of those injuries, like most occupational diseases, occur in developing countries where few workers are protected by health and safety legislation. Among the most vulnerable workers are migrants who are often employed in high risks jobs. It is not a workforce that needs another hazard.

World asbestos production has settled at about 2.3 million tonnes, nearly all of which is consumed in the developing world. The output is composed entirely of chrysotile. Even if one accepts that the latter is less toxic than amphiboles, it is roughly the equivalent of asbestos production in 1960, when Wagner revealed mesothelioma to the world. In some countries, particularly China, Kazakhstan, and Pakistan, levels of consumption have increased dramatically in the twenty-first century.[18] Nearly all of that chrysotile goes into the manufacture of asbestos-cement (a/c) sheets and pipes. With the passing of the corporate giants like Manville, the industry has been taken over by small indigenous companies. However, the debate about asbestos hazards is still conducted along the same old lines. Those defending asbestos emphasize, as ever, cost-benefit analysis, the low-risks associated with 'controlled' conditions, the 'safety' of chrysotile, and the unquantified hazards of substitutes. The players too have often remained the same. Canada remains an important influence in the industry, by providing financial support, a scientific model, and an example.

In the developing world, history has a way of repeating itself in almost a surreal fashion. In 1996, when the French prohibition on chrysotile imports confronted the Canadian industry with the prospect of an EU ban, as a publicity stunt a team of Quebec miners entered the French marathon. It was to demonstrate the supposed benefits of working with asbestos. Their efforts were applauded in the Canadian parliament.[19] A decade later, Zimbabwe's Shabanie mine announced that it would host an athletics meeting. The event, designed to publicize the safety of chrysotile, was to feature a number of sprint races and finish with a half-marathon

[17] J. LaDou, 'International Occupational Health', *International Journal of Hygiene and Environmental Health* 206 (August 2003), pp. 303–13.

[18] L. Kazan-Allen, *Killing the Future: Asbestos Use in Asia* (2007), p. 38.

[19] K. Sentes, 'Asbestos and the World Trade Organisation: A Case Study Challenging the Legitimacy of the WTO as a Forum for the Adjudication of Public Health and Environmental Issues' (University of Alberta MA, 2002), p. 28.

dubbed the 'Great Chrysotile Run'. Shabanie Athletics Club chairman, Lazurus Ambress, explained that his club had decided to use sport to highlight the safety of the mine and its products. 'The sponsorship for the race especially the half marathon has been good and prizes are fantastic', he said. 'We want to show the world that we can participate in athletics competitions because we are very fit and are not suffering from any asbestos-related diseases.'[20]

Not everything that Canada does is quite so humorous, though many of its policies might be considered as bizarre as anything staged in Robert Mugabe's Zimbabwe. In May 2006, the Chrysotile Institute held a conference in Montreal entitled 'Chrysotile at a Turning Point'. The Canadian asbestos industry now employed well under a thousand workers and the Chrysotile Institute itself was only sustained by a $250,000 federal subsidy. That payment in effect, meant that the government was paying the industry to lobby the legislature. Nevertheless, the conference attracted delegates from Mexico, Brazil, and India—all countries which import Canadian asbestos. The conference was designed to promote the use of asbestos in developing countries.[21] Anyone scanning the programme might be forgiven for feeling a sense of déjà vu at both the attendees and the agenda. It featured many old hands of the North American industry scene, such as Bob Pigg, Hans Weill, and not least, the conference spokesperson, Dr Jacques Dunnigan, a one-time adversary of Selikoff. Most of the arguments were familiar, too, not to say cranky: only heavy exposure causes adverse health effects, asbestos is safely bonded into cement products, and any disease is purely historical. No mention was made of the suffering of victims or compensation, though barbs were fired at personal-injury attorneys. The insouciance of the delegates towards asbestos was summed up by English presenter John Hoskins. When discussing pleural plaques, he stated that the only problem was that 'lawyers love them'.

The Secretary to the Minister for Natural Resources, Christian Paradis, gave the welcoming address. He told the audience that his government wanted to ensure the safety of exposed workers and dispel the myths surrounding asbestos. Paradis explained that the issue has been clouded by the amphiboles which are hazardous. The Canadian government did not market or promote the use of chrysotile, but through the Chrysotile

[20] 'Global Fight against Ban of White Asbestos Taken to the Sports Field', *The Herald* [Zimbabwe], 8 November 2006.

[21] International Conference on Chrysotile, Montreal, 23–24 May 2006. Posted at: http://www.chrysotile.com/en/conferences/default.aspx

Institute supported the safe use of that material. 'The Government of Canada', he said, 'has always based its position on the use of chrysotile on international scientific evidence'.[22] According to Paradis, Canada had played a leading role in raising awareness of risk and had actively helped developing countries to understand safe use. The latter is a paradoxical statement if chrysotile is supposed to be benign. Moreover, Paradis omitted to tell his audience that Canada, which is the only country to fund the export of asbestos, discourages its domestic consumption.[23] In Canada, the use of asbestos in many consumer products is prohibited, while regulations limit releases of asbestos from mines and mills.

Paradis explained that Canada would continue to resist the listing of chrysotile as a hazardous substance under the Rotterdam Convention. The Convention is the result of the dramatic growth in the international chemicals trade over the past three decades. That trade means that many countries in the developing world are unable to monitor imports let alone understand the health effects. In response, the UN fostered the Rotterdam Convention as a legally binding multilateral agreement to promote shared responsibilities in relation to importation of hazardous chemicals. By 2006, 73 countries were signatories and 111 were parties. The Convention promotes open exchange of information and calls on exporters of hazardous chemicals to use proper labelling, include directions on safe handling, and inform purchasers of any known restrictions or bans. The multilateral agreement covers materials which are banned or restricted for health or environmental reasons. The Convention does not prohibit trade, but merely sets guidelines in terms of knowledge and consent, with the final decision on a listing taken by a conference of all members. The thirty-nine chemicals currently listed include the pesticides 2,4,5-T, Aldrin, and Dieldrin. The list also includes five of the six types of asbestos. Chrysotile, which accounts for virtually all global use, is omitted.

The banning of chrysotile has been on the Convention's agenda since its creation. At a 2004 review, chrysotile satisfied all the criteria for listing. It was widely regarded as a carcinogen and more than two member states had imposed bans. The only difference from materials already listed was its widespread production and use. Canada and Russia, both exporters, organized the opposition. During the hearings the Russian Federation

[22] Paradis address to International Conference on Chrysotile, pp. 1–2.

[23] J. Brophy, 'The Public Health Disaster Canada Chooses to Ignore' in *Chrysotile Asbestos: Hazardous to Humans, Deadly to the Rotterdam Convention* (2006), pp. 17–20.

claim that chrysotile has not been proven carcinogenic was supported by Ukraine and Kazakhstan. In a brief statement, Canada simply endorsed controlled use at home and abroad. In opposing a listing, the Indonesian delegation argued: 'Inclusion of chrysotile will create a more complex national situation [for Indonesia].'[24] A similar response about a listing being tantamount to an international ban was repeated by Canada's other developing world supporters. The EC delegation remarked that the process for inclusion had been strictly followed and the need for a ban proven. It also noted that a failure to list chrysotile would strike at the heart of the Convention.[25]

At the October 2006 meeting in Geneva, Canada again led a group including Ukraine, Kazakhstan, Iran, Peru, and the Russian Federation in opposing the listing of chrysotile.[26] They argued that a listing would be tantamount to an international ban as it would force tighter domestic regulations on use. Once again the move to list chrysotile under the Rotterdam Convention (favoured by over 100 signatories) failed; the only decision taken was to postpone further consideration until 2008. Canada had all but destroyed the Convention.

On many important international issues Canada has played a progressive role, but in regard to asbestos, successive Canadian governments have kept the worst possible company. In April 2007, Canada joined Russia, Kazakhstan, Zimbabwe, Ukraine, Uzbekistan, India, Columbia, Brazil, and Mexico in sending representatives to an inaugural meeting of the Chrysotile International Alliance of Trade Unions, a new industry pressure group. After the event, an 'appeal' was made to Margaret Chan, Director-General of the WHO, to reconsider the WHO's policy on the elimination of ARDs. This request was apparently granted in May 2007, when the WHO suddenly included a let-out clause for chrysotile in its document 'Workers' Health: Global Plan of Action', which stated that a 'differentiated approach' should be adopted for asbestos regulation.

While the Chrysotile Institute was defending *controlled use*, labour in poor countries continued to be exposed to airborne fibre. This was especially true in the breaker yards of Bangladesh, China, and Pakistan, where each year hundreds of ships are turned into scrap by teams of men using simple tools.[27] At Alang in India (the largest of those yards) conditions

[24] L. Kazan-Allen, 'Chronological Record of the Contributions of National Delegations and Others', *Hazardous to Humans*, p. 30.

[25] Kazan-Allen, 'Chronological Record', p. 29.

[26] 'Asbestos Kept off Global List of Toxic Substances', *Reuters*, 13 October 2006.

[27] 'End of Life Ships—The Human Cost of Breaking Ships.' Posted at: http://www.fidh.org/IMG/pdf/shipbreaking2005a.pdf

remain hazardous. Although India banned the import of asbestos waste in 1998, that ban has not been applied to the yards where old ships full of asbestos lagging are scrapped (in conditions that offer an interesting insight into *controlled use*). In contrast, France has stringent laws on asbestos disposal, which make the option of sending its decommissioned ships overseas an enticing option.

After its heroics in the Gulf War, the aircraft carrier *Le Clemenceau* was decommissioned in 1997. Faced with local scrapping costs of £20 million, in 2005 France sold the ship to a Panama registered company SDIC, which planned to break up the carrier at Alang. The French claim that most of the lagging in *Le Clemenceau* had already been removed at the Toulon dockyards prior to its sale was strongly disputed. Ban Asbestos France and Greenpeace argued in court that France was trying to avoid its international obligations and in particular the Basel Convention forbidding the export of hazardous waste. According to the Basel Action Network: 'Their [the French] standards for handling asbestos are amongst the highest in the world. But instead of investing in safe removal and disposal of the asbestos on *Le Clemenceau,* they are trying to dupe the Indian Government, and dump their toxic wastes onto the poorest of the poor of the world.'[28] French courts never adjudicated on the legality of exporting a ship full of toxic waste, but dealt only with the technical question of whether or not the ship could leave French waters. Appeals by Greenpeace and Ban Asbestos were finally lost on 31 December 2005, when the Court cleared the way for the ship to depart for Alang.[29] The next day *Le Clemenceau* began its final voyage, but immediately hit choppy waters when the Indian courts began to question the actual quantity of asbestos in the carrier (a mere 200 tonnes according to the French; up to a thousand tonnes according to a Toulon company which had worked on the ship). Within a week, the Indian Supreme Court on hazardous waste ruled that any transnational movement of hazardous waste violated the Basel Convention, arguing: 'Why should we sacrifice our precious soil to bury some other country's junk? If a ship comes with 1,000,000 cobras, will we accept it just because some Indians can catch cobras?'[30] In February 2006, the French court ordered that *Le Clemenceau,* which by that stage had reached the Arabian Sea, be towed back to France. The

[28] Greenpeace Press Release, 'Greenpeace Stops Clemenceau Leaving European Territory', 12 January 2006. Posted at: http://www.greenpeaceweb.org/shipbreak/news117.asp

[29] Basel Action Network [BAN], 'France's Export of Decommissioned Aircraft Carrier *Clemenceau* in Violation of International, and National Law', 12 January 2006. Posted at: http://www.ban.org/pdfs/ban_clemenceau_summary_of_violations.pdf

[30] 'French Ship Must not Enter India', *The Hindu*, 7 January 2006.

disposal of *Le Clemenceau* had turned into a public relations disaster for the French government.

In their history of the US chemical industry, Gerald Markowitz and David Rosner make extensive use of documents gained through legal discovery. That material exposes the gap between the public statements on safety made by major US companies with what those companies actually knew about the risks facing their employees.[31] The stories Markowitz and Rosner tell of how the industry slowed down regulation and denied culpability are hauntingly familiar to anyone conversant with the behaviour of the asbestos companies, and often feature the same *dramatis personae.* There are corporations such as Union Carbide, General Motors, Eagle-Picher, W.R. Grace, and Johns-Manville. There are also a parallel set of professional associations that intertwine with public relations firms, such as Hill & Knowlton. And protecting the interests of its client are the ubiquitous corporate attorneys. Ranged against them are public health and regulatory bodies, such as OSHA, NIOSH, the EPA, and the FDA; a handful of public-spirited physicians such as Irving Selikoff; and the trade unions and activists.

National asbestos bans have usually come about as the result of protracted struggles between such groups. Opposing the use of asbestos is a growing network of trade unionists, scientists, lawyers, asbestos victims, politicians, physicians, government officials, environmentalists, and human rights activists. The industry has not been able to move to the developing world without inadvertently taking its opposition along. Campaigns against asbestos have been fought in Peru, Brazil, Malaysia, South Korea, India, Indonesia, Thailand, China, Vietnam, Pakistan, Lebanon, and Cyprus. Fernanda Giannasi, a fearless inspector with the Brazil Ministry of Labor and a founder of the Brazilian asbestos victims' group ABREA, has been subject to libel suits, administrative complaint, and death threats. Dr Tushar K. Joshi, an occupational physician in New Delhi, has been targeted by both the asbestos industry and government officials for advocating a ban. On 1 January 2005, all forms of asbestos were banned in the EU. By November of that year some forty countries, including many in Africa, the Middle East, and South America, had imposed bans. A central figure in the global effort, Laurie Kazan-Allen, coordinator of the International Ban Asbestos Secretariat, has described many of these struggles in her paper 'The Asbestos War'.[32] Japan's recent ban of asbestos

[31] G. Markowitz and D. Rosner, *Deceit and Denial: The Deadly Politics of Industrial Pollution* (2002).

[32] L. Kazan-Allen, 'The Asbestos War', *IJOEH* 9 (July/September 2003), pp. 173–93.

friction products and building materials may help shift the momentum in Asia, which to date has been a stronghold of asbestos use.

The involvement of these groups in the history of asbestos suggests that science is never sufficient to resolve occupational health problems. The most striking feature of the asbestos tragedy—and the most poignant—is that there has never been any shortage of information. What has been lacking are the social and political safeguards to enable that knowledge to be used for public benefit. Without those safeguards, data can be disputed or even invented, and risk assessment manipulated for commercial or political advantage. The US and British industries systematically used science against the poor and vulnerable to slow down regulation. They never used science to speed it up. Effective regulation requires a particular kind of political will or what Selikoff called the 'recruitment of constituencies'.[33] Selikoff rightfully believed that knowledge can only be effective if it is in the hands of the people who work with dangerous materials.

How is this knowledge to be released? Selikoff—who is one of the few individuals to emerge with credit from this story—pondered that question in the 1970s and offered his own 'incomplete and tentative' recommendations: dissemination of information by the gatekeepers of knowledge was first on his list, followed by avoidance of additional exposure, medical surveillance for asbestosis and a range of cancers, educational programmes especially with regard to smoking, and more research into high risk occupations, such as those in the shipyards.[34] These yardsticks provide a measure of how well the asbestos industry and government and scientists performed in reacting to the warnings about asbestos. Little comment is needed from us, except to add that throughout the history of asbestos the industry has consistently extolled the benefits of the mineral and trumpeted their own honesty and good faith. The history, however, shows a consistent pattern of duplicity and escalating injuries to individuals. In both the OECD states and the world's poorest countries, the asbestos industry has only thrived because the actual costs of that injury have been discounted or moved out of sight. If the disease burden among asbestos workers had been truly reflected in the price of asbestos products, then the global industry would have spiralled into decline in the 1950s.

[33] I.J. Selikoff, '*Twenty Lessons from Asbestos*', *EPA Journal* (May 1984).

[34] I.J. Selikoff and E.C. Hammond, 'Asbestos-Associated Disease in United States Shipyards', *CA: A Cancer Journal for Clinicians* 28 (1978), pp. 87–99.

Sources and Acknowledgements

Asbestos means headlines. That would have surprised the pioneers of the industry because, although asbestos was highly profitable and utilitarian, it was always relatively dull and unglamorous. Even the biggest companies struggled to make products such as gaskets, roofing slates, brake shoes, and ironing-board mats seem heroic. The industry was not widely photographed, it had few trade magazines, and its advertising campaigns, at least until the late twentieth century, were low key. Consequently, the literature on the industry before the 1960s is relatively sparse. Apart from the slowly growing medical literature, much of the information about asbestos is in technical texts, such as Jones (1897) and Summers (1919).

However, the public's knowledge of the industry was transformed after the 1960s, when the hazards of asbestos brought it under unprecedented scrutiny. First, a wave of litigation forced the industry to divulge a mountain of records; second, a growing number of published studies examined the history of the industry, particularly from the perspective of ARDs.

Our book is based primarily on the documents of the leading asbestos companies and on other asbestos archive collections. The industry did not divulge these documents willingly. They came into the public domain as the result of litigation in America, where so-called legal discovery gives plaintiffs extensive rights to examine records, and obliges defendants to place documents before the court. In America, unlike in Europe, these documents are available for consultation. After 1970, hundreds—indeed thousands—of American court cases made available a vast archive on the asbestos industry. Although the tidal wave of asbestos documents has been surpassed recently by the tsunami of tobacco archives, at the time no industry had been compelled to bare its soul to such an extent.

Not everything survives and not everything is transmitted to paper. The documents, too, consist of only what is 'produced' by the defendants. Nevertheless, up to the 1970s company directors, plant managers, physicians, technologists, and attorneys candidly committed to paper many of their thoughts and strategies, little realizing that decades later these would provide the 'smoking guns' for the courtroom. No one knows how many asbestos documents have become available over the last thirty years or so, but it runs into millions of pages. The picture that these pages conjure of the so-called free enterprise system is very different from the sanitized view that is popular with most business and medical historians, who still rely heavily on government and official repositories.

The plethora of documents in America can be contrasted with other countries, where many problems are encountered in looking at controversial industries. In Canada, it is almost impossible to obtain access to the holdings at the National Archives. The Quebec Records Concerns Act bars the transfer of certain documents out of the Province, even as photocopies. In the South African National Archives, no documents are available on the asbestos mines after 1963 and absolutely nothing on asbestos manufacture. In Australia, until recently very few company records were available. Little, too, was known about the British asbestos industry. The leading firms denied having any records, and documents produced in litigation are not in the public domain. In Europe, this enabled the leading companies to maintain a wall of secrecy.

This situation loosened in the UK after 1995, when T&N was sued by Chase Manhattan Bank in New York. This well-publicized case was facilitated by Chase's extensive discovery of T&N's files in Manchester, UK. Chase copied about a million pages of documents to prime its case. In 1996, Geoffrey Tweedale and his colleague David Jeremy purchased a complete microfilm copy of the Chase/T&N documents. This remarkable collection has been described in Tweedale (2000). Since these documents had been produced in an American court, they were not bound by the same censorship that operates in the UK. Not only did they shed an unprecedented light on T&N, but they also illuminated the global industry. They have proved invaluable for this study.

In America, we have been helped enormously by David Egilman of Brown University, Providence, Rhode Island. For many years, as an expert witness on occupational lung diseases, he has patiently collected and scanned every article, affidavit, and asbestos document that came his way. His entire collection was generously copied to us on DVDs. Not only did this hoard contain innumerable medical articles and published reports, but it also contained a vast array of key documents relating to the American and Canadian asbestos industries. These files gave us a window into the past on the actions of Johns-Manville and other leading companies, such as Grace, Raybestos-Manhattan, Pittsburgh-Corning, Owens-Illinois, and many others. They also uncovered the role of insurers, such as MetLife and Travelers, and proved particularly useful in disentangling the role of the industry's trade organizations, such as the ATI and the AIA/NA. Within this universe of records were also the Sumner Simpson papers, Anthony Lanza's correspondence, the Saranac Laboratory files, the minutes and commissioned reports of the Quebec Asbestos Mining Association, and many more. These scanned records were a springboard for the present study and a collection to which we turned repeatedly. However, they were not the only archive materials that we used.

The biggest asbestos documents collection in America is the Johns-Manville archive, which is based in Denver, Colorado, and is known as the Asbestos Claims Research Facility. We were unable to visit the ACRF because its primary purpose is to facilitate legal claims and even a day's visit can cost several hundred dollars. Its holdings are also vast (over 40,000 box files and over 7,000 rolls of microfilm). However, our disappointment at not being able to visit Denver was tempered by the help of Barry Castleman and Albert Donnay, who copied to us a large collection of Johns-Manville records, many of them from Denver. In addition, one of America's leading personal-injury attorneys, Motley Rice, gave us another electronic collection that featured many of the key discovery documents from the 1970s and 1980s. We also obtained for ourselves a complete run of John-Manville's annual reports.

Our understanding of the occupational health scene, both in the US and internationally, was helped greatly by gaining unprecedented access to the Irving Selikoff archive at Mount Sinai Hospital in New York City. For years Selikoff had avoided litigation and had fought to keep his papers away from the legal arena. After his death in 1992, these papers were still not generally made available. In 2006, however, Jock McCulloch was able to review these documents and they are a key source for this book. For this we thank the staff at Mount Sinai Hospital, especially Valerie Josephson, Philip Landrigan, and Stephen Levin.

In South Africa, the situation was more problematic. In 2004, as a stipulation of Cape Asbestos's settlement with its South African workers, plaintiffs' lawyers agreed to destroy all documentation relating to the legal action. Fortunately, the past is resilient and despite Cape's attempt to suppress the historical record, we have been able to obtain many documents on the South African mining industry. The most notable sources have come from the National Institute of Occupational Health (NIOH) in Johannesburg. We are grateful for the sustained help of J.C.A. Davies, David Reese, and Danuta Kielkowski. Richard Spoor and Richard Meeran, who have bravely run cases against Cape and now Anglo-American on behalf of miners, gave us access to important materials. The Concerned People Against Asbestos based at Prieska in the Northern Cape gave us access to their members and also to important materials documenting the group's struggle against a British conglomerate. Hetta Hager at the Mary Moffat Museum in Griekwastad provided important documents on the pay records of Cape Asbestos. We also benefited from a donation of documents by Laurie Flynn, who since the 1980s had tracked events in South Africa, and especially the role of Cape Asbestos. This collection of depositions, American discovery documents, and press cuttings, informed us about aspects of the Cape Asbestos story of which we had been unaware. In Kazakhstan, John Gilbert of ATEC, with the assistance of Peter Cullington and Sagila Kakabaeva, arranged a visit for Jock McCulloch and provided access to power stations and occupational health scientists.

Because most of these documents are not easily available, we have not tagged them with a specific location (unless they are located in a public archive). However, we hold a copy of every document cited in this book, either in hard copy or electronic form.

Running parallel with this litigation has been the publication of a steady stream of books and articles on the asbestos industry. Most of this literature dates from the 1980s and much of it is American. The authors are journalists, activists, and academics. The most important of the early writers on asbestos was Paul Brodeur, a staff writer for the *New Yorker*. A series of articles for the *New Yorker*, entitled 'The Magic Mineral' (1968), was widely reprinted; and another five-part series on asbestos became a paperback, *Expendable Americans* (1973). Brodeur followed this with *Outrageous Misconduct* (1985), which remains the most readable and humane book on asbestos. Brodeur has written about the background to these studies in his autobiography (1997). Another landmark publication was by Barry Castleman, an American environmental consultant and activist: *Asbestos: Medical and Legal Aspects*. First published in 1984, this book became a courtroom bible in the US and is now into its very much expanded fifth edition (2005, p. 894). Like Brodeur, Castleman writes in an understated fashion, but is appalled by the social and political system which has allowed the problems of asbestos to escalate. Besides responding to innumerable requests for information, his research has given us a map without which our own book would have been more difficult to write.

To these heavyweight treatments can be added David Ozonoff's shorter and more accessible account (1988). Other American academics who have shed light on the development of ARDs include David Rosner and Gerald

Markowitz. Their classic account of dust diseases (2nd edition 2006) is essential reading for anyone wishing to understand the politics of industrial disease in America. Robert Proctor also pursues a political theme in his history of cancer (1995), which contains a chapter on asbestos and offers parallels with tobacco. Proctor's *Nazi War on Cancer* (1999) includes an illuminating discussion of the German recognition of asbestos-related cancer during the 1940s and 1950s.

Official statistics on asbestos—both historical and contemporary—have been collected by the US Geological Survey. The data, patiently assembled by Robert Virta, is indispensable for an understanding of the economic history of asbestos (though his own interpretation is conservative). Numerous think tanks and interest groups have looked at asbestos, especially regarding the impact of litigation on the American economy. The opposite ends of this political debate can be sampled, for example, in the reports of the Manhattan Institute or Ralph Nader's Public Citizen. The Rand Corporation has produced one of the most widely cited reports (2005) on US litigation. It can be read in conjunction with *Asbestos: Think Again* (2004), produced by the Environmental Working Group in Washington. This report is notable for an online document gallery.

Asbestos continues to inspire journalists. Since 1999, Andrew Schneider and his colleagues at the *Seattle Post-Intelligencer* have written numerous reports (online at: http://seattlepi.nwsource.com), especially on vermiculite mining in Virginia and Montana. Their efforts resulted in a book on the Libby mining disaster (2004), a story also covered by Andrea Peacock (2003). Michael Bowker (2003) looks at Libby, besides recent developments, such as the World Trade Center attack. A particularly moving and well-researched account of asbestos and the shipyard community of Hampton Roads (Virginia) is by Bill Burke, a journalist for *The Virginian-Pilot* (2001). Similar reportage by Jessica Centers and Matthew Korade (2005) looks at the history of Cement Asbestos Products Company (Capco), based in Ragland (Alabama). A sobering coda to this story was the death of *Anniston Star* editor Randy Henderson (aged 53) from mesothelioma in 2006: he had spent two college summers working at Capco.

In contrast to the US, books on the Canadian asbestos industry are sparse. No comprehensive study has been published of the Canadian industry. Lloyd Tataryn (1979) contains a chapter on Thetford but no one has followed this with a comprehensive treatment of the industry. Suzanne LeBlanc (2003) has told the story of the Cassiar chrysotile mines in British Columbia, but ARDs receive only a brief treatment. More relevant for dust diseases is Esther Delisle's and Pierre K. Malouf's extended discussion (2004) of the key texts and documents relating to the 1949 asbestos strike in Quebec. Since the 1980s, the official voice of the Canadian asbestos industry has been the Asbestos Institute (now the Chrysotile Institute: http://www.chrysotile.com/), but its comforting view of asbestos has been challenged by Doug Smith (2000), who has described asbestos mining and manufacture in Canada as a national disgrace. Particularly damning have been Jim Brophy's studies (1999, 2007) of ARDs in the Chemical Valley of the Sarnia-Lambton area in Ontario. Canada's asbestos problems have also been highlighted in a CBC video/radio archive compilation, 'Asbestos: Magic Mineral or Deadly Dust?'.

The first book on asbestos and ARDs in the UK was written by Alan Dalton (1979) from a socialist perspective. The first academic study appeared in 1994 by Nick Wikeley, a professor of law, who wrote about asbestos compensation. Geoffrey Tweedale (2000) wrote the first study of an asbestos company. It was closely followed by Ronnie Johnston's and Arthur McIvor's account of the asbestos tragedy in Scotland (2000), which was notable for its oral history dimension. Morris Greenberg, an independent epidemiologist and once a Factory Inspector involved with asbestos, has produced a steady stream of carefully researched articles on the international asbestos industry.

In Australia, events at the James Hardie-owned Baryulgil mine in New South Wales, the Wittenoom crocidolite mine, and in the Latrobe Valley power stations provided the material for studies by Matt Peacock (1978), Ben Hills (1989), and George Wragg (1995). A more comprehensive account of the Australian asbestos industry was provided by Jock McCulloch (1986). More recently, the James Hardie scandal has been analysed in Gideon Haigh's 'secret history' of the company (2006).

In Continental Europe, for a long period much of the asbestos industry remained inscrutable. The only study of the most important company, Eternit, was by journalist Werner Catrina (1985). Within the last decade, however, Salvator Nay (1997) has provided an account of Eternit in Belgium; Guy Meria (2000) has studied Eternit's Corsican mine; and Odette Hardy-Hémery (2005) has analysed in detail the business history of Eternit's operations in France. An account of Eternit that focuses on its occupational health history is provided by Bob Ruers and N. Schouten (2006). An expanded version of this report (2006), which includes information by F. Iselin on Eternit in its homeland Switzerland, has been posted on the Swiss victims' group CAOVA's web site.

The only European country that has so far been covered in depth, apart from the UK, is France. The Collectif of Jussieu activists (1977) made the first inroads. Additional information on the 'social movement' against asbestos in France can be found in François Malye (1996, 2004), Roger Lenglet (1996), Caroline Faesch (2002), Emmanuel Henry (2007), and in several articles by Annie Thébaud-Mony.

The developing world is even less well covered in the literature. Journalist and documentary film producer Laurie Flynn wrote a pioneering study of mining in South Africa (1992), which included chapters on asbestos. Jock McCulloch (2000) described the political and medical history of South African asbestos in more detail, with the help of interviews and surviving records. No book-length studies have been written on the industry in any other developing nation, though several informative articles have appeared in the pages of the online journal, *International Journal of Occupational & Environmental Health* (IJOEH). In July–September 2003 and April–June 2004 the IJOEH published Special Issues on asbestos that included, inter alia, studies of Mexico, South Africa, Argentina, and several Asian countries. The 'Ban Asbestos' movement has also organized international congresses, in which discussions of asbestos in the developing world have been prominent. The proceedings are available in electronic versions as: *Annals of Global Asbestos Congress: Past Present and Future, Osasco, Brazil, 17–20 September 2000* (CD Proceedings, 2000); and *Annals of the Global Asbestos Congress, Tokyo, Japan, 19–21 November 2004* (CD Proceedings, 2005).

Victims' action groups provide a key contemporary narrative, illuminating aspects that are not covered in the official literature. Since May 1990, Laurie Kazan-Allen has published the *British Asbestos Newsletter* (http://www.lkaz.demon.co.uk/index.htm). In over sixty issues, BAN has given a running commentary on the struggle for compensation. Its sister organization, the Ban Asbestos Secretariat (http://www.ibasecretariat.org/), covers the same struggle from an international perspective and has links to groups in other countries (e.g. CAOVA, ABREA, and ANDEVA). All these sites provide articles, statistics, and documents on the international asbestos tragedy.

Since the scientific literature has been so polarized, it is perhaps not surprising that the historiography has suffered the same fate. Both Rachel Maines (2005) and Jacqueline Corn (2000) provide subtle defences of the industry. Maines furnishes many interesting details of the fire protection offered by asbestos and argues that the mineral was indispensable: the reader is not told, though, why society now thrives without it. Less subtle is British medical historian Peter Bartrip, who in authorized histories (2001, 2006) has written a comprehensive defence of the UK and US asbestos companies. He does not explain, though, why they are bankrupt.

Our book has been supported by the university institutions where we work. Geoffrey Tweedale would like to thank the Research Division at Manchester Metropolitan University Business School, who facilitated the writing of this book in various ways, and especially David Jeremy for his encouragement over many years. Without the unstinting support of Mary Titchmarsh, writing the book would have been far more arduous. Jock McCulloch would like to thank the Australian Research Council and all those in South Africa, the US, Canada, the UK, and Australia, who have been so generous during his travels.

Besides the individuals mentioned above, we would also like to thank for help and advice at various times: Jason Addy, Petr Brhel, Nigel Bryson, Anthony Coombs, Jim Fite, Arthur Frank, Morris Greenberg, Olaf Hagemeyer, Emmanuel Henry, Andy Higgison, Laurie Kazan-Allen, Richard Lemen, Katherine Lippel, James McNulty, Chris and Larry Madeksho, Salvator Nay, Louie Palu, Elihu Richter, David Rosner, Bob Ruers, David Sunderland, Nancy Tait, Annie Thébaud-Mony, Lars Vedsmand, Andrew Watterson, Nick Wikeley, and Kate Wilson. Finally, our editors at Oxford University Press—David Musson, Matthew Derbyshire, and Tanya Dean—must be thanked for guiding this book through to publication.

Media (TV/Radio)

BBC TV Horizon, 'Killer Dust', 20 January 1975.

CBC video/radio archive compilation, 'Asbestos: Magic Mineral or Deadly Dust?' (http://archives.cbc.ca/IDD-1-75-608/science_technology/asbestos/).

ITV World in Action, 'Dust to Dust', 16 November 1981. Transcript.

TSR [Switzerland], 'Mourir Amiante, en Silence', 28 September 2006.

Yorkshire TV, 'Alice—A Fight for Life', 20 July 1982.

Yorkshire TV, 'Too Close to Home', First Tuesday documentary, 6 December 1988.

Unpublished

Behrakis, P., Presentation, European Asbestos Conference: Policy, Health and Human Rights, Brussels, 22–23 September 2005.

Blinn, L., 'The History of the Saranac Laboratory 1885–1958', unpublished typescript, 1959.

Brophy, J., 'Carcinogens at Work', Conference on Everyday Carcinogens: Stopping Cancer Before It Starts', McMaster University, Hamilton, Ontario, 27 March 1999.

Browne, K., 'Chrysotile: Thresholds of Risk', Paper presented at an International Seminar on Safety in the Use of Chrysotile Asbestos: Basis for Scientifically-Based Regulatory Action, Havana, Cuba, 12–13 September 2000. Posted at: www.chrysotile.com/en/hltsfty/browne.htm

De, A., 'Petrology of Dikes Emplaced in the Ultramafic Rocks of South Eastern Quebec' (Princeton University PhD thesis, 1961).

Elmes, P., Testimony, *County of Anderson, Tennessee (Head of Education and Superintendent of Schools) v. US Gypsum and National Gypsum*, US District Court for the Eastern District of Tennessee Northern Division, 4 March 1985.

Grieve, I.M.D., 'Asbestosis' (Edinburgh University MD thesis, September 1927).

Hagan, G., 'James Hardie Industries 1880–1980' (Macquarie University BA thesis, 1980).

Kazan-Allen, L., 'A Comparative Review of European Asbestos Compensation' (2005). Unpublished typescript available on ANDEVA website: http://andeva.free.fr/

La Dou, J., 'An International Review of Occupational and Environmental Asbestos Issues: Funding of International Agencies', Paper presented to International Ban Asbestos Conference, Ottawa, 13 September 2003.

Little, A.D., 'Characterisation of the US Asbestos Papers Market' (Final Draft Report to Sores Inc. C-79231, May 1976).

Murray, R., Testimony, *Milton Jenkins v. Fibreboard Corp.*, Superior Court of the State of California for the County of Los Angeles, 12 July 1993.

Myers, J., 'Asbestos and Asbestos Related Disease in South Africa', Southern Africa Labour and Development Research Unit, Working Papers No. 28 (June 1980).

Peto, J., 'Asbestos Cancer Deaths in the UK: The Past, Present, and Future', Lane Lecture, Manchester University, 23 November 2005.

Roberts, J.E., 'What is the Price of 80 KG's? The Failure of the Detection of, and Compensation for, Asbestos-Related Disease: Social Exclusion in Sekhukhuneland' (University of Natal MA thesis, December 2000).

Schmidheiny, S., 'My Path—My Perspective (2nd edn., 2006). http://www.stephanschmidheiny.net

Slade, G.F., 'The Incidence of Respiratory Disability in Workers Employed in Asbestos Mining with Special Reference to the Type of Disability Caused by the Inhalation of Asbestos Dust' (University of Witwatersrand MD thesis, 1930).

Sentes, K., 'Asbestos and the World Trade Organisation: A Case Study Challenging the Legitimacy of the WTO as a Forum for the Adjudication of Public Health and Environmental Issues' (University of Alberta MA, 2002).

Wagner, J.C., 'The Pathology of Asbestosis in South Africa' (Department of Pathology, University of the Witwatersrand PhD, 1962).

Wagner, J.C., Deposition, *Claude Cimino v. Raymark Industries*. US District Court for the Eastern District of Texas, Beaumont Division Civil action No. B-86–0456-CA, pp. 1–245. Stafford Hotel, London, 30 May 1990.

Official Publications

Annual Report of [Barking] Medical Officer of Health (1945).

Beaudry R., Lagace G., and Jukau L., *Rapport Final: Comité d'Étude sur La Salubrité dans L'Industrie de L'Amiante* (Quebec: Le Comité, 1976).

Cirkel, F., *Asbestos: Its Occurrence, Exploitation and Uses* (Ottawa: Mines Branch Department of the Interior, 1905).

Department of Mines, *Report... for Year Ending 31 December 1978* (Pretoria: Government Printer, 1979)

Deriot, M.G. and Godefroy, M.J.-P., *Rapport [No 37]... sur... Le Bilan et Les Consequences de La Contamination par L'Amiante* (Senat: Session Ordinaire de, 2005–6), p. 83. Posted at: www.senat-fr/rap/r05-037-1/r05-037-1-mono.html

Doll, R. and Peto, J., *Asbestos: Effects on Health of Exposure to Asbestos* (HSE Books, 1985).

Dreessen, W.C., Dellavalle J.M., Edward T.I., Miller J.W., and Sayers R.R., A *Study of Asbestosis in the Asbestos Textile Industry. Public Health Bulletin No 241, August 1938* (Washington: Government Printing Office, 1938).

Glen, H.W. (ed.), Asbestos *Symposium Johannesburg, 1977* (Randburg: Department of Mines, 1978).

Hearing on Code of Fair Competition for the Asbestos Industry: Transcripts of the Hearing and Appendix (Washington, DC: National Recovery Administration, 19 October 1933).

Hall, A.L., *Asbestos in the Union of South Africa* (Pretoria: Department Mines and Industries Geological Survey, Memoir No. 12, 2nd edn., 1930).

Health and Safety Commission, *Asbestos. Vol. 1: Final Report of the Advisory Committee* (London: HMSO, 1979),

House of Commons Expenditure Committee (Trade and Industry Sub-Committee), Session 1973–4, *Fifth Report... on Wages and Conditions of African Workers Employed by British Firms in South Africa. HC 116: Minutes of Evidence and Memoranda*. HC 21 I–IV.

Institut National de Santé Publique du Quebec, *Report of the Epidemiology of Asbestos-Related Diseases in Quebec* (Quebec: July 2004). Posted at: http://www.inspq.qc.ca/pdf/publications/293-EpidemiologyAsbestos.pdf

International Programme on Chemical Safety, *Environmental Criteria 203: Chrysotile Asbestos* Geneva: WHO, 1998), p. 94. Posted at: www.inchem.org/documents/ehc/ehc/ehc203.htm

Merewether, E.R.A. and Price, C.W., *Report on the Effects of Asbestos Dust on the Lungs and Dust Suppression in the Asbestos Industry* (London: HMSO, 1930).

Minimum Requirements for Safety and Industrial Health in Contract Shipyards. US Navy Department (Washington: US Government Printing Office, 1943).

Ministry of Labour & Factory Inspectorate, *Annual Report...for 1947* (London: HMSO 1949).

Monopolies Commission, *Asbestos and Certain Asbestos Products* (London: HMSO, 1973).

MRC Institute for Environment and Health, *Chrysotile and its Substitutes: A Critical Evaluation* (Leicester: IEH, December 2000). Posted at: www.le.ac.uk/ieh/webpub/webpub.html

——*Fibrous Materials in the Environment: A Review of Asbestos and Man-Made Mineral Fibres* (Leicester: IEH, 1997).

New South Wales Government: Cabinet Office, *Report of the Special Commission of Inquiry into the Medical Research and Compensation Foundation* (September 2004). Posted at: www.cabinet.nsw.gov.au/publications

Noakes L., *Asbestos Supplement Department of Supply and Shipping Mineral Resources of Australia* (Summary Report No. 17, July 1945).

Official Transcript, *Tariff Board Inquiry Re. Asbestos Fibre* (Commonwealth of Australia, 2 June 1954).

Report of the Royal Commission on Matters of Health and Safety Arising from the Use of Asbestos in Ontario (Toronto: Queen's Printer, 3 vols, 1984).

Stander, E., and La Grange, J.J., *Asbestos* (Pretoria: The Government Printer, 1963).

Subcommittee on Environmental Health (Department of National Health and Welfare), *Report of the Asbestosis Working Group* (Ottawa, 15 February 1976).

US Public Health Service, *Objectives and General Plan for Occupational Health Study of the Asbestos Products* (Public Health Service: Industry Division of Public Health, 21 August, 1962).

Virta, R.L., *Worldwide Asbestos Supply and Consumption Trends from 1900 through 2003* (USGS, 2006). Posted at: http://pubs.usgs.gov

WHO/IARC, *IARC Monographs on the Evaluation of Carcinogenic Risks to Humans. Overall Evaluations of Carcinogenicity: An Updating of IARC Monographs vols 1–42, Supplement 7* (IARC: Lyon, 1987).

Wilson, N.H., *Notes on the Mining Industry of Southern Rhodesia* (Salisbury: Government Printer, 1932).

WTO, 'European Community—Measures Affecting Asbestos and Asbestos-Containing Products'. Report of the Panel, WT/DS135/R, 18 September 2000; Report of the Appellate Body, WT/DS135/AB/R, 12 March 2001.

Newspapers and Magazines

The Anniston Star
Asahi Shimbun
Basel Action Network
British Asbestos Newsletter

British Medical Journal
Business Standard
The Cape Times
Daily Telegraph
Economist
Financial Gazette [Harare]
Financial Times
Fortune
Greenpeace
The Guardian
The Herald (Harare)
The Hindu
Independent
Japan Times
Le Monde
Lloyds List
Mail & Guardian [Johannesburg]
New Statesman
New York Times
New Yorker
New Zealand Herald
Observer
Ottawa Gazette
Reuters
Star Tribune [Washington]
St Louis Dispatch
Seattle P-I
The Times
Time
The Virginian-Pilot
Wall Street Journal
Yomiuri Shimbun

Books/Published Reports/Pamphlets

AIA/NA, *Asbestos and Brake Linings* (Washington: AIA/NA, 1973).
AIC, *Asbestos and Your Health* (AIC, 1975).
AIC, *Asbestos—Safety and Control* (AIC, 1976).
AIC, *Mesothelioma* (AIC, 1976).
AIC, *Asbestos—Killer Dust or Miracle Fibre?* (AIC, 1976).
Albarado, R., *A Story Worth Telling: An Asbestos Tragedy* (Bloomington, IN: AuthorHouse, 2005).

Aldrich, M., *Safety First: Technology, Labor, and Business in the Building of American Work Safety 1970–1939* (Baltimore, MD: Johns Hopkins University Press, 1997).

Annals of the Global Asbestos Congress—Past, Present and Future (September 17–20, 2000, Osasco, Brazil, CD Proceedings).

Bartrip, P., *The Way from Dusty Death: Turner & Newall and the Regulation of Occupational Health in the British Asbestos Industry 1890s–1970* (London: Athlone, 2001).

——*Beyond the Factory Gates: Asbestos and Health in Twentieth Century America* (London: Continuum International Publishing Group, 2005).

Bennett, M.J., *The Asbestos Racket: An Environmental Parable* (Washington: Free Enterprise Press, 1991).

Biggs, J.L., 'Proposed Resolution Regarding the Need for Effective Asbestos Reform', American Academy of Actuaries Statement' (10 July 2003) (www.actuary.org/pdf/casualty/asbestos_10July03.pdf)

Blacksmith Institute, *The Hidden Tragedy: Pollution in the Developing World* (Blacksmith Institute, 2000) Posted at: http://www.blacksmithinstitute.org/hidden.pdf

Bowker, M., *Fatal Deception: The Untold Story of Asbestos. Why It Is Still Legal and Still Killing Us* (Emmaus, PA: Rodale, 2003).

Brodeur, P., *Expendable Americans: The Incredible Story of How Tens of Thousands of American Men and Women Die Each Year of Preventable Industrial Disease* (New York: Viking, 1974).

Brodeur, P., *Outrageous Misconduct: The Asbestos Industry on Trial* (Pantheon: New York, 1985).

Brodeur, P., *Secrets: A Writer in the Cold War* (Winchester, MA: Faber & Faber, 1997).

Canadian Asbestos Information Centre, *Asbestos, Health and Society Proceedings of the World Symposium on Asbestos* (Montreal: CAIC, 1982).

Carroll, B., *'A Very Good Business': One Hundred Years of James Hardie Industries Limited 1888–1988* (Sydney: James Hardie Industries, 2001).

Carroll, S.J., Hensler, D., Gross, J., Sloss, E.M., Schonlau, M., Abrahamse, A. and Ashwood, J.S., *Asbestos Litigation* (Santa Monica, CA: RAND, 2005).

Castleman, B.I., *Asbestos: Medical and Legal Aspects* (New York: Aspen Publishers, 5th edition, 2005).

Catrina, W., *Der Eternit-Report: Stephan Schmidheinys Schweres Erbe* (Zurich: Orell Füssli, 1989).

Cherniack, M., *The Hawk's Nest Incident: America's Worst Industrial Disaster* (New Haven, CT: Yale University Press, 1986).

Clapp, J., *Toxic Exports: The Transfer of Hazardous Wastes from Rich to Poor Countries* (Ithaca, NY: Cornell University Press, 2001).

Collectif Intersyndical Securite des Universites-Jussieu CFDT, CGT, FEN, *Danger! Amiante* (Paris: Éditions François Maspero, 1977).

Corn, J.K., *Environmental Public Health Policy for Asbestos in Schools: Unintended Consequences* (Boca Raton, FL: Lewis Publishers, 1999).

Dalton, A., *Asbestos Killer Dust: A Worker/Community Guide: How to Fight the Hazards of Asbestos and Its Substitutes* (London: BSSRS Publications, 1979).
Delisle, E., and Malouf, P.K., *Le Quatuor D'Asbestos: Autour de la Grève de L'Amiante* (Montreal: Les Éditions Varia, 2004).
Environmental Working Group, *Asbestos Think Again* (Washington, DC: EWG, 2004). Posted at: www.ewg.org/reports/asbestos
European Forum of the Insurance Against Accidents at Work and Occupational Diseases, *Asbestos-Related Occupational Diseases in Europe: Recognition, Figures, Specific Systems* (Brussels: Eurogip, 2006).
Faesch, C., *Salariés de L'Amiante, Employés de L'Indifférence* (Villeurbanne: Éditions Golias, 2002).
Fletcher, M., *Digging People Up For Coal* (Melbourne: Melbourne University Press, 2002).
Flynn, L., *Studded with Diamonds and Paved with Gold: Miners, Mining Companies and Human Rights in Southern Africa* (London: Bloomsbury, 1992).
Glantz, S., Slade, J., Bero, L.A., Hanauer, P., and Barnes, D.E. (eds.), *The Cigarette Papers* (Berkeley, CA: University of California Press, 1996).
Grove, M.J., *Asbestos Cancer: One Man's Experience* (Hanover, MA: Christopher Publishing House, 1995).
Haigh, G., *Asbestos House: The Secret History of James Hardie Industries* (Melbourne: Scribe, 2006).
James Hardie Asbestos Ltd, *Hardie Ferodo 1000: A James Hardie Group and Activity Report* (1978).
Hardy-Hemery, O., *Eternit et L'Amiante 1922–2000: Aux Sources du Profit, une Industrie du Risque* (Villeneuve d'Ascq: Presses Universitaires du Septentrion, 2005).
Harvey, C., and Rollinson, M., *Asbestos in Schools: A Guide for School Administrators, Teachers and Parents* (New York: Praeger Publishers, 1987).
Health Effects Institute, *Asbestos in Commercial Buildings: A Literature Review and Synthesis of Current Knowledge* (Boston, MA: HEI, 1991).
Henry, E., *Amiante: Un Scandal Improbable. Sociologie d'un Problème Public* (Rennes: Presses Universitaires de Rennes, 2007).
Hills, B., *Blue Murder* (Melbourne: Sun Books, 1989).
IARC Monographs on the *Evaluation of the Carcinogenic Risk of Chemicals to Humans* (IARC Monographs, Lyon, September 1979).
Johnston, R., and McIvor, A., *Lethal Work: A History of the Asbestos Tragedy in Scotland* (Glasgow: Tuckwell Press, 2000).
Jones, R., *Asbestos and Asbestic* (London: Crosby, Lockwood & Son, 1897).
Knuckey, G., *Chrysotile Arizona 1914 to 1945* (Tuscon, AZ: Wheatmark, 2007).
Kember, L., *Lean on Me: Cancer Through a Carer's Eyes* (Beechboro, W. Australia: L. Kember Productions, 2003).
Kluger, R., *Ashes to Ashes* (New York: Vintage Books, 1997).
Lanza, A. (ed.), *Silicosis and Asbestosis* (London: Oxford University Press, 1938).

LeBlanc, S., *Cassiar: A Jewel in the Wilderness* (Prince George, BC: Caitlin Press, 2003).

Kazan-Allen, L., *Killing the Future: Asbestos Use in Asia* (London: International Ban Asbestos Secretariat, 2007), p. 38. Posted at: http://www.lkaz.demon.co.uk/ktf_web_fin.pdf

Landrigan, P.J., and Kazemi, H. (eds.), 'The Third Wave of Asbestos Disease: Exposure to Asbestos in Place. Public Health Control', *Annals of New York Academy of Sciences* 643 (1991), pp. 1–628.

Leneghan, J., *Victims Twice Over: A Report into How Members of Clydeside Action on Asbestos are Disabled by Lung Disease and Further Handicapped by Medical and Social Services* (Glasgow: Clydeside Action on Asbestos, 1994).

Lenglet, R., *L'Affaire de L'Amiante* (Paris: La Decouverte, 1996).

Levine, M., *Surviving Cancer: One Woman's Story and Her Inspiring Program for Anyone Facing a Cancer Diagnosis* (New York: Broadway, 2001).

London Hazards Centre, *The Asbestos Hazards Handbook* (London: LHC, 1995).

——*Rising from the Dust: Building a Support Network on Asbestos and Environmental Disease in Barking and Dagenham* (2004). Posted at: www.lhc.org.uk

McCulloch, J., *Asbestos: Its Human Cost* (St Lucia: University of Queensland Press, 1986).

——*Asbestos Blues: Labour, Capital, Physicians and the State in South Africa* (Oxford: James Currey, 2002).

McGiffen, S.P. (ed.), *The Polluter Pays: Notes from the International Conference on Asbestos Held in Amsterdam in May 2004* (Amsterdam: Comité Asbestslachtoffers, 2005).

McKessock, B., *Mesothelioma: The Story of an Illness* (Glasgow: Argyll Publishing, 1995).

Maines, R., *Asbestos and Fire: Technological Trade-offs and the Body at Risk* (New Brunswick, NJ: Rutgers University Press, 2005).

Malye, F., *Amiante: Le Dossier de L'Air Contaminé* (Paris: Le Pré aux Clercs, 1996).

Manhattan Institute, *Trial Lawyers Inc: A Report on the Lawsuit Industry in America* (New York, 2003). Posted at: www.triallawyersinc.com

Markowitz, G., and Rosner, D., *Deceit and Denial: The Deadly Politics of Industrial Pollution* (Berkeley, CA: University of California Press, 2002).

Meria, G., *L'Aventure Industrielle De L'Amiante en Corse* (Ajaccio: Alain Piazzola, 2003).

National Asbestos Summit, *Final Proceedings, National Asbestos Summit,* 24–26 November 1998 (Johannesburg, Esselen Park: The National Parliamentary Portfolio Committee on Environment & Tourism, 1998).

Nay, S., *Mortel Amiante* (Brussels: Editions Evo, 1997).

Peacock, A., *Libby, Montana: Asbestos and the Deadly Silence of an American Corporation* (Boulder, CO: Johnson Books, 2003).

Peacock, M., *Asbestos: Work as a Health Hazard* (Sydney: Australian Broadcasting Commission with Hodder & Stoughton, 1978).

Proctor, R.N., *Cancer Wars: How Politics Shapes What We Know and Don't Know About Cancer* (New York: Basic Books, 1995).
——*The Nazi War on Cancer* (Princeton, NJ: Princeton University Press, 1999).
Public Citizen, *Federal Asbestos Legislation: The Winners Are...* (Washington: Congress Watch, May 2005).
——*Asbestos Cases in the Courts: No Logjam* (Washington, DC, Public Citizen, 2006).
Raybestos-Manhattan Inc., *Annual Report.*
Roselli, M., *Die Asbestlüge* (Zurich: Rotpankverlag, 2007).
Rosner, D., and Markowitz, G., *Deadly Dust: Silicosis and the On-Going Struggle to Protect Workers' Health* (Ann Arbor, MI: Michigan University Press, 2nd edn., 2006).
Ross, J.G., *Chrysotile Asbestos in Canada* (Ottawa: F.A. Acland, 1931).
Rossi, N., *From This Day Forward: A True Love Story* (New York: Times Books, 1983).
Ruers, R.F., and Schouten, N., *The Tragedy of Asbestos* (Rotterdam: Socialist Party, 2005), p. 13. Posted at: http://www.international.sp.nl/publications/asbestos.pdf
Ruers, R.F. et al., Eternit le Blanchiment de l' Amiente Sale: Les Consequesces Tragiques de 100 Ans d' Amiante - Ciment (CAOVA, 2006). Posted at: http://caova.ch
Safe Buildings Alliance, *Asbestos in Buildings: What Owners and Managers Should Know* (Washington, DC, 1989).
Salfellner, H., *Franz Kafka and Prague* (Prague: Vitalis, 3rd edn., 2003).
Schneider, A., and McCumber, D., *An Air that Kills: How the Asbestos Poisoning of Libby, Montana, Uncovered a National Scandal* (New York: G.P. Putnam's Sons, 2004).
Selikoff, I.J., and Lee, D.H.K., *Asbestos and Disease* (New York: Academic Press, 1978).
Sellers, C.C., *From Industrial Disease to Environmental Health Science* (Chapel Hill, NC: University of North Carolina Press, 1997).
Shapiro, H.A. (ed.), *Pneumoconiosis Proceedings of the International Conference Johannesburg 1969* (Cape Town: Oxford University Press, 1970).
Skinner, H.C.W., Ross, M., and Frondel, C., *Asbestos and Other Fibrous Minerals* (New York: Oxford University Press, 1988).
Smith, D., *Consulted to Death: How Canada's Workplace Health and Safety System Fails Workers* (Winnipeg: Arbeiter Ring Publishing, 2000).
Stiglitz, J., Orszag, J.M., and Orszag, P.R., *The Impact of Asbestos Liabilities on Workers in Bankrupt Firms* (Sebago Associates: Commissioned by American Insurance Association, 2002).
Summers, A.L., *Asbestos and the Asbestos Industry: The World's Most Wonderful Mineral and Other Fireproof Materials* (London: Pitman, 1919).
Tait, N., *Asbestos Kills* (London: SPAID, 1976).
Tataryn, L., *Dying for a Living: The Politics of Industrial Death* (Ottawa: Deneau & Greenberg, 1979).
Turner & Newall, *Annual Reports.*

Tweedale, G., *Magic Mineral to Killer Dust: Turner & Newall and the Asbestos Hazard* (Oxford: Oxford University Press, 2nd edn., 2001).

Vircondelet, A., *Mortel Amiante* (Paris: Editions Anne Carriere, 1998).

Whipple, H. E. (ed.), 'Biological Effects of Asbestos', *Annals of the New York Academy of Sciences* 132 (31 December 1965), i, pp. 1–765.

Wikeley, N.J., *Compensation for Industrial Disease* (Aldershot: Dartmouth Publishing, 1993).

Winfield, M., Coumans, C., Kuyek, J.N., Meloche, F., and Taylor, A., *Looking Beneath the Surface: an Assessment of the Value of Public Support for the Metal Mining Industry in Canada* (Canada: The Pembina Institute and MiningWatch Canada, 2002). Posted at: http://www.miningwatch.ca/updir/belowthesurface-eng.pdf

Wragg, G., *The Asbestos Time Bomb* (Sydney: Catalyst Press, 1995).

Zegart, D., *Civil Warriors: The Legal Siege on the Tobacco Industry* (New York, 2000).

Articles

Abelson, P.H., 'The Asbestos Removal Fiasco', *Science* 247 (2 March 1990), p. 1017.

Aguilar-Madrid, G., Juárez-Férez, C.A., Vasquez-Grameix, J.H., and Hernéndez-Avila, M., 'Occupational and Environmental Rights Violation in the Asbestos Industry in Mexico', *Annals of the Global Asbestos Congress — Past, Present and Future* (Osasco, Brazil, 17–20 September 2000).

Alleman, J.E., and Mossman, B.T., 'Asbestos Revisited', *Scientific American* (July 1997), pp. 54–7.

Annals of the Global Asbestos Congress, Tokyo, Japan, 19–21 November 2004 (CD Proceedings, 2005).

Anon, 'The Manville Bankruptcy: Treating Mass Tort Claims in Chapter 11 Proceedings', *Harvard Law Review* 96 (March 1983), pp. 1121–42.

'Asbestos Outlook', *The South African Mining and Engineering Journal* (October 1974), pp. 18–34.

Baur, X., and Czuppon, A.B., 'Regulation and Compensation of Asbestos Diseases in Germany', in G.A. Peters and B.J. Peters (eds.), *Sourcebook on Asbestos Diseases: Volume 15* (Charlottesville, VA: Lexis Law Publishing, 1997), pp. 405–19.

Beckett, C., 'An Epidemiologist at Work: The Personal Papers of Sir Richard Doll', *Medical History* 46 (2002), pp. 403–21.

Berry, M., 'Mesothelioma Incidence and Community Asbestos Exposure', *Environmental Research* 75 (1997), pp. 34–40.

Best, R., 'Liability for Asbestos-Related Disease in England and Germany', *German Law Journal* 4 (1 July 2003), pp. 661–83.

'Biological Effects of Asbestos: Report of the Advisory Committee on Asbestos Cancers to the Director of the International Agency for Research on Cancer, Lyon, 5–6 October 1972', *Annals of Occupational Hygiene* 16 (1973), pp. 9–17.

Boffetta, P., and Nyberg, F., 'Contribution of Environmental Factors to Cancer Risk', *British Medical Bulletin* 68 (2003), pp. 71–94.

Bohme, S.R., Zorabadeian, J., and Egilman, D.S., 'Maximising Profit and Endangering Health: Corporate Strategies to Avoid Litigation and Regulation', *International Journal of Occupational and Environmental Health* 11 (2005), pp. 338–48.

Borow, M., Livornese, L.L., and Schalet, N., 'Mesothelioma and Its Association with Asbestosis', *Journal of the American Medical Association* 201 (July–September 1967), pp. 587–91.

Braun, D.C., and Truan, T.D., 'An Epidemiological Study of Lung Cancer in Asbestos Miners', *AMA Archives of Industrial Health* 17 (June 1958), pp. 634–53.

Braun, L., Green, A., Manseau, M., Singhal, R., Kisting, S., and Jacobs, N., 'Scientific Controversy and Asbestos: Making Disease Invisible', *International Journal of Occupational and Environmental Health* 9 (July–September 2003), pp. 194–205.

Breslow, L., Hoaglin, L., Rasmussen, G., and Abrams, H.K., 'Occupations and Cigarette Smoking as Factors in Lung Cancer', *American Journal of Public Health* 44 (1954), pp. 171–81.

Brhel, P., 'Occupational Respiratory Diseases in the Czech Republic', *Industrial Health* 41 (2003), pp. 121–3.

Brodeur, P., 'Annals of Industry: Casualties of the Workplace. III Some Conflicts of Interest', *New Yorker*, 12 November 1973.

——*A Reporter at Large: The Magic Mineral* (*The New Yorker*, 1968).

Brophy, J., and Parent, M., 'Documenting the Asbestos Story in Sarnia', *New Solutions* 9 (1999), pp. 297–316.

——'The Public Health Disaster Canada Chooses to Ignore', in *Chrysotile Asbestos: Hazardous to Humans, Deadly to the Rotterdam Convention* (London: Building & Woodworkers International and International Ban Asbestos Secretariat, 2006), pp. 17–20. Posted at: www.lkaz.demon.co.uk/chrys_hazard_rott_conv_06.pdf

Browne, K., and Wagner, J.C., 'Environmental Exposure to Amphibole-Asbestos and Mesothelioma', in R.P. Nolan, A.M. Langer, M. Ross, F.J. Wicks, and R.F. Martin (eds.), *The Health Effects of Chrysotile Asbestos: Contribution of Science to Risk-Management Decisions* (Ottawa: Mineralogical Association of Canada, 2001), pp. 21–8.

Burdorf, A., Dahhan, M., and Swuste, P., 'Occupational Characteristics of Cases with Asbestos-Related Diseases in the Netherlands', *Annals of Occupational Hygiene* 47 (2003), pp. 485–92.

——————'Pleural Mesothelioma in Women Associated with Environmental Exposure to Asbestos', *Ned Tijdschr Geneeskd* 148 (28 August 2004), pp. 1727–31.

Burke, B., 'Shipbuilding's Deadly Legacy', Virginian-Pilot, 6–10 May 2001. Posted at: http://hamptonroads.com/pilotonline/special/asbestos/indext.html

Canadian Medical Association Journal, Editorial, 'A Ban on Asbestos: Is Now the Time?', *CMAJ* 164 (20 February 2001), p. 453.

Cape Asbestos Company Ltd, *Asbestos* (The magazine), Diamond Jubilee, 7–8 July 1953.

Collegium Ramazzini, 'D.T.K. Joshi and Asbestos in India: A Message from the Collegium Ramazzini', *American Journal of Industrial Medicine* 45 (2004), pp. 125–8.

Castleman, B., 'The Export of Hazardous Factories to Developing Nations', *Indian Journal of History of Science* 9 (1979), pp. 569–606.

——'Implications of the World Trade Organisation Verdict for Public Health and Global Trade', in *Annals of the Global Asbestos Congress—Past, Present and Future* (Osasco, Brazil, 17–20 September 2000).

——'The Manipulation of "Scientifie Organisations": Controversies at International Organisations over Asbestos Industry Influence', *Annals of the Global Asbestos Congress – Past, Present and Future* (CD-ROM, 17–20 September 2000, Osasco, Brazil).

——'Doll's 1955 Study on Cancer from Asbestos', *American Journal of Industrial Medicine* 39 (2001), pp. 237–40.

——'WTO Confidential: The Case of Asbestos', *Indian Journal of History of Science* 32 (2002), pp. 489–510.

——'Controlled Use of Asbestos', *International Journal of Occupational and Environmental Health* 9 (July–September 2003), pp. 294–8.

——and Lemen, R.A., 'The Manipulation of International Scientific Organizations', *International Journal of Occupational and Environmental Health* 4 (1998), pp. 53–5.

Centers, J., and Korade, M., 'Special Report: Cement Asbestos Products Company (Capco)', *The Anniston Star*, 27–30 March 2005. Posted at: http://www.annistonstar.com/showcase/2005/as-specialreport-0327-0-5c28k0245.htm

Churg, A., 'Chrysotile, Tremolite, and Malignant Mesothelioma in Man', *Chest* 93 (1988), pp. 621–8.

——Wiggs, B., Depaoli, L., Kampe, B., and Stevens, B., 'Lung Asbestos Content in Chrysotile Workers with Mesothelioma', *American Review of Respiratory Disease* 130 (1984), pp. 1042–5.

Comba, P., Merler, E., and Pasetto, R., 'Asbestos-Related Diseases in Italy: Epidemiological Evidences and Public Health Issues', *International Journal of Occupational and Environmental Health* 11 (2005), pp. 36–44.

Cooke, W.E., 'Fibrosis of the Lungs Due to the Inhalation of Asbestos Dust', *BMJ* (26 July 1924), ii, p. 147.

Corn, J., and Starr, J., 'Historical Perspective on Asbestos: Policies and Protective Measures in World War II Shipbuilding', *American Journal of Industrial Medicine* 11 (1987), pp. 359–73.

Dalquen, P., Dabbert, A.F., and Hinz, I., 'The Epidemiology of Pulmonary Mesothelioma: A Preliminary Report on 119 Cases from the Hamburg Area', *German Medical Monthly* 15 (February 1970), pp. 89–95.

Davies, J.C.A., Williams, B.G., Debeila, M.A., and Davies, D.A., 'Asbestos-Related Lung Disease Among Women in the Northern Province of South Africa', *The South African Journal of Science*, 97 (March/April 2001), pp. 87–92.

Degiovanni, D., 'History of the Asbestos Ban in Casale Monferrato', in *Annals of Global Asbestos Congress: Past Present and Future* (Osasco, Brazil, 17–20 September 2000).

Dement, J.M., Harris, R.L., Symons, M.J., and Shy, C.M., 'Exposures and Mortality Among Chrysotile Asbestos Workers', Part II. Mortality, *American Journal of Industrial Medicine* 4 (1983), pp. 421–33.

Dodson, R.F., Atkinson, M.A.L., and Levin, J.L., 'Asbestos Fiber Length as Related to Potential Pathogenicity: A Critical Review', *American Journal of Industrial Medicine* 44 (2003), pp. 291–7.

Doll, R., 'Mortality from Lung Cancer in Asbestos Workers', *British Journal of Industrial Medicine* 12 (1955), pp. 81–6.

——'Occupational Cancer: A Hazard for Epidemiologists', *International Journal of Epidemiology* 14 (March 1985), pp. 22–31.

Drinker, P., 'Engineering Methods in the Control of Silicosis', in Kuechle, B. (ed.), *Second Symposium on Silicosis . . . at Saranac Lake, 3–7 June 1935* (Saranac Laboratory: Unofficial published transcript, 1935), pp. 176–81.

Drucker, E., Nagin, D., Michaels, D., Lacher, M., and Zoloth, S., 'Exposure of Sheet-Metal Workers to Asbestos During the Construction and Renovation of Commercial Buildings in New York City', *Annals of the New York Academy of Sciences* 502 (1987), pp. 230–44.

Du Plessis, H., 'Asbestos's Sorrowful Legacy: A Photo-essay', *International Journal of Occupational and Environmental Health* 9 (July–September 2003), pp. 236–43.

Egilman, D., 'The Asbestos TLV: Early Evidence of Inadequacy', *American Journal of Industrial Medicine* 30 (1996), pp. 369–70.

——and Reinert, A., Letter re. 'Corruption of Previously Published Asbestos Research', *Archives of Environmental Health* 55 (January/February 2000), pp. 75–6.

——Fehnel, C., and Bohme, S.R., 'Exposing the "Myth" of ABC, "Anything But Chrysotile": A Critique of the Canadian Asbestos Mining Industry and McGill University Chrysotile Studies', *American Journal of Industrial Medicine* 44 (2003), pp. 540–57.

——Tweedale, G., McCulloch, J., Kovarik, W., Castleman, B., Longo, W., Levin, S., and Bohme, S.R., 'P.J.W. Bartrip's Attack on Irving J. Selikoff', *American Journal of Industrial Medicine* 46 (2004), pp. 151–5.

Ellman, P., 'Pulmonary Asbestosis: Its Clinical, Radiological and Pathological Features, and Associated Risk of Tuberculous Infection', *Journal of Industrial Hygiene* 15 (1933), pp. 165–83.

Elmes, P.C., 'Current Information on the Health Risk of Asbestos', *Royal Society of Health Journal* 96 (1976), pp. 248–52.

Enterline, P.E., 'Asbestos and Cancer: The International Lag', *American Review of Respiratory Disease* 118 (1978), pp. 975–8.

——'Changing Attitudes and Opinions Regarding Asbestos and Cancer 1934–1965', *American Journal of Industrial Medicine* 20 (1991), pp. 685–700.

Environmental Working Group, *Asbestos: Think Again* (Washington, DC, 2004). Posted at: www.ewg.org/reports/asbestos

Fabianova, E., Szeszenia-Dabrowska, N., Kjaerheim, K., and Boffetta, P., 'Occupational Cancer in Central European Countries', *Environmental Health Perspectives* 107 Supp. 2 (May 1999), pp. 279–82.

Flanagan, J., and Whitson, T., 'Asbestos Victims Support Groups in England', *International Journal of Occupational and Environmental Health* 10 (April/June 2004), pp. 177–9.

Fletcher, D.E., 'Asbestos-Related Chest Diseases in Joiners', *Proceedings of Royal Society of Medicine* 64 (1971), pp. 837–8.

Fleischer, W., Viles, F., Gade, R., and Drinker, P., 'A Health Survey of Pipe Covering Operations in Constructing Naval Vessels', *Journal of Industrial Hygiene and Toxicology* 28 (January 1946), pp. 9–16.

Flynn, L., *Asbestos: The Dust that Kills in the Name of Profit* (London: Socialist Worker, 1974).

Fumento, M., 'The Asbestos Rip-Off', *American Spectator* (October 1989). Posted at: www.fumento.com/asbest.html

Furuya, S., Natori Y., and Ikeda, R., 'Asbestos in Japan', *International Journal of Occupational and Environmental Health* 9 (July/September 2003), pp. 260–5.

Gardner, L., 'General Technique of Dust Exposure', in *Silicosis: Records of the International Conference held at Johannesburg 13–27th August 1930* (London: ILO, 1930), pp. 53–4.

Garfinkel, L., 'Asbestos: Historical Perspective', *CA: A Cancer Journal for Clinicians* 34 (1984), pp. 44–7.

Gennaro, V., Ugolini, D., Viarengo, P., Benfatto, L., Bianchelli, M., Lanzzarotto, A., Montarano, F., and Puntoni, R., 'Incidence of Pleural Mesothelioma in Liguria Region, Italy (1996–2002)', *European Journal of Cancer* 47 (November 2005), pp. 2709–14.

Gerin-Lajoie, J., 'Financial History of the Asbestos Industry', in Pierre Trudeau (ed.), *The Asbestos Strike* (Toronto: James Lewis & Samuel, 1974), pp. 83–104.

Giannasi, F., 'Globalisation from Below: Building an Anti-asbestos Movement in Brazil', in *Annals of the Global Asbestos Congress—Past, Present and Future* (Osasco, Brazil, 17–20 September 2000).

Gibbs, G.W., Valic, F., and Browne, K. (eds.), 'Health Risks Associated with Chrysotile Asbestos', Report of Workshop in Jersey, Channel Islands, 14–17 November 1993, *Annals of Occupational Hygiene* 38 (1994), pp. 399–646.

Gilson, J.C., 'Asbestos Cancer: Past and Future Hazards', *Proceedings Royal Society of Medicine* 66 (1973), pp. 395–403.

Gloag, D., 'Can Society Live with Asbestos?', *BMJ* 284 (12 June 1982), pp. 1728–9.

Gloyne, S.R., 'Pneumoconiosis: A Histological Survey of Necropsy Material in 1205 Cases', *Lancet* 260 (14 April 1951), i, pp. 810–14.

Goldberg, M., Goldberg, S., and Luce, D., 'Regional Differences in the Compensation of Pleural Mesothelioma as Occupational Disease in France, 1986–1993', *Revue Épidémiologie Santé Publique* 47 (1999), pp. 421–31.

Greenberg, M., 'Classical Syndromes in Occupational Medicine: The Montague Murray Case', *American Journal of Industrial Medicine* 3 (1982), pp. 351–6.

——'A Study of Lung Cancer Mortality in Asbestos Workers: Doll, 1955', *American Journal of Industrial Medicine* 36 (1999), pp. 331–47.

——'Re. Call for an International Ban on Asbestos: Trust Me I'm a Doctor', *American Journal of Industrial Medicine* 37 (2000), pp. 232–4.

——'Biological Effects of Asbestos: New York Academy of Sciences 1964', *American Journal of Industrial Medicine* 43 (2003), pp. 543–52.

——'The British Approach to Standard Setting: 1898–2000', *American Journal of Industrial Medicine* 46 (2004), pp. 534–41.

——'A Report on the Health of Asbestos, Quebec Miners 1940', *American Journal of Industrial Medicine* 38 (2005), pp. 230–7.

——'Revising the British Occupational Hygiene Society Asbestos Standard: 1968–1982', *American Journal of Industrial Medicine* 49 (2006), pp. 577–604.

——and Lloyd Davies, T.A., 'Mesothelioma Register 1967–68', *British Journal of Industrial Medicine* 31 (1974), pp. 91–104.

Hagemeyer, O., Otten, H., and Kraus, T., 'Asbestos Consumption, Asbestos Exposure and Asbestos-Related Occupational Diseases in Germany', *International Archives of Occupational and Environmental Health* 79 (September 2006), pp. 613–20.

Hammond, E.C., Selikoff, I.J., and Seidman, H., 'Asbestos Exposure, Cigarette Smoking, and Death Rates', *Annals of the New York Academy of Sciences* 330 (1979), pp. 473–90.

Hansen, J., Rasmussen, T.R., Omland, O., and Ølsen, J.H., 'Registration of Selected Cases of Occupational Cancer (1994–2002) with the Danish National Board of Industrial Injuries', *Ugeskr Laeger* 169 (30 April 2007), pp. 1674–8.

Harrison, P.T.C., Levy, L.S., Patrick, G., Pigott, G.H., and Smith, L.L., 'Comparative Hazards of Chrysotile Asbestos and Its Substitutes: A European Perspective', *Environmental Health Perspectives* 1999, 107, pp. 607–11.

Heenan, J., 'Graceful Maneuvering: Corporate Avoidance of Liability through Bankruptcy and Corporate Law', *Vermont Journal of Environmental Law* (2003). Online journal published at: www.vjel.org/roscoe/roscoe03a.html

Hills, B., 'The James Hardie Story: Asbestos Victims' Claims Evaded by Manufacturer', *International Journal of Occupational and Environmental Health* 11 (2005), pp. 212–14.

Hills, D.W., 'Economics of Dust Control', in H.E. Whipple (ed.), 'Biological Effects of Asbestos', *Annals of the New York Academy of Sciences* 132 (31 December 1965), i, pp. 322–37.

Hinne, K., Opening Address at European Asbestos Conference, Dresden, 2003. Posted at: www.hvbg.de/e/asbest/konfrep/konfrep/index.html

Holt, P., 'Asbestos-Related Diseases: Neighbourhood and Domestic Hazard', in Holt (ed.), *Inhaled Dust and Disease* (New York: John Wiley & Sons, 1987), pp. 157–66.

Huff, J., 'Industry Influence on Occupational and Environmental Public Health', *International Journal of Occupational and Environmental Health* 13 (2007), pp. 107–17.

Huncharek, M., 'Asbestos and Cancer: Epidemiological and Public Health Controversies', *Cancer Investigation* 12 (1994), pp. 214–22.

Ilgren, E.B., and Wagner, J.C., 'Background Incidence of Mesothelioma: Animal and Human Evidence', *Regulatory Toxicology and Pharmacology* 13 (1991), pp. 133–49.

International Conference on Chrysotile, Montreal, 23–24 May 2006. Proceedings. Posted at: http://www.chrysotile.com/en/conferences/default.aspx

International Union Against Cancer, 'Report and Recommendations of the Working Group on Asbestos and Cancer', *Annals of the New York Academy of Sciences* 132 (1965), pp. 706–21.

Jacques, L.C., 'Shipboard Asbestos Use: An Historical Perspective', in George A. Peters and B. Peters (eds.) *Current Asbestos Issues: Legal, Medical and Technical Research. Volume 10: Sourcebook on Asbestos Diseases* (New Hampshire: Butterworths, 1994).

Jasanoff, S., and Perese, D., 'Welfare State or Welfare Court: Asbestos Litigation in Comparative Perspective', *Journal of Law and Policy* 12 (2004), pp. 619–39.

Joachim-Woitowitz, H., 'Asbestos-Related Occupational Diseases—The Current Situation', European Asbestos Conference, Dresden, 2003. Posted at: www.asbestkonferenze2003.de/

Johns-Manville Corporation, *Annual Reports*.

Johnson, W.G., and Heler, E., 'Compensation for Death from Asbestos', *Industrial and Labor Relations Review* 37 (July 1984), pp. 529–40.

Kashansky, S.V., Shcherbakov, S.V., and Kogan, F., 'Dust Levels in Workplace Air (A Retrospective View of "Uralasbest")', in G.A. Peters and B. Peters (eds.), *Current Asbestos Issues: Sourcebook on Asbestos Disease Vol. 15* (New Haven, CA: Butterworths, 1994), pp. 337–54.

——Domnin, S., Kochelayev, V., Monakhov, D., and Kogan, F., 'Retrospective View of Airborne Dust Levels in Workplace of a Chrysotile Mine in Ural, Russia', *Industrial Health* 39 (2001), pp. 51–6.

Kazan-Allen, L., 'The Asbestos War', *International Journal of Occupational and Environmental Health* 9 (July/September 2003), pp. 173–93. Posted at: http://www.ijoeh.com/archive.html

——'Canadian Asbestos: A Global Concern', *International Journal of Occupational and Environmental Health* 10 (April/June 2004), pp. 121–43.

Kilburn, K.H., Warshaw, R., and Thornton, J.C., 'Asbestos Diseases and Pulmonary Symptoms and Signs in Shipyard Workers and Their Families in Los Angeles', *Archives of Internal Medicine* 146 (November 1986), pp. 2213–20.

Kjellstrom, T.E., 'The Epidemic of Asbestos-Related Diseases in New Zealand', *International Journal of Occupational and Environmental Health* 10 (April/June 2004), pp. 212–19.

Knox, J.F., Doll, R.S., and Hill, I.D., 'Cohort Analysis of Changes in Incidence of Bronchial Carcinoma in a Textile Asbestos Factory', *Annals of the New York Academy of Sciences* 132 (December 1965), i, pp. 527–35.

——Holmes, S., Doll, R.S., and Hill, I.D., 'Mortality from Lung Cancer and Other Causes among Workers in an Asbestos Textile Factory', *British Journal of Industrial Medicine* 25 (1968), pp. 293–303.

Koniak, S.P., 'Feasting While the Widows Weep: Georgine v. Anchem Products Inc.', *Cornell Law Review* 80 (May 1995), pp. 1045–58.

Kotelchuck, D., 'Asbestos Research', *Health Advisory Center*, No. 61 (Nov/Dec 1974), pp. 1–6, 20–7.

G. Krishna, 'Asbestos Kills Europeans, Australians and Japanese But Not Indians', *Toxicslink* (6 January 2006). Posted at: toxicslink.org

Kroll-Smith, S., Brown, P., and Gunter, V.J. (eds.), *Illness and the Environment: A Reader in Contested Medicine* (New York: New York University Press, 2000).

Kuehn, R.R., 'The Suppression of Environmental Science', *American Journal of Law and Medicine* 30 (2004), pp. 333–69.

LaDou, J., 'International Occupational Health', *International Journal of Hygiene and Environmental Health* 206 (August 2003), pp. 303–13.

——'The Asbestos Cancer Epidemic', *Environmental Health Perspectives* 112 (March 2003), pp. 285–90.

Landrigan, P.J., 'Asbestos—Still a Carcinogen', *New England Journal of Medicine* 338 (28 May 1998), pp. 1618–19.

——Lioy, P.J., Thurston, G., Berkowitz, G., et al., 'Health and Environmental Consequences of the World Trade Center Disaster', *Environmental Health Perspectives* 112 (May 2004), pp. 731–9.

Lanza, A., McConnell, W.J., and Fehnel, J.W., 'Effects of the Inhalation of Asbestos Dust on the Lungs of Asbestos Workers', *Public Health Reports* 50 (January 1935), pp. 1–12.

Lanza, A.J., and Vane, R.J., 'Industrial Dusts and the Mortality from Pulmonary Disease', *American Review of Tuberculosis* 39 (1939), pp. 419–38.

LeDoux, B., 'Asbestosis at East Broughton—A Village of Three Thousand Suffocates in Dust', *Le Devoir*, 12 January 1949.

Lee, D.H.K., and Selikoff, I.J., 'Historical Background to the Asbestos Problem', *Environmental Research* 18 (April 1979), ii, pp. 300–14.

Leigh, J., and Driscoll, T., 'Malignant Mesothelioma in Australia, 1945–2002', *International Journal of Occupational and Environmental Health* 9 (July/September 2003), pp. 206–17.

Leigh, J.P., and Robbins, J.A., 'Occupational Disease and Workers' Compensation: Coverage, Costs and Consequences', *The Milbank Quarterly* 82 (2004), pp. 689–721.

Lemen, R.A., 'Challenge for the 21st Century—A Global Ban on Asbestos', in *Annals of the Global Asbestos Congress — Past, Present and Future* (17–20 September 2000, Osasco, Brazil).

——'Chrysotile Asbestos as a Cause of Mesothelioma: Application of the Hill Causation Model', *International Journal of Occupational and Environmental Health* 10 (2004), pp. 233–9.

Liddell, F.D.K., 'Magic, Menace, Myth and Malice', *Annals of Occupational Hygiene* 41 (1997), pp. 1–12.

——McDonald, A.D., and McDonald, J.C., 'Dust Exposure and Lung Cancer in Quebec Chrysotile Miners and Millers', *Annals of Occupational Hygiene* 42 (1998), pp. 7–20.

Lieben, J., and Pistawka, H., 'Mesothelioma and Asbestos Exposure', *Archives of Environmental Health* 14 (April 1967), pp. 559–63.

Lilienfeld, D.E., 'The Silence: The Asbestos Industry and Early Occupational Cancer Research—A Case Study', *American Journal of Public Health* 81 (June 1991), pp. 791–800.

Lynch, J.R., and Ayer, H.E., 'Measurement of Asbestos Exposure', *Journal of Occupational Medicine* 10 (1968), pp. 21–4.

McLaughlin, A.I.G., 'The Prevention of Dust Diseases', *Lancet* (1953), ii, p. 49.

McCulloch, J., 'The Discovery of Mesothelioma on South Africa's Asbestos Fields' *Social History of Medicine* 16 (2003), pp. 419–36.

——'Women Mining Asbestos in South Africa, 1893–1980', *Journal of Southern African Studies* 29 (2003), pp. 411–30.

——'Dust, Disease and Labour at Havelock Mine', *Journal of Southern African Studies* 31 (June 2005), pp. 251–66.

——'Beating the Odds: The Quest for Justice by South African Asbestos Mining Communities', *Review of African Political Economy* 32 (March 2005), pp. 63–77.

——'The Mine at Wittenoom: Blue Asbestos, Labour and Occupational Disease', *Labor History* 47 (February 2006), pp. 1–19.

——and Tweedale, G., 'Shooting the Messenger: The Vilification of Irving Selikoff', *International Journal of Health Services* 37 (2007), pp. 619–34.

————'Science is Not Sufficient: Irving J. Selikoff and the Asbestos Tragedy', *New Solutions* 17 (2007), pp. 293–310.

McDonald, A.D., and McDonald, J.C., 'Mesothelioma after Crocidolite Exposure during Gas Mask Manufacture', *Environmental Research* 17 (1978), pp. 340–6.

——Case, B.W., Churg, A., Dufresne, A., Gibbs, G.W., Sebastien, P., and McDonald, D., 'Mesothelioma in Quebec Chrysotile Miners and Millers: Epidemiology and Aetiology', *Annals of Occupational Hygiene* 41(1997), pp. 707–19.

McDonald, J.C., 'Unfinished Business: The Asbestos Textiles Mystery', *Annals of Occupational Hygiene* 42 (1998), pp. 3–5.

——and McDonald, A.D., 'Chrysotile, Tremolite and Carcinogenicity', *Annals of Occupational Hygiene* 41(1997), pp. 699–705.

————Gibbs, G.W., Siemiatycki, J., and Rossiter, C.E., 'Mortality in the Chrysotile Asbestos Mines and Mills of Quebec', *Archives of Environmental Health* 22 (June 1971), pp. 677–86.

Manaouil, C., Graser, M., and Jardie, O., 'Compensation of Asbestos Victims in France', *Medicine and Law* 25 (September 2006), pp. 435–3.

'Management by Morgan', *Fortune* (Magazine) 9 (1934), pp. 82–9.

Markowitz, G., and Rosner, D., 'The Limits of Thresholds: Silica and the Politics of Science', *American Journal of Public Health* 85 (1995), pp. 253–62.

Magnani, C., et al., 'Multicentric Study on Malignant Pleural Mesothelioma and Non-Occupational Exposure to Asbestos', *British Journal of Cancer* 83 (2000), pp. 104–11.

Maltoni, C., 'The Long-Lasting Legacy of Industrial Carcinogens: The Lesson of Asbestos', in E. Bingham and D.P. Rall (eds.), 'Preventive Strategies for Living in a Chemical World: A Symposium in Honor of Irving J. Selikoff', *Annals of the New York Academy of Sciences* 837 (1997), pp. 570–86.

Marchand, P.E., 'The Discovery of Mesothelioma in the Northwestern Cape Province in the Republic of South Africa', *American Journal of Industrial Medicine* 19 (1991), pp. 241–6.

Martonik, J.F., Nash, E., and Grossman, E., 'The History of OSHA's Asbestos Rulemakings and Some Distinctive Approaches that They Introduced for Regulating Occupational Exposure to Toxic Substances', *American Industrial Hygiene Association Journal* 62 (March/April 2001), pp. 208–17.

Meeran, R., 'The Boots Mesothelioma Cases', in G.A. Peters and B. Peters (eds.), *Current Asbestos Issues: Legal, Medical and Technical Research. Volume 10: Sourcebook on Asbestos Diseases* (New Hampshire: Butterworths, 1994), pp. 273–84.

——'Cape Plc: South African Mineworkers' Quest for Justice', *International Journal of Occupational and Environmental Health* 9 (September 2003), pp. 218–27.

Merewether, E.R.A., 'A Memorandum on Asbestosis', *Tubercle* 15 (1933–34), pp. 69–81, 109–18, 152–9.

Merler, E., 'A Cross-Sectional Study on Asbestos Workers Carried Out in Italy in 1940: A Forgotten Study', *American Journal of Industrial Medicine* 33 (1998), pp. 90–3.

——Ercolanelli, M., et al., 'On the Italian Migrants to Australia who Worked at the Crocidolite Mine at Wittenoom Gorge, Western Australia', in A. Grieco, S. Iavicoli, and G. Berlinguer (eds.), *Contributions to the History of Occupational and Environmental Prevention* (Amsterdam: Elsevier Science B.V, 1999), pp. 295–6.

Merler, E., Vineis, P., Alhaique, D., and Milligi, L., 'Occupational Cancer in Italy', *Environmental Health Perspectives Supplements* (S2) 107 (May 1999), pp. 259–71.

Michaels, D., 'Manufactured Uncertainty: Protecting Public Health in the Age of Contested Science and Product Defense', *Annals of the New York Academy of Sciences* 1076 (September 2006), pp. 149–62.

——and Monforton, C., 'Manufacturing Uncertainty: Contested Science and the Protection of the Public's Health and Environment', *American Journal of Public Health* 95 (2005), Supp 1, pp. 39–48.

Miller, A., 'Mesothelioma in Household Members of Asbestos-Exposed Workers: 32 United States Cases Since 1990', *American Journal of Industrial Medicine* 47 (2005), pp. 458–62.

Moatamed, F., Lockey, J.E., and Parry, W.T., 'Fiber Contamination of Vermiculites: A Potential Occupational and Environmental Health Hazard', *Environmental Research* 41 (1986), pp. 207–18.

Montanaro, F., Vitto, V., Lagattolla, N., Lazzarotto, A., Bianchelli, M., Puntoni, R., and Gennaro, V., 'Occupational Exposure to Asbestos and Recognition of Pleural Mesothelioma as Occupational Disease in the Province of Genoa', *Epidemiologia e Prevenzione* 25 (March/April 2001), pp. 71–6.

Mossman, B.T., and Gee, J.B.L., 'Asbestos-Related Diseases', *NEJM* 320 (29 June 1989), pp. 1721–30.

——Bignon, J., Corn, M., Seaton, A., and Gee, J.B.L., 'Asbestos: Scientific Developments and Implications for Public Policy', *Science* 247 (19 January 1990), pp. 294–301.

Muchlinski, P., 'Corporations in International Litigation: Problems of Jurisdiction and the United Kingdom Asbestos Cases', *International and Comparative Law Quarterly* 50 (January 2001), pp. 1–25.

Murray, T.H., 'Regulating Asbestos: Ethics, Politics, and the Values of Science', in R. Bayer (ed.), *The Health and Safety of Workers* (New York: Oxford University Press, 1988), pp. 271–92.

Nay, S.Y., 'Asbestos in Belgium: Use and Abuse', *International Journal of Occupational and Environmental Health* 9 (July–September 2003), pp. 287–93.

Newhouse, M.L., 'Asbestos in the Workplace and the Community', *Annals of Occupational Hygiene* 16 (1973), pp. 97–102.

——and Thompson, H., 'Mesothelioma of Pleura and Peritoneum Following Exposure to Asbestos in the London Area', *British Journal of Industrial Medicine* 22 (1965), 261–9.

Nicholson, W.J., 'Comparative Dose-Response Relationships of Asbestos-Fiber Types: Magnitudes and Uncertainties', in P.J. Landrigan and H. Kazemi (eds.), 'The Third Wave of Asbestos Disease: Exposure to Asbestos in Place. Public Health Control', *Annals of the New York Academy of Sciences* 643 (1991), pp. 74–84.

——Selikoff, I.J., Seidman, H., Lilis, R., and Formby, P., 'Long-Term Mortality in Experience of Chrysotile Miners and Millers in Thetford Mines, Quebec', *Annals of the New York Academy of Sciences* 330 (1979), pp. 11–21.

——Perkel, G., and Selikoff, I.J., 'Occupational Exposure to Asbestos: Population at Risk and Projected Mortality 1980–2030', *American Journal of Industrial Medicine* 3 (1982), pp. 259–311.

Oliver, L.C., 'Asbestos in Buildings: Management and Related Health Effects', in M.A. Mehlman and A. Upton, *Advances in Modern Toxicology. Vol. 22: The Identification and Control of Environmental and Occupational Diseases: Asbestos and Cancers* (Princeton: Princeton Scientific Publishing Co. 1994), pp. 174–88.

Ozonoff, Denial 'Failed Warnings: Asbestos-Related Disease and Industrial Medicine', in R. Bayer (ed.), *The Health and Safety of Workers* (New York: Oxford University Press, 1988), pp. 139–218.

Paek, D., 'Asbestos Problems in Korea: History and Current Situation', *Annals of the Global Asbestos Congress—Past, Present and Future* (Osasco, Brazil, 17–20 September 2000).

Pandita, S., 'Banning Asbestos in Asia', *International Journal of Occupational and Environmental Health* 12 (2006), pp. 248–53.

Parloff, R., 'The $200 Billion Miscarriage of Justice', *Fortune* 145 (4 March 2002), pp. 154–8, 162, 164.

Pedley, F., 'Asbestosis', *Canadian Public Health Journal* 21 (1930), pp. 576–7.

——'Asbestosis', *Canadian Medical Association Journal* 22 (1930), pp. 253–4.

Peto, J., Hodgson, J., Matthews, F.E., and Jones, J.R, 'Continuing Increase in Mesothelioma Mortality in Britain', *The Lancet* 345 (4 March 1995), pp. 535–9.

——Decarli, A., La Vecchia, C., Levi, F., and Negri, I., 'The European Mesothelioma Epidemic'. *British Journal of Cancer* 79 (1999), pp. 666–72.

'The Quebec Asbestos Cohort', *Annals of Occupational Hygiene* 41 (1997), p. 1.

Reitze, W.B., Nicholson, W.J., Holaday, D.A., and Selikoff, I.J., 'Application of Sprayed Inorganic Fiber Containing Asbestos: Occupational Health Hazards', *American Industrial Hygiene Association Journal* 33 (March 1972), pp. 178–91.

Richter, E.D., and Laster, R., 'The Precautionary Principle, Epidemiology and the Ethics of Delay', *Human and Ecological Risk Assessment* 11 (February 2005), pp. 17–27.

Rivlin, K., and Potts, J.D., 'Not So Fast: The Sealed Air Asbestos Settlement and Methods of Risk Management in the Acquisition of Companies with Asbestos Liabilities', *New York University Environmental Law Journal* 11 (2003), pp. 626–61.

Robinson, B.W.S., Musk, A.W., and Lake, R.A., 'Malignant Mesothelioma', *The Lancet* 366 (30 July 2005), pp. 397–408.

Roggli, V.L., Pratt, P.C., and Brody, A.R., 'Asbestos Fiber Type in Malignant Mesothelioma: An Analytical Scanning Electron Microscope Study of 94 Cases', *American Journal of Industrial Medicine* 23 (1993), pp. 605–14.

Rosenberg, D., 'The Dusting of America: A Story of Asbestos—Carnage, Cover-Up, and Litigation', *Harvard Law Review* 99 (1986), pp. 1693–706.

Ross, M., 'The Schoolroom Asbestos Abatement Program: A Public Policy Debacle', *Environmental Geology* 26 (October 1995), pp. 182–8.

Sawyer, R.N., 'Asbestos Exposure in a Yale Building: Analysis and Resolution', *Environmental Research* 13 (1977), pp. 146–69.

——and de Melo, J., 'Institutional Knowledge, Common Knowledge: Occupational Asbestos Diseases and Gender', *Annals of the Global Asbestos Congress—Past, Present and Future* (Osasco, Brazil, 17–20 September 2000).

Scavone, L., Giannasi, F., and Thébaud-Mony, A., 'Asbestos Diseases in Brazil and the Building of Counter-Powers: A Study in Health, Work, and Gender', *Annals of the Global Asbestos Congress—Past, Present and Future* (Osasco, Brazil, 17–20 September 2000).

Schepers, G.W.H., 'Chronology of Asbestos Cancer Discoveries: Experimental Studies of the Saranac Laboratory', *American Journal of Industrial Medicine* 27 (1995), pp. 593–606.

Sebastien, P., Billion-Galland, M.A., Dufour, G., and Bignon, J., 'Measurement of Asbestos Air Pollution Inside Buildings Sprayed with Asbestos' (EPA 560/13–80–026, August, 1980).

Seidman, H., Selikoff, I.J., and Hammond, E.C., 'Short-Term Asbestos Work Exposure and Long-Term Observation', *Annals of the New York Academy of Sciences* 330 (1979), pp. 61–89.

Selikoff, I.J., 'Partnership for Prevention—The Insulation Industry Hygiene Research Program', *Industrial Medicine* 39 (April 1970), pp. 21–5.

——'Twenty Lessons from Asbestos', *EPA Journal* (May 1984).

——and Greenberg, M., 'A Landmark Case in Asbestosis', *Journal of the American Medical Association* 265 (20 February 1991), pp. 898–901.

——and Hammond, E.C., 'Asbestos-Associated Disease in United States Shipyards', *CA: A Cancer Journal for Clinicians* 28 (1978), pp. 87–99.

——Churg, J., and Hammond, E.C., 'Asbestos Exposure and Neoplasia', *Journal of the American Medical Association* 188 (1964), pp. 22–6.

——Hammond, E.C., and Churg, J., 'Asbestos Exposure, Smoking, and Neoplasia', *Journal of the American Medical Association* 204 (1968), pp. 106–12.

————and Seidman, H., 'Mortality Experience of Insulation Workers in the United States and Canada, 1943–1976', *Annals of the New York Academy of Sciences* 330 (1979), pp. 91–116.

Sharp, A., and Hardt, J., *Five Deaths a Day: Workplace Fatalities in Canada 1993–2005* (Ottawa: Centre for the Study of Living Standards, 2006). Posted at: http://www.csls.ca/

Shcherbakov, S., Kashansky, S., Domnin, S.G., and Kogan, F., 'The Health Effects of Mining and Milling Chrysotile: The Russian Experience', in R.P. Nolan, A. Langer, M. Ross, F. Wicks, and R.F. Martin (eds.), *The Health Effects of Chrysotile Asbestos: Contributions of Science to Risk-Management Decisions* (Montreal: Mineralogical Association of Canada, 2001).

Smith, A.H., and Wright, C.C., 'Chrysotile Asbestos is the Main Cause of Pleural Mesothelioma', *American Journal of Industrial Medicine* 30 (1996), pp. 252–66.

Smith, C., *'Don't Ask, Don't Tell'*, in *Chrysotile Asbestos: Hazardous to Humans, Deadly to the Rotterdam Convention* (London: Building and Woodworkers International and International Ban Asbestos Secretariat, 2006), pp. 11–13, 17–20. Posted at: www.lkaz.demon.co.uk/chrys_hazard_rott_conv_06.pdf

Smith, W.E., 'Surveys of Some Current British and European Studies of Occupational Tumor Problems', *Archives of Industrial Hygiene and Occupational Medicine* 5 (1952), pp. 242–62.

Stanton, M.F., Layard, M., Tegeris, E., Miller, E., May, M., Morgan, E., and Smith, A., 'Relation of Particle Dimension to Carcinogenicity in Amphibole Asbestos and Other Fibrous Materials', *Journal of the National Cancer Institute* 67 (1981), pp. 965–75.

Stayner, L.T., Dankovic, D., and Lemen, R.A., 'Occupational Exposure to Chrysotile Asbestos and Cancer Risk: A Review of the Amphibole Hypothesis', *American Journal of Public Health* 86 (1996), pp. 179–86.

Stein, R.C., Kitajeska, J.Y., Kirkham, J.B., Tait N., Sinha, G., and Rudd, R.M., 'Pleural Mesothelioma Resulting from Exposure to Amosite Asbestos in a Building', *Respiratory Medicine* 83 (1989), pp. 237–9.

Stellman, S.D., 'Issues of Causality in the History of Occupational Epidemiology', *Sozial-und Praventivmedizin* 48 (2003), pp. 151–61.

Stone, R., 'No Meeting of Minds on Asbestos', *Science* 254 (16 November 1991), pp. 928–31.

Storey, R., and Lewchuck, W., 'From Dust to DUST to Dust: Asbestos and the Struggle for Worker Health and Safety at Bendix Automotive', *Labour/Le Travail* [*Journal of Canadian Labour Studies*], 45 (2000), pp. 103–40.

Suzuki, Y., and Kohyama, N., 'Translocation of Inhaled Asbestos Fibres from the Lung to Other Tissue', *American Journal of Industrial Medicine* 19 (1991), pp. 701–4.

——and Yuen, S.R., 'Asbestos Tissue Burden Study on Human Malignant Mesothelioma', *Industrial Health* 39 (2001), pp. 150–60.

———'Asbestos Fibers Contributing to the Induction of Human Malignant Mesothelioma', *Annals of New York Academy of Sciences* 982 (2002), pp. 160–76.

Sweeney, E., 'Asbestos Diseases Foundation of Australia, Inc.', paper presented at Global Asbestos Congress, Osasco, Brazil, 17–20 September 2000.

Swetonic, M.W., 'Death of the Asbestos Industry', in J. Gottschalk (ed.), *Crisis Response: Inside Stories of Managing Image under Siege* (Detroit, IL: Gail Research Inc., 1993), pp. 289–308.

Swuste, P., Burdoff, A., and Ruers, B., 'Asbestos, Asbestos-Related Diseases, and Compensation Claims in the Netherlands', *International Journal of Occupational and Environmental Health* 10 (April/June 2004), pp. 159–65.

Talcott, J.A., Thurber, W., Kantor, A.F., Gaensler, E.A., Danahy, J.F., Antman, K.H., and Li, F.P., 'Asbestos-Associated Diseases in a Cohort of Cigarette-Filter Workers', *NEJM* 321 (2 November 1989), pp. 1220–3.

Talent, J.M., Harrison, W.O., Solomon, A., and Webster, I., 'A Survey of Black Mineworkers of the Cape Crocidolite Mines', in J.C. Wagner (ed.), *Biological Effects of Mineral Fibres*. Vol. 2 (Lyon: IARC Scientific Publications, No. 30, 1980), pp. 723–9.

Terracini, B., 'A Precautionary Programme for Workers Who Have Been Exposed to Asbestos?', *Annals of the Global Asbestos Congress—Past, Present and Future* (Osasco, Brazil, 17–20 September 2000).

Thébaud-Mony, A., 'Justice for Asbestos Victims and the Politics of Compensation: The French Experience', *International Journal of Occupational and Environmental Health* 9 (July/September 2003), pp. 280–6.

Thomson, J.G., 'Asbestos and the Urban Dweller', in H.E. Whipple (ed.), 'Biological Effects of Asbestos', *Annals of the New York Academy of Sciences* 132 (31 December 1965), i, pp. 196–214.

Tossavainen, A., 'Epidemiological Trends for Asbestos-Related Cancers', *Annals of the Global Asbestos Congress—Past, Present and Future* (Osasco, Brazil, 17–20 September 2000).

——'Asbestos, Asbestosis and Cancer: The Helsinki Criteria for Diagnosis and Attribution'. *Scandinavian Journal of Work and Environmental Health* 23 (1997), pp. 311–16.

Toxicslink, 'Asbestos: Fibre of Subterfuge', No. 11 (2001) Posted at toxicslink.org

Tweedale, G., 'Management Strategies for Health: J.W. Roberts and the Armley Asbestos Tragedy, 1920–1958', *Journal of Industrial History* 2 (1999), pp. 72–95.

——'Sprayed "Limpet" Asbestos: Technical, Commercial, and Regulatory Aspects', in G. Peters and B.J. Peters (eds.), *Sourcebook on Asbestos Diseases* Vol. 20 (Charlottesville, VA: Lexis Law Publishing, 1999), pp. 79–109.

——'Sources in the History of Occupational Health: The Turner & Newall Archive', *Social History of Medicine* 13 (2000), pp. 515–34.

——'Science or Public Relations? The Inside Story of the Asbestosis Research Council', *American Journal of Industrial Medicine* 38 (December 2000), pp. 723–34.

——'Asbestos and Its Lethal Legacy', *Nature Reviews Cancer* 2 (April 2002), pp. 311–15.

——'What You See Depends on Where You Sit: The Rochdale Asbestos Cancer Studies and the Politics of Epidemiology', *International Journal of Occupational and Environmental Health* 13 (January/March 2007), pp. 70–9.

——'Hero or Villain? Sir Richard Doll and Occupational Cancer', *International Journal of Occupational and Environmental Health* 13 (2007), pp. 233–5.

——and Flynn, L., 'Piercing the Corporate Veil: Cape Industries and Multinational Corporate Liability for a Toxic Hazard, 1950–2004', *Enterprise and Society* 8 (June 2007), pp. 268–96.

——and McCulloch J., 'Chrysophiles versus Chrysophobes: The White Asbestos Controversy, 1950s-2004', *Isis* 95 (2004), pp. 239–59.

——and Warren, R., 'Chapter 11 and Asbestos: Encouraging Private Enterprise or Conspiring to Avoid Liability?', *Journal of Business Ethics* 55 (November 2004), pp. 31–42.

Uno, R., 'Asbestosis: My Pain from Asbestosis', *Annals of the Global Asbestos Congress, Tokyo, Japan, 19–21 November 2004* (CD Proceedings, 2005).

Vianna, N.J., and Polan, A.K., 'Non-Occupational Exposure to Asbestos and Malignant Mesothelioma in Females', *The Lancet*, i, 20 May 1978, pp. 1061–3.

Vigliani, E.C., *Studio Sull' Asbestosi Nelle Manifatture Di Amianto* (Cirie': Stabilmento Tipografico Giovanni Capella, 1940).

Wagner J.C., 'Mesothelioma and Mineral Fibers' [Charles Mott Prize paper], *Cancer* 57 (15 May 1986), pp. 1905–11.

——'Statement by Dr. J. Christopher Wagner concerning some of the conclusions of the Literature Review Panel, August 27th 1991', in *The Health Effects Institute: Asbestos in Commercial Buildings: A Literature Review and Synthesis of Current Knowledge* (Boston, MA: HEI, 1991), pp. S2–1 to S2–3.

——and Berry, G., 'Mesothelioma in Rats Following Inoculation with Asbestos', *British Journal of Cancer* 23 (1969), pp. 578–81.

——Sleggs, C.A., and Marchand, P., 'Diffuse Pleural Mesotheliomas and Asbestos Exposure in the North-Western Cape Province', *British Journal of Industrial Medicine* 17 (1960), pp. 260–71.

——Berry, G., and Timbrell, V., 'Mesotheliomas in Rats Following the Intra-Pleural Inoculation' with Asbestos', H.A. Shapiro (ed.), *Pneumoconiosis: Proceedings of an International Conference, Johannesburg, 1969* (Cape Town: Oxford University Press, 1970).

————Skidmore, J.W., and Timbrell, V., 'The Effects of the Inhalation of Asbestos in Rats', *British Journal of Cancer* 29 (1974), pp. 252–69.

——————and Pooley, F.D., 'The Comparative Effect of Three Chrysotiles by Injection and Inhalation in Rats', in IARC, *Biological Effects of Mineral Fibres. IARC Scientific Publications No. 30* (Lyon: IARC, 1980), pp. 363–73.

————and Pooley, F.D., 'Carcinogenesis and Mineral Fibres', *British Medical Bulletin* 36 (1980), pp. 53–6.

Wagstaff, A., 'Richard Doll: Science Will Always Win in the End', *Cancer World* (December 2004), pp. 28–34.

Waller, B. and Marrett, L. (eds.), *The Occupational Cancer Research and Surveillance Project* (Toronto: Cancer Care Ontario and WSIB, 2006), p. 23. Posted at: www.cancercare.on.ca/documents/Occupreport2006.pdf

Waterman, Y.R.K., and Peters, M.G.P., 'The Dutch Institute for Asbestos Victims', *International Journal of Occupational and Environmental Health* 10 (April/June 2004), pp. 166–76.

Watterson, A., Gorman, T., Malcolm, C., Robinson, M., and Beck, M., 'The Economic Costs of Health Service Treatments for Asbestos-Related Mesothelioma Deaths', *Annals of the New York Academy of Sciences* 1076 (September 2006), pp. 871–88.

Webster, I., 'Methods by which Mesothelioma can be Diagnosed', in H.W. Glen (ed.), *Asbestos Symposium* (Johannesburg: Department of Mines, 1978), pp. 3–8.

Western Australian Cancer Registry, *Cancer in Western Australia* (Perth: Department of Health, 2005).

White, M.J., 'Asbestos and the Future of Mass Torts', *Journal of Economic Perspectives* 18 (Spring 2004), pp. 183–204.

Wicks, F.J., Royal Ontario Museum, Canadian Society of Exploration Geophysics Conference abstract, 2000. Posted at: http://www.cseg.ca/conferences/2000/2000abstracts/786.PDF

Wyers, H., 'Asbestosis', *Postgraduate Medical Journal* 25 (December 1949), pp. 631–8.

Index

Figures and tables are indexed in bold, e.g. 172**t**

Au:
chec
rang
reflo
line
folio
264
"def

Ingram Content Group UK Ltd.
Milton Keynes UK
UKHW020427070423
419654UK00005B/160